Foundations of
Information Integration Theory

Titles in Information Integration Theory
by Norman H. Anderson

FOUNDATIONS OF INFORMATION INTEGRATION THEORY
METHODS OF INFORMATION INTEGRATION THEORY (in press)

Foundations of
Information Integration Theory

Norman H. Anderson

Department of Psychology
University of California, San Diego
La Jolla, California

ACADEMIC PRESS   1981

A Subsidiary of Harcourt Brace Jovanovich, Publishers

New York   London   Toronto   Sydney   San Francisco

ACADEMIC PRESS, INC.
111 Fifth Avenue, New York, New York 10003

United Kingdom Edition published by
ACADEMIC PRESS, INC. (LONDON) LTD.
24/28 Oval Road, London NW1 7DX

Library of Congress Cataloging in Publication Data

Anderson, Norman Henry, Date.
    Foundations of information integration theory.

    (Foundations, methodology, and application of
information integration theory)
    Bibliography: p.
    Includes index.
    1. Human information processing. 2. Human information
processing—Mathematical models. 3. Human information
processing—Experiments. I. Title. II. Series:
Anderson, Norman Henry, Date. Foundations, methodology,
and application of information integration theory.
1. [DNLM: 1. Models, Psychological. 2. Psychology,
Experimental. BF 181 A548f]
BF441.A54   153s   [153]   80-1769
ISBN 0-12-058101-9   AACR2

Contents

2 Basic Experiments on Information Integration

3 Meaning Constancy in Person Perception

4 Special Problems of Information Integration

5 Measurement Theory

Preface

The intended subtitle of this book, *A Case History in Experimental Science*, was dropped for simplicity, but it does in fact help to describe the nature of this work. Information integration theory has developed through experimental analysis, and its nature has been determined largely by the outcomes of these experimental investigations. The work began in 1959–1960, and most of the basic experiments were completed by 1965. The emotions of that early period cannot be recaptured, but the discussions of Chapters 2–4 will suggest something of the difficulties that had to be resolved. Each new experiment was half expected to demolish the developing formulation. By 1965, however, it was clear that a simple, unified account of a complex of data had emerged.

Even today this outcome seems surprising. Person perception, which was the main substantive area of the initial work, is manifestly complex and not manifestly promising for quantitatively oriented investigation. However, this quantitative approach had considerable success in some very demanding tests, qualitative and quantitative. Both the initial results and the theoretical interpretation have stood the test of time. Little emendation has been required, even in the many generalizations to other areas.

The subsequent 15 years have brought many important developments. On the technical side are the linear fan analysis, accommodation to the difficulties of differential weighting, and improved methods for monotone analysis. On the substantive side are conceptual and experimental extensions to virtually every area of cognitive psychology. These developments, which represent the contributions of my students and associates at the University of California, San Diego, have transformed a local model into a unified, general theory.

Chapter 1 gives an overview of basic concepts and methods accompanied by a cross section of applications, each chosen to illustrate some substantive issue. This chapter also provides the theoretical background needed for the later chapters in the volume. Because of the central role of psychological measurement in this work, the final chapter presents a critique of measurement theory in psychology.

Chapters 2–4 present the foundation work on the personality adjective task. The problems considered in these chapters concern basic phenomena of information integration. Primary attention is on the foundation experiments, which have not been superseded but retain their original interest and value. However, later work is also included to round out the treatment of certain topics. In this respect, a reasonably complete review is given of certain issues, such as meaning constancy and the two-memory hypothesis. Although this experimental work is largely restricted to the personality adjective task, similar results have been found in many areas, as illustrated in Chapter 1. Thus, the methods and theory developed in the initial work appear to have substantial generality.

The companion *Methods* volume goes into detail on experimental procedures and methods of data analysis. It is intended for investigators who wish to pursue research in this area. Additional volumes are in progress that will cover developments in other substantive areas, including social cognition, decision making, developmental psychology, psychophysics, and psycholinguistics.

Acknowledgments

This book is dedicated to my students, whose work has formed the foundation of information integration theory. First among them are Ann Norman Jacobson and Anita Lampel; their help in the initial investigations of person perception was beyond price. Stephen Hubert and Ralph Stewart also did important work during this initial stage, and James Anderson provided valuable technical assistance.

Nearly all my students at the University of California, San Diego, have contributed to the field of person perception, but they have mainly been concerned with exploring new areas. The theoretical developments in decision theory were initiated by James Shanteau, in psychophysics by David Weiss, and in psycholinguistics by Gregg Oden. A foundation of applications in developmental psychology has been provided by Clifford Butzin, Diane Cuneo, and Manuel Leon. Other extensions have been made by John Clavadetscher to perceptual illusions, by Arthur Farkas to equity theory, by Cheryl Graesser to group dynamics, by Michael Klitzner to motivation, by Barbara Sawyers to attitude theory, by John Verdi to moral judgment, by James Zalinski to statistical analysis, and by Lola Lopes to many problems. The work of Donald Blankenship and JoAnn Kahn was also important. It is to these men and women that integration theory owes its breadth.

The theoretical development is also much indebted to work by colleagues at other institutions. Foremost are Martin Kaplan and Clyde Hendrick, who made basic and enduring contributions in the early period, and Michael Birnbaum, who has done cogent work in many areas. Other contributions have been made by Richard Bogartz, Edward

Carterette, Samuel Himmelfarb, Wilfried Hommers, Irwin Levin, Jordan Louviere, Dominic Massaro, Kent Norman, Thomas Ostrom, Allen Parducci, Charles Schmidt, Joseph Sidowski, Ramadhar Singh, Friedrich Wilkening, and Robert Wyer. In addition, I am grateful to Edward Alf, Reid Hastie, Miriam Rodin, and Lennart Sjöberg for many helpful comments and criticisms, and to Eileen Beier and Frank Logan for intellectual stimulation.

Among my teachers, I wish to record my appreciation of Cletus Burke, David Grant, Carl Hovland, and, especially, Paul Halmos, whose life and teaching went hand in hand. In addition, my heartfelt gratitude goes to four teachers at Milaca High School: Dale Dougherty, Gordon Mork, Nels Tosseland, and Leslie Westin.

I take this opportunity to express my deepest personal appreciation to Lucille Kirsch, Eileen Beier, Ann Norman Jacobson, Joan Prentice, Mary Pendery, and my beloved Margaret. For their love and understanding I am profoundly grateful. For friendships that have meant much to me in difficult times, I am grateful to Frank Allen, Robert Doan, Herb and Lee Kanner, Melvin Kniseley, Richard Miller, and Louise Warner. And to James Alexander, Lucille Kirsch, and Mary Pendery, I owe more than words can say.

Besides these personal debts are various institutional obligations that are pleasant to record. The National Science Foundation has supported this work since 1962, and I appreciate the backing of Henry Odbert and Kelly Shaver at critical junctures. My graduate training at the University of Wisconsin turned me into an experimentalist, and a postdoctoral fellowship at Yale University sponsored by the Social Science Research Council was important in widening my psychological outlook. Also notable was a year spent at the Center for Advanced Study in the Behavioral Sciences, that among other things, provided a healthy separation from experimental work and started me on the long path to this book. Above all, I am indebted to the University of Chicago, where, as an undergraduate, I found worlds of which I had never dreamed.

Cognitive Algebra

This chapter has two main purposes. The first is to present the basic ideas and concepts of information integration theory. The second is to show, by experimental example, how these ideas and concepts manifest themselves across many areas of psychology.

Primary emphasis is on experimental analysis. This emphasis reflects the inductive nature of information integration theory. One main business of the chapter, therefore, is to demonstrate the empirical validity of the integration-theoretical approach. These diverse experimental applications provide an empirical foundation for the theory as well as a means to illustrate its nature in operation. More general comments on the nature and qualities of the theory are presented in the final section of this chapter.

Cognitive algebra has been important in this theoretical development. Many tasks of information integration have been found to obey simple algebraic rules, collectively called *cognitive algebra*. The development of general methods that can reveal the operation of these algebraic rules constitutes a major contribution of information integration theory. The experimental studies that have affirmed the operation of these rules constitute the main foundation of the theory. These algebraic rules provide a new capability for cognitive analysis: as a foundation for psychological measurement and, more important, as a means to delineate processing structure.

This chapter is also intended to provide adequate preparation for reading the later empirical chapters. Those who wish to pursue research on information integration will need to study the more technical material of the companion volume, *Methods of Information Integration Theory* (Anderson, in press). The basic ideas and methods are simple, however, and they can readily be applied to diverse experimental areas. Accordingly, this introductory chapter includes sufficient material to understand the basic experiments of Chapter 2 and the special topics of Chapters 3 and 4.

1.1 Information Integration Theory

1.1.1 BASIC CONCEPTS

Integration theory has developed around four interlocking concepts: *stimulus integration, stimulus valuation, cognitive algebra,* and *functional measurement.* These concepts will be outlined briefly here and in more detail in the following sections.

Integration. Stimulus integration is the central concept. Thought and behavior typically depend on the joint action of multiple stimuli. Single causes are seldom sufficient for understanding or prediction. Multiple causation is the rule.

Integration theory has central concern with the study of stimulus integration. Two kinds of questions arise. Given the effective stimuli, how are they combined or integrated to produce the response? Given the response, what were the effective stimuli? The first question, that of synthesis, constitutes the central theoretical concern. The study of synthesis also provides a useful, novel approach to the second question, that of analysis.

Valuation. Stimuli may be considered at two levels, physical and psychological. Physical stimuli are observables that can potentially be controlled in experimental studies. However, they are distant, partial causes of thought and behavior. Integration theory is primarily concerned with stimuli at the psychological level, for these are the immediate causes of thought and behavior. The chain of processing that transforms the physical stimulus into its psychological counterpart is represented by the valuation operation.

The role of the valuation operation may be underscored by individual differences. A word or glance will have different meanings for different persons. Even a simple color plate will not appear the same to color-weak and normal persons. The concept of valuation takes cognizance of the fundamental importance of representing individual differences within the theory.

Cognitive Algebra. The work in this research program has shown that stimulus integration often obeys simple algebraic rules. The human organism frequently appears to be averaging, subtracting, or multiplying the stimulus information to arrive at a response. Because of the ubiquity of these algebraic rules, they are generically termed cognitive algebra.

Functional Measurement. The fourth basic concept is functional measurement. Implicit in the notion of cognitive algebra is a numerical representation of the stimuli. To say that two stimuli are averaged or multiplied seems to presuppose numerical values. Accordingly, the study of any algebraic rule is integrally bound up with measurement of psychological values. A conceptual foundation as well as practical techniques have been provided by functional measurement theory.

The Zeitgeist. A view of the organism as an integrator of stimulus information is time-honored in traditional perception and forms a basic paradigm of modern judgment theory. Valuation is a general name for a class of processes that represent the role of sensory systems in simple sensation and the role of semantic memory in understanding words and phrases. Integration and valuation are basic concepts that have been widely and intensively studied in every area of psychology.

Algebraic models for stimulus integration have also been suggested by many investigators. These have generally been isolated applications that were developed in connection with particular substantive problems. When considered in the aggregate, however, it is surprising how often such models have arisen, and in what diverse areas.

Despite this wide interest, previous progress on algebraic models was very limited. The essential difficulty has been the lack of a theory of measurement. Without a measurement capability, adequate tests of the models are not generally possible. Functional measurement, based on the guiding principle of using the model itself as the base and frame for measurement, provided a resolution to this difficulty.

Functional measurement itself is heavily indebted to contributions by others. On the conceptual side, there is probably no single idea in this approach that has not been considered by many previous workers. On the technical side, functional measurement methodology has incorporated many ideas and techniques from statistical theory. In both conceptual and technical aspects, functional measurement is part of the Zeitgeist.

The program of research on information integration theory is thus part of the ongoing tradition of experimental science. Within its own domain, it seems fair to say that it has carried previous work forward in a significant way. Functional measurement itself provides a unification of ideas and methods that constitutes a general theory of psychological measurement, one that has the essential virtue of empirical validity. The more general theory of stimulus integration includes, among other results, an extended network of experimental studies that have placed cognitive algebra on a solid empirical foundation. This work has provided a cogent approach and new tools, both conceptual and technical, to many traditional problems of valuation and integration.

1.1.2 FUNCTIONAL MEASUREMENT DIAGRAM

How the four basic concepts interlock can be seen in the functional measurement diagram of Figure 1.1. The sense of this diagram is straightforward. Physical stimuli S impinge on the organism and are processed by valuation functions V into their psychological values s. These psychological stimuli are combined by the integration function I into an implicit response r. This implicit response is then externalized by the response function M to become the observable response R.

Thus, the path from the observable stimulus S to the observable response R is represented by three linked functions. These are

Valuation function: $V(S) = s;$ (1a)

Integration function: $I\{s\} = r;$ (1b)

Response function: $M(r) = R.$ (1c)

In this notation, the observable stimuli and response are denoted by uppercase letters S and R, whereas lowercase letters s and r are used to denote their unobservable, subjective counterparts.

The integration function or psychological law has primary interest in integration theory. Its determination requires the solution of three interlocked problems, corresponding to the three laws in Figure 1.1. These problems are

1. Measuring the psychological values of the stimuli
2. Measuring the psychological value of the response
3. Determining the psychological law or integration function I

As the diagram implies, therefore, two distinct problems of measurement are involved in the determination of the integration function.

All three problems are interlocked, and functional measurement provides a joint solution to all three. The integration function implicitly contains the measurement scales for stimuli and response. Moreover, it provides the structural frame for obtaining these scales.

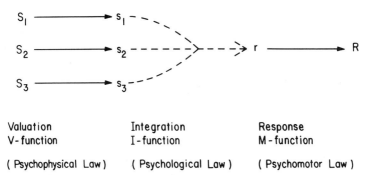

FIGURE 1.1. *Functional measurement diagram. Chain of three linked functions,* **V, I,** *and* **M,** *lead from observerable stimulus field,* {S_i}, *to observable response,* R. Valuation function *or psychophysical law,* **V,** *maps physical stimuli,* S_i, *into subjective counterparts,* s_i. Integration function *or psychological law,* **I,** *maps subjective stimulus field,* {s_i}, *into implicit response,* r. Response function *or psychomotor law,* **M,** *maps implicit response,* r, *into observable response,* R.

This proposition may seem doubtful because all three problems deal with unobservable entities. Their existence and definition are inferential. Nevertheless, a rigorous, objective foundation for these three unobservables can be constructed from the observable stimuli and response. That is actually quite easy to do.

1.1.3 VALUATION

Valuation refers to the processes that extract the information from the physical stimulus. Two examples will bring out some important aspects of valuation. In the first example, with a sensory stimulus, a sound is given, to be judged on loudness or on pitch. In the second example, with a verbal stimulus, a man is described as happy-go-lucky, to be judged on his suitability as a picnic guest or as a statistical clerk.

In each case, the task instructions set some dimension of judgment. Relative to this dimension, each stimulus has some value. This value may be an immediate sensory effect, as with the sound, or a semantic inference, as with the words.

These scale values are not enough. A concept of weight is also necessary for many integration tasks. This represents the relative salience or importance of each stimulus in the overall response. Thus, a two-parameter, weight–value representation of the stimulus is required.

The valuation operation denotes the chain of processes that lead from the physical stimulus to its psychological representation. As the two examples show, the valuation operation depends on the dimension of judgment. A loud, low sound will have different scale values for loudness and for pitch. The scale value of happy-go-lucky will be positive for the picnic guest, negative for the statistical clerk.

In general, therefore, the weight–value representation is not a constant of the stimulus. Quite the contrary, it depends very sensitively on the prevailing dimension of judgment and also on the momentary motivational state of the organism.

The valuation operation obviously represents an involved chain of processing. For the sound, two major component chains may be distinguished. The first is a sensory-perceptual chain that leads to the conscious percept. The second chain is more cognitive, leading from this percept to the implicit judgment. Similarly, in the example from social perception, an initial decoding chain leads to the recognition of the word, and this is followed by a chain of inference that processes the implications for the given dimension of judgment.

It is tempting to think that the dimension of judgment affects only the second major component in the valuation chain. The conscious percept of the sound is presumably the same regardless of whether loudness or pitch is to be judged. Similarly, the meaning of the trait word is presumably constant, even though its implications have different values for the two dimensions of judgment (guest or clerk). In general, however, the dimension of judgment can also affect the processing in the first stage; for example, by selective attention.

Some way to avoid the necessity of a link-by-link attack on the valuation operation is desirable. Integration theory does that by analysis at a molar level in which valuation is represented by the **V** operator with molar effects represented by the weight–value parameters. This molar approach passes by important questions about molecular process. That is a limitation, in one respect, but also a great advantage, for it can yield results that are exact and valid at the molar level entirely independent of process details (Sections 1.1.5, 1.6.5, 1.8.2).

Three final points about the valuation operation may be mentioned briefly. First, valuation has been discussed as though it were independent of integration, but that will not always be true. For example, redundancy between stimuli will affect their valuation, but redundancy only exists in consequence of an attempted or implicit integration. Again, in a feature or set-theory representation, the integration may precede and mediate the valuation.

Second, present consideration is mainly restricted to situations with a dimensional response. The utility of this dimensional approach justifies its limitations. However, that should not obscure the interest and importance of numerous nondimensional responses, judgments of timbre of the sound, for example, or predictions about the happy-go-lucky person's probable reply to some question.

Third, it is a basic tenet of integration theory that stimuli must be considered in terms of their meaning and value to the individual. If two persons hear the same message, they may disagree about what was actually said. If they agree about what was actually said, they may still disagree about the implications. And even if they agree about the implications, they may nevertheless disagree about their desirability. Valuation processes may be similar across individuals, but their outcomes will differ. Integration theory can operate within the value system of the individual.

1.1.4 INTEGRATION

Virtually all thought and behavior is multiply caused, the resultant of numerous coacting factors. Depth perception is a classic example, for it depends on integration of a field of cues, including interposition, relative size, perspective, hue, and texture. The evolution of depth in painting reflects the gradual discovery and mastery of such informational cues to depth.

Social perception provides a corresponding example in the concept of deservingness. What a man deserves depends not simply on what he accomplished, but also on how his accomplishment relates to his abilities and on other factors such as his need, how hard he tried, how agreeable he was, and even on his physical appearance. These and other cues are integrated into the deservingness judgments that operate in social distributions of status symbols and material goods and determine social feelings of fairness and unfairness.

Multiple causation may be viewed in two related aspects, synthesis and analysis. Synthesis studies the response to a complex stimulus field, perceptual or social in the two cited examples. Synthesis corresponds to the integration function that represents how the effective stimuli combine to produce the response. Analysis is inverse to synthesis and seeks to dissect a given response into its causal components.

Problems of multiple causation can be extremely difficult. When several factors are at work, each pushing in its own direction, their combined effect is not generally predictable without the aid of quantitative analysis. Such analysis, moreover, must generally be in terms of the psychological values of the individual. Without such quantitative capability, many basic problems of multiple causation can hardly be touched.

Integration theory makes a direct attack on multiple causation. The central problem, that of determining the nature of the integration function, is the main focus of the theory. At the same time, functional measurement contributes to analysis for it dissects the observed response into its functional components.

The efficacy of this approach owes much to the fortunate fact that stimulus integration often obeys algebraic models. A glance forward at Sections 1.3 and 1.5 will show the great diversity of applications. These algebraic models are sufficiently common to indicate the existence of a general cognitive algebra of multiple causation. This has led to a new way of thinking about many problems in psychology (Section 1.8).

1.1.5 COGNITIVE UNITIZATION

Many stimuli have complex structure. A prose passage can be broken down into paragraphs, the paragraphs into sentences, the sentences into functional verbal units, these units into words, and the words themselves into features in semantic memory. Nor does analysis end there, for a further processing chain leads back through the perceptual–sensory system to the physical stimulus itself.

Different investigators begin analysis at different levels in the processing chain: at the level of letter–word recognition, for example, or at the level of ideas. There seems to be an implicit presumption that different levels will all have psychological validity, that each level can be represented in terms of functional units. That would certainly be desirable. It would allow simultaneous attacks at different levels rather than requiring a start at one level with a unique set of primitive units. However, choice of level has generally been guided only by common sense. That is quite reasonable as far as it goes, but it leaves open questions as to how far each level has psychological reality and how well analysis at one level will cohere with analysis at adjacent levels.

The concepts of valuation and integration may help elucidate this problem of cognitive units. Integration itself can be considered as a unitizing process. Recognition of words, for example, results from an integration of various cues, both focal and contextual. At higher levels, similarly, understanding meanings also results from integration of diverse cues. Moreover, the integration and valuation operations are interlinked: integration operations at one level constitute valuation operations with respect to the next level.

In this view, the processing chain is considered as a sequence of integration operations. At each level, integration has a simplifying effect, reducing a compound or complex stimulus field to a unitary resultant. By virtue of this simplification, integration at the next level can operate across a broader stimulus field.

Cognitive unitization has fundamental importance because valid analyses are possible at a molar level without immediate need to consider molecular detail. Moreover, this molar approach can provide boundary conditions for molecular analysis. When an algebraic model holds, it can be used to dissect an observed response into functional components and to measure each component on validated scales. These functional weights and scale values are complete and exact summaries of the molecular processing. Any molecular theory must obey these boundary conditions. They provide exact, quantitative constraints on molecular analysis that might otherwise be unavailable.

1.1.6 RESPONSE FUNCTION

A distinction between the observed response and an internal or implicit response is common throughout psychology. Accordingly, the concept of an output function or response function is required. This is denoted by $\mathbf{M}$ in the diagram of Figure 1.1, so that $R = \mathbf{M}(r)$. Thus, $\mathbf{M}$ represents the transformation from the internal state to the response device—verbal, motor, or other—imposed by the investigator.

Among the various questions than can be asked concerning the response function, one has attracted particular attention and controversy. This is the question of the "equal-interval" scale. Even the meaning of this term has caused confusion. Its usual sense seems to be that equal numerical intervals on the observable response scale must correspond to equal units of the psychological quality that is presumably being measured. This definition originates by analogy to physical measurement. Length, for example, can be measured by physical addition of unit lengths: that is what a meter stick accomplishes. Similarly, measurement of mass can be viewed as the physical addition of unit masses.

However, attempts to establish additive units have not been very satisfactory in psychology. Nor has any other approach to establishing the equal-interval property been generally accepted. Indeed, comparison of intervals at different locations on the underlying psychological scale may not even make sense.

Functional measurement provides a simple conceptual definition. The essential property is that $\mathbf{M}$ be linear; that is, that R be a linear function of r. If R is a linear scale in this functional measurement sense, then it will serve as an interval scale in the customary sense. However, the definition in terms of linearity has two advantages. First, it makes explicit that a measurement scale is a functional relation between two variables, in this case R and r. Second, it bypasses the quagmire of "equal intervals" and places measurement on a functional base. Since the term *interval scale* seems to perpetuate confusion, the term *linear scale* seems preferable and it will be used here.

The main problem, of course, is not definitional, but practical. How can it be proved that the observable R is or is not a linear function of the unobservable r? A general answer is provided by functional measurement as shown in the next section.

The rating response has played a major role in this research program and requires separate discussion. Rating scales have been almost universally condemned in measurement theory. Among other problems, rating numbers are really just words: to treat them as true numbers surely requires justification. Traditional measurement theory has seen no way to justify the rating method and, accordingly, has turned to paired comparisons or rank-order data. Even the dominant school that advocated numerical response joined in criticism of the rating method. Because of these objections to the rating method, its present use requires careful justification. This justification will appear later in the theoretical and empirical development of functional measurement methodology (see also Chapter 5 and Anderson [1979a, in press]).

The linear character of the rating scale is considered to reflect a general-purpose metric sense. Children and adults readily give quantified judgments of a variety of concepts, both perceptual and social in nature. It is assumed that these diverse concepts attain their quantification on the same internal metric.

The alternative of a separate metric for each concept seems uneconomical. Separate metrics would be reasonable for a limited number of concepts, especially basic sensory qualities such as loudness and heaviness. In person perception, however, every trait adjective defines a dimension, and there are innumerable others. It is hard to believe that all these judgment dimensions have separate internal scales.

There may, of course, be a limited number of distinct internal metrics. One argument for parsimony is that children as young as 3 years are able to use the graphic rating scale in a true linear manner. That suggests a developmental origin from reaching movements involved in spatial exploration by the child. The linearity of the rating scale may thus be a direct reflection of a veridical internal representation of local space.

For present purposes, the general-purpose metric can be viewed as a response capability. Accordingly, it must be calibrated to the stimulus range of each particular task. That is the function of the stimulus end anchors (Appendix A), which define the zero and unit of the scale for each task. The phenomenological or psychological quality being judged may be considered as a dimensional constant that is attached to the metric for the given task.

1.1.7 FUNCTIONAL MEASUREMENT

The guiding principle of functional measurement is that measurement scales are derivative from substantive theory. "The logic of the present scaling technique consists in using the postulated behavior laws to induce a scaling on the dependent variable [Anderson, 1962b, p. 410]." In terms of the diagram of Figure 1.1, *behavior law* corresponds to the psychological law or integration function, **I**. *Dependent variable* refers to the overt response, R, that is to be transformed to a linear scale; that is, to be a linear function of the underlying response, r.

The fundamental element is the integration function. Its mathematical form carries implicit scales of stimulus and response variables. This functional form provides the structural frame for the scaling itself as well as the validational base.

The term *functional measurement* derives from this fundamental property of the integration function. The term *functional* is also appropriate in other respects. In particular, the stimulus values provided by this method are those that were functional in the very thought and behavior under study.

A general theory of measurement must be able to work with monotone (ordinal) response scales. Observed response measures will not in general be linear, but they will often be monotone functions of the underlying response variable. The essential idea for analysis of monotone response scales is simple (Anderson, 1962b, p. 404). Since the observed response is a monotone scale, some monotone transformation will make it a linear scale. If the integration function is valid, then the desired transformation can be computed because it is the one that makes the data fit the function form. In a proper sense, therefore, functional measures may be called *scale free* because they depend solely on the algebraic form of the integration function.

The feasibility of this approach depends on many-variable analysis. Because the integration function depends on two or more variables, it allows determination of the monotone transformation and still leaves degrees of freedom to test whether the transformed data do fit the function. If the postulated integration function is not valid, of course, then the data will not in general pass this test. This test of goodness of fit is essential; it provides the validational criterion of the integration function and hence also of the derived scales.

Many technical problems arise in implementing monotone analysis (Section 1.3.8). These problems can be greatly simplified if the observed response scale can be linearized by experimental procedures rather than by statistical computation. That is why numerical response scales have been emphasized in the experimental work. This research program owes much to the fortunate capability of the human organism to use a linear response function.

Traditionally, measurement has been treated as a methodological preliminary to substantive inquiry. If valid scales of the stimulus variables could once be obtained, they would greatly facilitate the study of the psychological law. To begin with stimulus scaling is thus natural and obvious, seemingly like the approach used in physics. However, it has not worked very well in psychology. Functional measurement reverses the traditional approach and makes measurement theory an organic component of substantive investigation.

In principle this approach is straightforward. However, the final question is not one of logical principle, but of empirical fact. The true foundation of measurement lies in substantive theory. Only if these algebraic models of stimulus integration are empirically valid can this approach have value or meaning. The essence of functional measurement thus resides in the experimental applications that follow.

1.2 Adding-Type Models

1.2.1 EXPERIMENTAL PREVIEW

To those who speak its language, Figure 1.2 tells a remarkable story. These data are from an experiment on person perception in which subjects received pairs of adjectives that described an unknown person. The subjects' task was to form an impression of the person and to judge how likable the person would probably be. Each data point represents the judged likableness of one person, rated on a 1–20 scale. There were nine person descriptions altogether, specified by the 3 × 3 matrix of adjectives given in Figure 1.2. The nine data points for each subject, F. F. and R. H., are plotted as a set of three curves.

Figure 1.2 begins its story by saying that both subjects obeyed an adding-type rule. Likableness for each person description can be represented as a sum or an average of the values of the separate adjectives that describe the person. Integration of the adjective information into an overall impression is undoubtedly a complex process, yet it obeys a simple, exact algebraic rule.

Furthermore, the subjective values of the three row adjectives are represented by the elevations of the three curves. For subject F. F., therefore, *unsophisticated* lies about halfway in value between *ungrateful* and *level-headed*. Subject R. H. has a different system of values; for him *unsophisticated* is nearly equal in value to *level-headed*. In other words, the data provide direct readings on linear (equal-interval) scales of the functional values of the stimulus adjectives—and this is done within the personal value system of each individual subject.

In addition, Figure 1.2 speaks to the character of the rating response. It says that these responses are not mere number-words; they provide

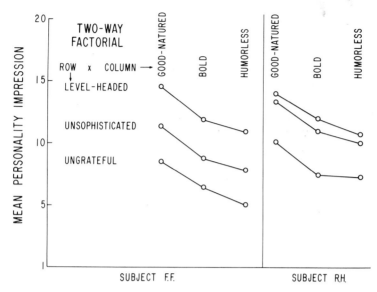

FIGURE 1.2. *Parallelism pattern supports adding-type rule in person perception. Subjects judge likableness of hypothetical persons described by two trait adjectives from indicated Row × Column design, with row adjectives of* level-headed, unsophisticated, *and* ungrateful *and column adjectives of* good-natured, bold, *and* humorless. *Each of these 3 × 3 = 9 person descriptions corresponds to one data point. Data averaged over third adjective for simplicity (see Section 2.2.1). (After Anderson, 1962a.)*

true quantification of the person impression. Thus, the response scale is also a linear (equal-interval) scale.

There is another implication of Figure 1.2, one that many have found disturbing. It says that each adjective has a fixed value and meaning. The meaning of *unsophisticated*, for example, remains the same regardless of whether the person is described as *good-natured, bold,* or *humorless.* This implication represents the hypothesis of meaning constancy for adjective descriptions.

In short, Figure 1.2 speaks to three issues. It tells how the information in the trait adjectives is integrated to form the person impression, it validates the rating response as a true numerical scale, and it provides validated scales of adjective values for each individual subject. How Figure 1.2 can speak so well will be shown in the following functional measurement analysis for adding-type models.

1.2.2 PARALLELISM ANALYSIS

The essential idea of parallelism analysis is simple. Suppose that two or more stimulus variables are thought to add together to yield the observed response. To test this hypothesis, manipulate the stimulus variables in a factorial design. If the hypothesis is true, then the factorial plot of the response data will exhibit a pattern of parallelism—which is the sign of an adding-type operation. Experimental diagnosis of adding-type operations can thus be straightforward because the pattern in the factorial plot speaks directly to the validity of the hypothesis.

Factorial Design. Factorial design, which is basic in functional measurement, involves joint manipulation of two or more stimulus variables. The variables are called *factors,* and the chosen values of the variables are called the *levels* of the factors. A two-variable (or two-way) factorial design can be represented as a rectangular, Row × Column matrix as illustrated in Figure 1.3. The levels of the row factor are physical stimuli denoted by S_{Ai}; the levels of the column factor are physical stimuli denoted by S_{Bj}. The experimental conditions are pairs of physical stimuli. Cell ij of the design matrix thus corresponds to the experimental condition (S_{Ai}, S_{Bj}). The subject's overt response to this stimulus pair is denoted by R_{ij}.

A *factorial plot* is obtained by placing the column stimuli, say, on the horizontal axis at any convenient spacing. The observed responses in each row of the design are then plotted as a separate curve. Examples are given in Figure 1.2 and in Figures 1.4–1.23. Under suitable conditions, the pattern in this factorial plot will reveal the nature of the integration operation.

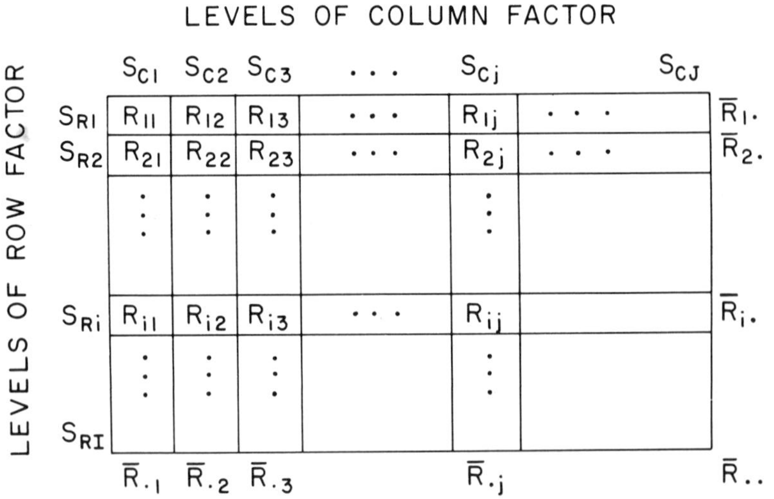

FIGURE 1.3.   *Factorial, Row × Column design.* S_{Ai} *and* S_{Bj} *represent stimulus levels of row and column variables, respectively. Cell entries* R_{ij} *denote observed response to stimulus combination* (S_{Ai}, S_{Bj}) *in cell* ij.

Adding Model. In the adding model to be considered here, the subject's implicit response is assumed to be a sum of the subjective values of the given stimuli. In cell ij of the factorial design, the subjective values of the physical stimuli, S_{Ai} and S_{Bj}, are denoted by corresponding lowercase letters, s_{Ai} and s_{Bj}. Similarly, the implicit value of the overt response, R_{ij}, is denoted by r_{ij}. Weight parameters and response variability may be ignored for the present since they do not affect the essential logic. Accordingly, the model may be written

$$r_{ij} = s_{Ai} + s_{Bj}. \tag{2}$$

In terms of the functional measurement diagram of Figure 1.1, the essential assumption is that the integration function, $\mathbf{I}$, is additive. Thus, the model assumes additivity at the subjective level.

Because all three terms in the model are unobservable, it may seem difficult or impossible to test. Fortunately, there is a straightforward test, one that requires nothing more than the matrix of observed responses, R_{ij}. If the underlying integration obeys an adding rule, then an additive pattern will manifest itself in the observed data. In this way, the model yields an observable implication that provides a validational test. The simplest and most useful case is covered by the following parallelism theorem.

Parallelism Theorem. *If the adding model of Eq. (2) is true, and if the observable response is a linear (equal-interval) scale, then (a) the factorial data plot will form a set of parallel curves; and (b) the row means of the factorial design will be estimates of the subjective values of the row stimuli on validated linear (equal-interval) scales and similarly for the column means.*

Proof of Parallelism. Proof of the first conclusion of the parallelism theorem is straightforward. The assumption that the overt response is a linear scale of the implicit response can be written

$$R_{ij} = C_0 + C_1 r_{ij}, \tag{3}$$

where C_0 and C_1 are constants that correspond to the zero and unit of the overt response scale. Consider rows 1 and 2. In column j, the theoretical responses in rows 1 and 2, respectively, are

$$R_{1j} = C_0 + C_1 r_{1j} = C_0 + C_1(s_{A1} + s_{Bj}),$$
$$R_{2j} = C_0 + C_1 r_{2j} = C_0 + C_1(s_{A2} + s_{Bj}). \tag{4}$$

Upon subtraction, the common terms in C_0 and s_{Bj} cancel. Thus, the differences reduce to

$$R_{1j} - R_{2j} = C_1(s_{A1} - s_{A2}). \tag{5}$$

But the expression on the right of Eq. (5) is constant, the same for each column j. Algebraically, the entries in rows 1 and 2 have a constant difference in every column. Geometrically, that means that these two rows of data will plot as two parallel curves. The same holds for any other two rows and that proves the first conclusion of the theorem.

Comment. This parallelism theorem provides a remarkably simple and precise way to test the model. Observed parallelism supports the basic model; it also supports the other assumption of the theorem, namely, that the observed response is a linear scale. If either assumption is incorrect, then parallelism will not in general be obtained. There is, of course, a logical possibility that nonlinearity in the response scale just balances nonadditivity in the integration rule to yield net parallelism (Section 1.2.6). Subject to this qualification, however, observed parallelism supports both assumptions. Observed parallelism thus accomplishes three simultaneous goals.

1. It supports the adding model.
2. It supports the linearity of the response scale.
3. It provides linear scales of the stimulus variables.

1.2.3 STIMULUS SCALING AND MEASUREMENT THEORY

No stimulus values have been needed so far. The analysis rests entirely on the pattern in the observed responses. This approach may seem puzzling because an obvious way to test the model is to begin by trying to measure the stimulus values. Curiously enough, this direct scaling approach has many pitfalls (Anderson, in press).

In the functional measurement approach, the stimulus values are derivative from the model itself. That is the second conclusion of the parallelism theorem and it will now be proved. Let I be the number of rows so that the mean of the entries in column j, denoted by $\bar{R}_{.j}$, has the expression

$$\bar{R}_{.j} = (1/I) \sum_{i=1}^{I} R_{ij}.$$

From Eq. (3), the theoretical value of this row mean is

$$\bar{R}_{.j} = (1/I)\sum_i [C_0 + C_1(s_{Ai} + s_{Bj})]$$
$$= (1/I)\sum_i C_0 + C_1(1/I)\sum_i s_{Ai} + C_1(1/I)\sum_i s_{Bj}$$
$$= C_0 + C_1 \bar{s}_A + C_1 s_{Bj}, \tag{6}$$

where $\bar{s}_A$ is the mean of the s_{Ai}.

But $C_0 + C_1\bar{s}_A$ is a constant, say, C_0'. Hence Eq. (6) reduces to

$$\bar{R}_{.j} = C_0' + C_1 s_{Bj}. \tag{7}$$

Thus, the column means are a linear function of the column stimulus values, and conversely. The same holds for the row means, by symmetry, which proves the second conclusion of the theorem. Under the conditions of the theorem, therefore, the marginal means of the factorial data table provide linear scales of subjective value. Thus, the integration model provides a validational base for the stimulus scales.

1.2.4 RESPONSE SCALING AND MEASUREMENT THEORY

Two distinct problems of measurement are involved in testing the adding model. One is the problem of stimulus scaling that has just been discussed. The other, more fundamental, problem is that of response scaling.

The parallelism theorem, as stated above, includes the assumption that the overt response is a linear scale; that is, a linear function of the implicit response. Unless this assumption is valid, this simple form of the parallelism theorem is not applicable. It is a primary problem, therefore, to obtain a linear response measure. Two facets of this problem require discussion.

Numerical Ratings. The experiment of Figure 1.2 used a numerical rating response. The interpretation of these data in Section 1.2.1 depends on the assumption that the rating method yielded a linear response measure.

However, rating scales have been severely criticized by most workers in measurement theory. One criticism is that the ratings are merely English words, that it is unwarranted to use them as though they are true numbers. Thus, the psychological difference between 18 and 19, near the end of the scale, may be quite different from the psychological difference between 11 and 12, near the center of the scale. These criticisms of numerical response measures are serious and must be answered.

There is a simple answer. Part of the answer is found in the parallelism theorem itself. Unless the rating scale is a linear response measure, the data will not in general exhibit parallelism even though the model is true. Observed parallelism therefore provides simultaneous support for the model—and for the assumption of response linearity.

The other, more important, part of the answer is in the experiments. They exhibit parallelism, as shown in Figure 1.2 and in many other applications noted in Section 1.3. This empirical foundation provides the essential validation for the rating response.

That rating scales are subject to various biases deserves emphasis. The development of procedures to minimize bias and promote response linearity has been of central concern in this research program. This methodological problem is important because possession of a linear response scale greatly simplifies analysis and interpretation.

Nonlinear Response. Not all response measures will be linear. Behavioral measures, in contrast to judgmental responses, may often be nonlinear. Ratings themselves are generally linear only with certain precautions.

It is a fortunate fact that a linear response measure is not actually necessary. If the model is correct, then it can be used to transform a nonlinear observed response into a true linear scale. "The basic idea of the present method is to use the theoretical equations directly in the scaling procedure [Anderson, 1962b, p. 408]."

This functional measurement logic has been outlined in Section 1.1.6 and will be illustrated in the study of grayness bisection in Section 1.3.8. All that is necessary, therefore, is to have a valid integration function. That function constitutes the base and frame for measurement on both the stimulus and the response sides.

1.2.5 INDEPENDENCE ASSUMPTION

The parallelism theorem contains an important, implicit assumption that requires explicit discussion. This is the independence assumption that the scale value of each stimulus is constant, independent of what other stimuli it is combined with. This is a standard assumption in model analysis, and it is usually treated as part of the model itself. Properly speaking, however, it should be kept separate because the model operation could apply even though the stimuli interacted to change one another's values. There could be interaction effects in the valuation operation, in other words, even though the integration operation remained the same.

As a preliminary comment, it may be re-emphasized that independence is always relative to some task or dimension of judgment. In a person perception task like that of Figure 1.2, the trait *unsophisticated* would obviously have different importance and value for various occupations, say, for a diplomat and for an engineer. In general, the parameters of any stimulus will vary with the dimension of judgment (Section 1.1.3), an obvious condition that any theory must allow for. Once the dimension of judgment is fixed, however, then the independence assumption becomes applicable.

The role of the independence assumption can be seen in Eqs. (4). Suppose there was stimulus interaction so that the scale value of each column stimulus varied depending on which row stimulus it was combined with. Then s_{Bj} could really be different in the two equations and so would not cancel in the subtraction. Equation (5) would not follow and parallelism would not obtain. Thus, the independence assumption is essential for the parallelism test.

Conversely, observed parallelism supports the independence assumption. Since stimulus interaction would in general cause the response differences in Eq. (5) to vary across columns, the lack of such variation is evidence against interaction.

Independence will not always hold. One stimulus may qualify another stimulus or be inconsistent with it, thereby changing its meaning and scale value. Or an added stimulus may be largely redundant, with concomitant reduction in its weight parameter. These and other kinds of interactions would be expected to cause predictable deviations from parallelism, reflecting the specific relations among the stimuli.

In the experiment on person perception of Figure 1.2, the independence assumption states, for example, that the scale value of *unsophisticated* is the same in the *good-natured* and *unsophisticated* person as in the *humorless* and *unsophisticated* person. This assumption has been widely doubted on strong introspective grounds that seem to imply adjective interaction (Chapter 3). But if the adjectives did interact to change one another's meanings and values, then parallelism would not be expected. The observed parallelism argues against the interaction hypothesis, therefore, and for the hypothesis of meaning constancy.

In utility theory, on the other hand, the independence assumption has generally been accepted. It is normatively rational and seems intuitively reasonable, at least for a fairly general class of situations. Utility theory, as a consequence, has been dominated by the additive hypothesis. In the functional measurement tests, however, subadditivity has been frequent (Section 1.7.1).

As these two examples indicate, intuition is an unreliable guide to stimulus interaction. Interaction is important, but it involves difficult problems of analysis. One comment may help point up the usefulness of the independence assumption in the study of interaction. To say that two stimuli interact almost necessarily implies that their resultant is different from what it would otherwise have been; that is, had there been no interaction. But that implies a no-interaction reference standard that is not generally available. Interactive interpretations often rest on unwarranted or invalid assumptions about the no-interaction reference. Model analysis can play a useful conceptual role in this problem as illustrated in the discussion of implicit additivity in Section 1.6.6.

1.2.6 COMMENTS AND QUALIFICATIONS

Although the logic of the parallelism theorem is simple, various complications can arise in applying it. A few brief comments may be helpful here.

Inverse Inference. Empirical inference typically has an inverse direction, from consequence to premise. The logical form is: "If A, then B; B; therefore some support for A." This logic provides no certainty, and that is the basis for the commonplace observation that science never "proves" anything. Two aspects of inverse inference that relate to the parallelism theorem deserve mention.

Model–Scale Tradeoff. Both assumptions of the parallelism theorem (additivity and response linearity) are necessary. If either one is violated, then parallelism will not in general obtain. It is possible, however, that both assumptions are violated but trade off to produce net parallelism. That may not seem too likely, but it is always a logical possibility and it can be a real problem in certain situations. The case of the Height + Width rule in children's judgment of area of rectangles illustrates this problem (Section 1.3.6).

Alternative Models. A related question is whether some other model could also predict parallelism. In fact, the averaging model has a linear form and yields parallelism under the equal weighting condition. However, the averaging model is not properly an exception to the parallelism theorem, as will be seen in Section 1.6. As long as the response scale is linear, any alternative model must also have an adding form.

Adding Model with Weights. A more general model than Eq. (2), which also predicts parallelism, could be written as

$$r_{ij} = w_{Ai}s_{Ai} + w_{Bj}s_{Bj}, \qquad (8)$$

where w_{Ai} and w_{Bj} are weight or importance parameters. However, each weight parameter is confounded with the unit of its stimulus scale and so is not in general identifiable. This point deserves emphasis since many attempts to compare importance of different variables are not meaningful (Section 1.6.6).

Failure of Parallelism. When parallelism is obtained, interpretation is reasonably straightforward. When parallelism is not obtained, however, interpretation is difficult. The deviation from parallelism could have been produced by nonlinear biases in the response, by nonlinearity in the integration operation, or by violations of the independence assumption of no stimulus interaction. Indeed, all three factors could be jointly at work, a situation that would be difficult to sort out.

An instance arose in the negativity effect that first appeared as a deviation from parallelism in a study of person perception (Anderson, 1965a). One interpretation was in terms of an averaging model, with more extreme negative adjectives having greater weight. However, the deviation could also reflect a bias in the rating response, in which case it could hardly be considered a substantive phenomenon. Later work has supported the averaging interpretation, but at the time the interpretation had to be left open.

As this example suggests, no general rule can be given for the interpretation of deviations from parallelism. They raise the same problems as deviations from any other theoretical hypothesis, and any interpretation depends heavily on collateral information that may be available. In particular, collateral evidence to support response linearity can be valuable, even indispensable in certain situations.

Generality. Even when parallelism is obtained, its generality may be uncertain. It may be limited to the particular choice of stimulus materials, to details of task instructions, or to other circumstances peculiar to the given experiment. This uncertainty can become acute when an unexpected result is obtained (e.g., Section 1.3.6).

No experiment goes very far alone. Replication, both literal and generalized, is important. No less important is the value of an interconnected network of data and theory as a base for confidence (Section 1.8.2).

1.2.7 GOODNESS OF FIT

Response variability, which has so far been neglected, is a serious problem in model analysis. Even if the assumptions of the parallelism theorem hold, observed data will not be perfectly parallel. Accordingly, it is necessary to test goodness of fit; that is, to assess whether the observed deviations from parallelism may reasonably be attributed to prevailing response variability or to failure of the model assumptions.

Inspection of the factorial graph provides an informal but often useful test of goodness of fit. The key idea is that the factorial graph in any but the smallest designs can provide a visual index of prevailing response variability. To illustrate, glance at the left panel of Figure 1.4. The four solid curves not only exhibit near-parallelism, but also relatively small pointwise fluctuations from parallelism. With due caution, these pointwise fluctuations may be taken as an index of prevailing response variability. It hardly needs a formal statistical test, therefore, to recognize that the crossover of the dashed curve must be reliable.

Although the graphical test of parallelism is quite useful, some more objective test of fit is generally necessary. Fortunately, the ordinary analysis of variance provides a straightforward method. Parallelism is the graphical equivalent of zero interaction term in the analysis of variance. If the parallelism theorem applies, then this statistical interaction is zero in principle and should be nonsignificant in practice. Distribution-free tests may be used in the same way (Anderson, in press).

Most of the experiments discussed in this book will be presented by graphical analyses, and little statistical detail will be given. It should be emphasized, however, that this research program has been much concerned with developing and reporting adequate statistical analyses. Principles and methods are given in Anderson (in press), and detailed analyses can be found in the original experimental articles. The need for such analyses is illustrated in the next section.

1.2.8 WEAK INFERENCE

Much work on algebraic models of judgment has relied on statistical analysis that seems to test the models but does not really do so. This "weak inference" methodology has been especially prevalent in work with linear models. One example will be given here to illustrate the problem and to highlight the need for proper tests of goodness of fit.

The right panel of Figure 1.4 presents a standard scatterplot of predicted values obtained from a linear model, plotted as a function of the actual observed values on the horizontal. If the linear model fitted perfectly, and if response variability was zero, then the points would lie exactly on the diagonal line. Since the points cluster very close to the line, it would seem that the linear model fits very well. This visual impression agrees with the exceptionally high correlation of .993 between predicted and observed. Judged by these correlation–scatterplot statistics, the linear model seems to do extremely well.

But, in fact, the linear model does very ill. The actual data are plotted in the left panel of Figure 1.4 as a factorial graph. A linear model cannot account for these data because it requires parallelism. The four solid curves are nearly parallel, to be sure, but the dashed curve shows a crossover interaction. The linear model is thus in serious error. The factorial graph reveals the inadequacy of the linear model, and the

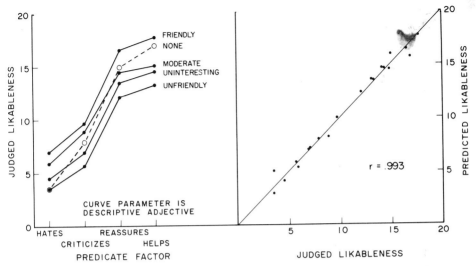

FIGURE 1.4. *Weak inference and strong inference tests of linear model. Right panel plots predictions from linear model on vertical axis as a function of the observed responses on the horizontal axis. Tight scatterplot and high correlation seem to give strong support to the linear model. But factorial plot of these same observed responses in left panel show marked crossover interaction, contrary to the linear model, which requires noncrossing curves. Thus, factorial plot provides strong inference test of linear model whose incorrectness is hidden by weak inference tests with correlation and scatterplot. (See also Sections 1.3.1 and 1.6.1.)*

analysis of variance showed that the nonparallelism was highly significant.

This example is one of many that tell the same story (see Notes 2.1.1b, 5.4.1b; Anderson, 1969c; Anderson & Shanteau, 1977). Correlation and scatterplot statistics obscure theoretical discrepancies and make bad models look good. These "weak inference" statistics are not adequate for model analysis.

Deviations from a model need not be fatal or even serious. They could reflect residual biases in the rating response, small stimulus interactions of no great importance, or the operation of some further process not included in the model. There is no routine recipe for deciding whether deviations are serious, but it is important to bring whatever deviations there are out into the open.

1.3 Experimental Studies of Adding-Type Models

Eight experimental applications of integration theory are discussed briefly in this section. They give empirical body and reality to the theoretical conception of the two previous sections. Each application is intended to highlight one or two issues, theoretical and methodological, that arise fairly generally in the integration-theoretical approach. More complete presentations are given in later chapters, but these summaries illustrate both the breadth of integration theory and the network of evidence on which it rests.

Although nearly all these models have been considered by previous workers, it seems fair to say that the present applications generally represent the first empirically adequate tests. Much previous work has relied on weak inference methods that are not adequate to test the models (Section 1.2.8). More appropriate statistical analyses have also been made, but using objective, rather than subjective, values of the stimuli. These tests have generally failed, as with the subjective probability models of Section 1.3.3, perhaps because the objective values were incorrect. The correctness of the model itself is thus left uncertain. The present applications became possible with the developments presented in Section 1.2.

These applications have often affirmed the correctness of previous conjectures but have even more often led to significant theoretical change. In the area of attitudes, for example, the popular adding model gave way to a more complex averaging formulation. For probability judgments, the formal model was supported, but the analysis led to a descriptive view that was conceptually quite different from the original normative view. And for judgments of area, the analysis revealed a model that was entirely unexpected but which appears to have a basic role in quantitative judgments.

1.3.1 PSYCHOLINGUISTICS

Person perception and psycholinguistics have much in common. Processing information about other persons often involves delicate webs of semantic inference and semantic integration. The experiment of Figure 1.2 on person perception can thus be viewed equally well as a study of semantic inference and integration. Indeed, the results had useful psycholinguistic implications for they supported the hypothesis of meaning constancy and also showed how meaning could be measured at the individual level. One further experimental example is noted here.

Adjective–Predicate Integration. Subjects received sentences of the form

The [adjective] *man* [verbs] *people.*

They judged how likable the man was expected to be. The experimental design can be seen in the left panel of Figure 1.4. Four verbs, listed on the horizontal, and four trait adjectives, listed by the four solid curves, were combined in a 4 × 4 factorial design to yield 16 sentences.

Each sentence contains two pieces of information about the man, the adjective trait, and the predicate action. The theoretical hypothesis was that this information would be integrated by an averaging rule. The near-parallelism of the four solid curves supports this averaging hypothesis (see also Figure 1.20).

This outcome was not surprising since adjective and predicate are informationally equivalent for the given judgment. The sentence task is thus similar to the personality adjective task studied in Figure 1.2. Other problems involving integration of information from connected discourse may perhaps be studied in similar fashion.

Semantic Inference. The integration model provides a useful capability for analyzing semantic inference. In the present sentence task, for example, judgments could be obtained on any of a myriad of other response dimensions, including honesty, impulsiveness, personal happiness, etc. Each such judgment dimension requires a semantic inference or valuation operation to arrive at the functional values of the given stimuli. Integration theory provides the first firm theoretical base for exact measurement of these semantic inferences. Moreover, it can operate at the level of the individual subject and the individual semantic unit, providing a powerful tool for the study of semantic relations and cognitive structure (Oden & Anderson, 1974).

Context Effects. Integration theory provides a straightforward attack on context effects by considering context stimuli to be integrated along with the focal stimulus. In some tasks, integration of context stimuli may be expected to obey the present averaging rule, and the previous comments on semantic inference would apply directly. Other tasks will require other models, such as Oden's (1978b) ratio model of semantic ambiguity. There is no restriction on the nature of the context stimuli; they may be linguistic cues such as intonation and inferences from background knowledge or nonlinguistic cues such as facial expressions and intentions. Establishing an integration model enables investigators to dissect and measure exact effects of subtle, intangible contextual cues.

1.3.2 ATTITUDE THEORY

Much behavior seems to be directed by attitudes; that is, by dispositions to react with characteristic judgments and toward characteristic goals across a variety of situations. Various people have characteristically different attitudes toward behaviorism and phenomenology, for example, and toward the men who symbolize these movements. Such attitudes and attitudinal judgments have many similarities with person perception and are expected to exhibit similar integration processes.

Attitudes toward United States Presidents. Subjects read biographical paragraphs about the lives and deeds of various United States presidents and judged them on statesmanship. These paragraphs were combined according to the 2 × 3 design represented in Figure 1.5. The two rows of the design represent positive or negative paragraphs; similarly, the three columns of the design correspond to paragraphs with values listed on the horizontal.

The main feature of these data is the parallelism of the curves. By the parallelism theorem, these data support an adding-type model for attitudes. Such parallelism could be produced by either a true adding process or by an averaging process, and the attitude literature has devoted some attention to deciding between them. Critical tests that eliminate the adding hypothesis and support the averaging hypothesis are given in Section 1.6 and in Chapter 2.

Cognitive Unitization. Attitudinal judgments typically rest on integration of large amounts of information. Even these short president paragraphs require complex information processing. Stated facts about slavery, for example, have to be understood in their historical context and evaluated on the given dimension of judgment. Inferences from these

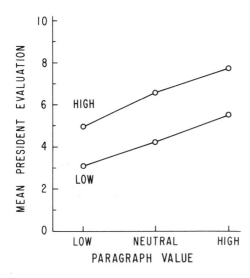

FIGURE 1.5. *Parallelism supports averaging rule for attitudes toward United States presidents. Subjects judge presidents on statesmanship, given four paragraphs of biographical information, one pair for the row factor, another pair for the column factor. (From Anderson, 1973c. [Copyright © 1973 by the American Psychological Association. Reprinted by permission.])*

facts will generally be needed, and the information within the paragraph may interact in various complex ways.

The integration model can treat this informational complex as a molar unit. No matter how interactive the paragraph processing, no matter how intricate the valuation and inference chain, the resultant is theoretically represented in the simple, molar w–s form (Section 1.6.5). With respect to the attitudinal judgment, therefore, the paragraph may theoretically be treated as a cognitive unit. The success of the parallelism prediction both depends on and supports this assumption of cognitive unitization.

Power. The stimulus materials used in this experiment illustrate an important problem of method. Each subject served in all experimental conditions, but the given information described a different United States president in each condition. This within-subject design was possible because of a collection of 220 paragraphs about 17 presidents that had been constructed for this research. The utility of this within-subject design may be indicated by a power comparison. To obtain the same statistical power for testing deviations from parallelism would have required 510 subjects in between-subject design instead of the 48 subjects used in the actual within-subject design.

1.3.3 SUBJECTIVE PROBABILITY

The uncertainty of life has commanded attention in all ages but especially in our own time as concepts of probability enter more deeply into common thought. Mathematical probability theory provides a normative or actuarial framework for probability calculation and yields simple algebraic models for probabilities of certain compound events. Various writers have speculated that human thinking would obey these same algebraic models.

Empirical tests generally showed discrepancies from the normative probability models, but their meaning was uncertain. Two possible sources of discrepancy had to do with measurement; the models would fail if the response scale was nonlinear or if the subjective probabilities differed from the objective values. More basic than these two measurement problems, the form of the psychological integration rule might differ from the normative form. The interpretation of any observed discrepancy was problematical, therefore, because it could result from any one cause or from all three acting jointly.

Nevertheless, it was still possible to hope that judgments of probability would exhibit the algebraic form of the normative models—if the two measurement problems could be handled. Functional measurement provided the needed measurement capability for the first adequate tests of a cognitive algebra of probability (Lopes, 1976b; Wyer, 1975a).

Test of a Probability Averaging Model. One experimental test is shown in Figure 1.6. The subject saw two urns, each with a specified proportion of red and white beads. One urn was picked with specified probability, and one bead was drawn from that urn. The subject estimated the probability that that bead would be white.

The indicated model from mathematical probability theory is

$$\text{Prob(White)} = \text{Prob(Urn A)Prob(White}|\text{Urn A)}$$
$$+ [1 - \text{Prob(Urn A)] Prob (White}|\text{Urn B)},$$

where Prob(White|Urn A) is the conditional probability of drawing a white bead, given that Urn A is chosen. The question is whether subjects' judgments will exhibit the same algebraic structure as the model. Thus, the experimental test must allow for subjective values of all the terms in this model.

Such a test is readily obtained. If Prob(Urn A) is held constant, then the model implies that Prob(White) is a weighted average of the two conditional probabilities. These two conditional probabilities can be independently varied by treating the red:white proportions of the two urns as factors in a two-way design. If the model is correct, the data should exhibit parallelism.

The success of the probability averaging model seems clear in Figure 1.6 for the three curves are very nearly parallel. Other parts of this same experiment gave some support to the subjective forms of the adding and multiplying rules for disjunctions and conjunctions of two independent events (see also Figure 1.23). A cognitive algebra of probability thus appears to have some validity, although how far this will extend to more complex tasks is unknown.

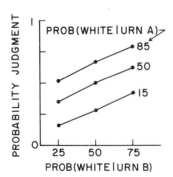

FIGURE 1.6. *Parallelism supports probability averaging rule for subjective probabilities; see also Figure 1.23. (From Anderson, 1975b.)*

Normative Theory and Descriptive Theory. When human judgment exhibits the form prescribed by a normative model, it is tempting to feel that there must be a relation. This view can be seen in the Bayesian movement in psychological decision theory and in the axiomatic approaches that begin with some set of rational assumptions about human judgment.

In the present view, these normative approaches are largely irrelevant to psychological theory. Cognitive algebra includes general purpose integration rules that are keyed in by aspects of task structure that have little or nothing to do with the normative concepts (see also Sections 1.5.2, 1.8.1). This point is well demonstrated by the two preceding and three following empirical applications. Although the rationalist approach can be undeniably attractive, it constitutes an inappropriate conceptual framework for cognitive analysis.

1.3.4 DECISION THEORY

Preference judgments are ubiquitous in daily life. Decisions about which car or house to buy, which research idea to pursue, how to spend the next vacation, all involve preference judgments among two or more alternatives. An obvious hypothesis is that preference between two alternatives obeys the subtracting rule

$$\text{Preference} = \text{Value}_1 - \text{Value}_2.$$

Numerical response data are of main concern here, in contrast to most work on preference, which has studied close decisions using probabilistic choice data. Although a few earlier statistical articles had considered the use of numerical response methods, the first extensive experimental study seems to be that described next.

Experimental Test. In this test of the subtracting rule for preference, subjects were presented with two food items and rated how much they preferred one over the other. Each subject served in all conditions of three 3 × 4 designs that are indicated in Figure 1.7. The left panel presents data for sandwich preferences, and the center panel presents data for beverage preferences. The observed parallelism supports the subtracting rule for sandwiches and beverages both. The same holds for preferences between sandwich–beverage lunches in the right panel.

Numerical preference judgments have several advantages over the customary choice data. They are more informative because they take account of degree of preference. They are not limited to probabilistic, threshold choices, but can be used to quantify suprathreshold preference. And they make practicable the use of within-subject design, al-

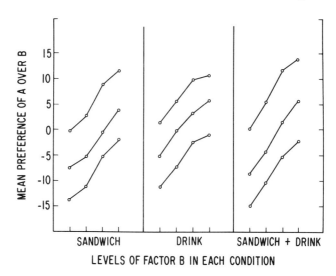

FIGURE 1.7. *Parallelism supports differencing model for preference judgments. Subjects rate preference of item A (row factor) over item B (column factor) for sandwiches, drinks, and sandwich-drink lunches. (After Shanteau & Anderson, 1969.)*

most a necessity for handling the wide individual differences that prevail in many preference studies.

Individual and Group Data. The treatment of individual differences in this experiment may be of interest. Since food preferences are idiosyncratic, the foods were preselected on rough graded categories for each individual subject. The main purpose of this preselection was to increase the power of the interaction test of the model, which depends on having substantial main effects. If each individual satisfies the parallelism property, then so will the mean group data. Even though the specific foods differ across individuals, the factorial plot of the mean data provides a valid test of parallelism.

Parallelism in the group data does not guarantee parallelism in the individual data, however, since idiosyncratic deviations could cancel in the group average. In this early experiment, the individual tests showed statistically significant discrepancies from parallelism for about one subject in four. Detailed scrutiny suggested that these discrepancies resulted in part from number preferences, but otherwise no meaningful pattern was found. Although these discrepancies did not seem to require any serious qualification of the model, this frequency of individual discrepancies is not satisfactory.

Difference Judgments. Despite the long popularity of the subtracting model for judgments of preference and difference, little is known about the validity of the model. The bulk of the work has relied on choice or rank order data that have not provided satisfactory tests of goodness of fit. The method of paired comparisons, for example, typically requires pooling choices across individuals with different preference orders (Section 5.3). Whatever meaning such tests may have, they do not speak to the essential question of individual behavior.

Among studies that have used numerical response measures, only a few, mainly in psychophysical judgment, have reported individual tests. These studies generally agree with the present experiment. The group test often works well, whereas the individuals frequently show moderate but significant deviations. Deviations from the model need not be serious. The high statistical power obtained with numerical response measures means that number preferences or other nuisance nonlinearity in the response could be significant. Again, the deviations might reflect real but minor idiosyncratic interactions among the stimuli. The deviations may be serious, however, and deserve more intensive study, both to help define the domain of model validity and to help improve experimental procedures.

1.3.5 DESERVINGNESS

Deservingness is a central concept in social life. On the positive side, deservingness underlies judgments about what is fair and constitutes one determinant of gratitude, reward, and reciprocity bargaining. On the negative side, deservingness ramifies into feelings of blame and resentment and actions toward getting even. These feelings and actions are very real and present a variety of interesting problems. This section presents one result from a developmental study of cognitive algebra.

Children's Judgments of Deservingness. Subjects were children in four age groups, from 4 to 8 years of age. They played Santa Claus, giving a "fair share" of toys to hypothetical children who were described by their need (how many toys they already had) and by their achievement (how many dishes they had washed for their mother). These variables formed a 2 × 3, Achievement × Need factorial design. The hypothesis was that achievement and need are added or averaged to determine deservingness.

$$\text{Deservingness} = \text{Achievement} + \text{Need}.$$

The mean fair share judgments, shown in Figure 1.8, have two features of interest. First, the slopes of the curves show that the children are

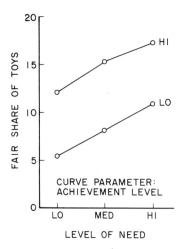

FIGURE 1.8. *Parallelism supports fair-shares model for judgments by subject children who play Santa Claus and distribute toys to story children, given information about how much work the recipients did and their level of need. (After Anderson & Butzin, 1978 [Experiment 2, Part 3].)*

sensitive to need, while the vertical elevations show the same for achievement. Second, the parallelism supports the hypothesis that these two kinds of information are integrated by an adding-type rule.

A surprising outcome of this experiment was the lack of any age trend. It had been expected that younger children would place less weight on the more subjective factor of need and perhaps also show deviations from parallelism. In fact, all four age groups showed essentially the same pattern of data. These results indicate, therefore, that complex and sensitive social–cognitive capacities appear at rather early ages.

Information Integration by Children. The integration-theoretical approach has many advantages for developmental analysis. On the technical side, the use of numerical response and factorial design provides far greater analytical power than can be obtained with the choice data characteristic of the Piagetian approach. Thus, the well-known Piagetian claim about "centration," that preoperational children are typically unable to integrate diverse pieces of information, turns out to be largely a methodological artifact.

On the conceptual side, the study of integration tasks is providing new insight on the development of cognition. Overall, these results reveal considerably greater and more varied cognitive abilities in young children than had been recognized in traditional views. Other developmental studies of cognitive algebra are given by Anderson (1980), Butzin (1978), Cuneo (1978, 1980, 1981), Leon (1976, 1980), Gupta (1978), Hommers (1980), Kun, Parsons, and Ruble (1974), Singh, Sidana, and

Saluja (1978), Singh, Sidana, and Srivastava (1978), Surber (1977), Verge and Bogartz (1978), and Wilkening (1979, 1980, 1981).

1.3.6 HEIGHT + WIDTH RULE IN CHILDREN'S JUDGMENTS OF AREA

Five-year-olds judge area of rectangles by an adding, Height + Width rule (Anderson & Cuneo, 1978a). This is puzzling because direct response to perceptual area should mirror the normative, physical rule— Area = Height × Width.

Experimental Study of Area Judgments. In a typical experiment in this series, children judged how happy a hungry little girl would be with a given amount of cookie to eat. The cookies were rectangular, with height and width varied in factorial design. Judgments were made on a graphical, face-happiness rating scale designed for use with young children.

The data for 5-year-olds from the initial experiment are shown in Figure 1.9. This pattern of parallelism points to the operation of a Height + Width integration rule. It is as though the children are adding height and width in arriving at their judgment.

The expected pattern was a linear fan, of course, corresponding to the

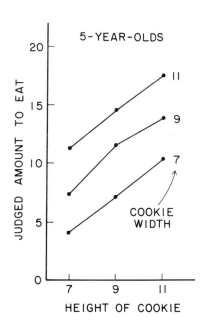

FIGURE 1.9. *Parallelism implies that 5-year-old children judge area of rectangle by Height + Width rule. (After Anderson & Cuneo, 1978a.)*

physical multiplying rule for area. Even if 5-year-olds were not very good at area perception, even if they had difficulty using the rating scale, some semblance of linear fan pattern should appear. Older age groups did show linear fans, but the data for 5-year-olds were neatly parallel.

The natural reaction to this parallelism is that something is wrong. However, a series of eight experiments showed that the parallelism was not due to nonlinearity in the rating response, nor to a law of diminishing returns for the happiness judgment, nor to lack of statistical power, nor to averaging artifacts, nor to specific stimulus conditions. Overall, the Height + Width rule appears to be both robust and valid (see also Anderson & Cuneo, 1978b; Bogartz, 1978).

General-Purpose Integration Rules. The Height + Width rule is thought to result from the operation of a general-purpose integration rule. Young children lack adult concepts for area and other quantities. They recognize, however, that a "how much" judgment is required and seize on cues that are meaningful and relevant to them. These cues are then integrated by a general-purpose adding-type rule.

If this interpretation is correct, then similar adding-type rules should appear in other tasks in which a multiplying rule would be expected on physical grounds. This implication has been supported by findings of a Length + Density rule for judged numerosity of a row of beads (Cuneo, 1978, 1981), a Diameter + Height rule for judgments of cone volume (Wilkening, 1980), and of a Subjective Probability + Utility rule for a roulette game (Anderson, 1980). The appearance of these adding rules as early as 3 years of age in Cuneo's work suggests that cognitive algebra reflects basic and natural modes of information integration.

1.3.7 PSYCHOPHYSICS

A curious psychological fact is that a pound of lead feels heavier than a pound of feathers. This is a strong contrast illusion and makes a sure-fire classroom demonstration. It is caused by the visual appearance; a larger object (the feathers) feels lighter than a smaller object (the lead) of the same gram weight.

Weight–Size Integration. The weight–size illusion presents interesting problems in stimulus integration. How are the two cues, physical weight and visual appearance, integrated into the unitary perception of heaviness? This question was answered by varying the two cues in the factorial design of Figure 1.10. The observed parallelism supports an adding-type rule. Symbolically,

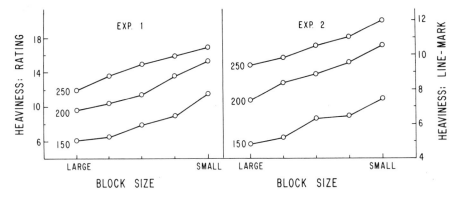

FIGURE 1.10. *Parallelism supports adding-type rule for weight–size illusion. Subjects lift and judge heaviness of cubical blocks in 3 × 5, Gram Weight × Block Size design. Gram weight is listed as curve parameter, with block size on horizontal axis. Verbal rating response is in left panel, graphic rating response in right panel. (After Anderson, 1970a.)*

Subjective Heaviness = Weight + Appearance,

where the two terms on the right represent the subjective values of the two stimulus cues.

This outcome may seem counterintuitive. In previous work, the most popular hypothesis had been that subjects were really judging density. However, the density hypothesis corresponds to the dividing rule, Weight ÷ Size. That would imply a linear fan pattern (Section 1.4), not the observed parallelism. Functional measurement thus provided a simple diagnosis of the operative integration rule.

The Psychophysical Law. There is a further benefit from this experiment: It yields the psychophysical law for weight. The parallelism theorem implies that the vertical elevations of the three curves provide a linear scale of subjective heaviness of gram weight. The distance between the bottom and middle curves is greater than the distance between the middle and top curves. This means that the first 50-gram increment adds more heaviness than the second, a law of diminishing returns that has been supported in later, more extensive work.

The special value of this approach is that it provides a validational base for the psychophysical function. By virtue of the parallelism theorem, observed parallelism provides validational support for the subjective heaviness values (Section 1.2). In this way, functional measurement provides a solution to the problem of the psychophysical law posed by Fechner more than a century ago (see also Chapter 5).

Stimulus Invariance. A followup experiment included both the weight–size task and a weight-averaging task in which subjects lifted two unseen weights and made an intuitive judgment of their average heaviness. The weight–size data again exhibited parallelism and a law of diminishing returns. The weight-averaging data also exhibited parallelism, which is hardly surprising, of course, because the instructions prescribe an averaging rule. The weight-averaging task is important, however, because it provides a second functional scale of heaviness, one that may be compared to the heaviness scale from the weight–size task. This comparison showed that both tasks yielded equivalent functional scales of heaviness (Anderson, 1974a, Figure 3; see also Anderson, 1974c). This cross-task stimulus invariance provides interlocking support for both models stronger than obtained from testing each one separately.

Moreover, stimulus invariance also carries important clues about information processing. The integration operation has a different locus in the two tasks, coming before the conscious heaviness sensation in the weight–size task but after the conscious heaviness sensations in the weight-averaging task. Presumably, therefore, the sensory representations of the physical weights have different cognitive loci and so they need not be equivalent. Their observed equivalence thus points to an interesting feature of the cognitive economy and illustrates the potential of integration tasks to help map out internal flow of information (Anderson, 1975a, pp. 477–478).

Ratings versus Magnitude Estimation. In the method of magnitude estimation, the subject is instructed to assign numbers to stimuli in proportion to their apparent magnitude (Section 5.4.1). This method does not constrain the response numbers to lie within a fixed range, as does the rating method. Otherwise, the two methods are rather similar, and subjects find them both natural and easy to use.

But these two methods yield quite different results. For example, magnitude estimation yields a psychophysical function for gram weight that is bowed upwards, exactly opposite to the law of diminishing returns implied by Figure 1.10. With a magnitude estimation response, moreover, parallelism is not obtained in the weight–size experiment. How is the sharp difference between these two plausible methods of obtaining numerical judgments to be adjudicated?

This issue can be resolved by functional measurement methodology. It is neutral in the controversy and provides a validational basis for discriminating between ratings and magnitude estimation (Anderson, 1970b, p. 166). Both have equal opportunity to satisfy the parallelism theorem. When ratings succeed and magnitude estimation fails, the presumption is that ratings are valid and magnitude estimation is invalid.

One experiment may not go very far, but this one experiment is buttressed by many others. Some of these obtain equivalent psychophysical functions from other integration tasks, cross-task validation on the stimulus side. Numerous others have successfully used the rating method in studies of algebraic integration models in a variety of areas, cross-task validation on the response side. Since magnitude estimation does not seem to possess any comparable validational base, it seems appropriate to conclude that it is indeed biased and invalid (Section 5.4).

1.3.8 MONOTONE ANALYSIS AND THE PROBLEM OF BISECTION

The century-old problem of bisection has considerable interest in psychophysical judgment and also provides a good illustration of functional measurement with monotone response measures. This approach makes it possible to dispense with the special assumptions used by previous workers while providing substantially more powerful tests of the basic model (Anderson, 1977a; Carterette & Anderson, 1979; Weiss, 1975).

Bisection Model. The present experiment studied bisection of grayness. The subject was presented with two stimulus chips, each varied in grayness from black to white in a 5 × 6, Left Chip × Right Chip design. The subject's task was to select a third, "response" chip that lay midway in grayness between the two stimulus chips.

The standard hypothesis is that the subject is equating the two sense distances

$$s_1 - r = r - s_2, \tag{9}$$

where s_1 and s_2 are the subjective grayness values of the two stimulus chips, and r is the subjective grayness of the response chip. Solving for r yields the average

$$r = \tfrac{1}{2}(s_1 + s_2). \tag{10}$$

The great attraction of this bisection task is that it deals directly with perceptions; no arbitrary verbal response measure intrudes.

But this bisection model has long resisted analysis precisely because it deals solely in nonobservables. All three terms in Eq. (10), including the response term, r, are subjective. That is as it should be, for the bisection occurs in the subjective metric. The investigator, however, must make

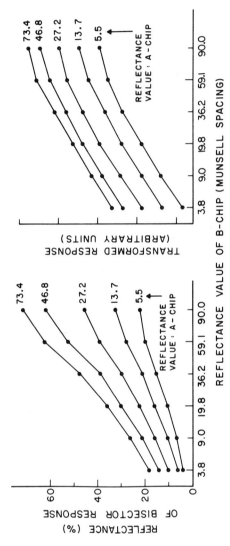

FIGURE 1.11. *Monotone analysis provides parallelism test of bisection model in psychophysics. Subject chooses response chip to lie midway in grayness between left and right stimulus chips. Left panel plots mean physical reflectance value of response chip; right panel plots mean response value after monotone transformation. (After Anderson, 1976c.)*

do with the observable response, R, measured on the physical reflectance scale.

Physical reflectance certainly does not obey the parallelism theorem: that can be seen in the left panel of Figure 1.11. That is as expected, of course, for physical measures are not generally linear scales of subjective value. The observable response data, therefore, do not speak directly to the question of whether bisection involves an equating of sense distances or some other process.

Nevertheless, there is a way to use the observed data to attain the desired end following the logic given in Section 1.1.7. Since subjective grayness is a monotone function of physical reflectance, some monotone transformation of R will yield r. If the bisection model is correct, then the desired monotone transformation can be determined, for it is the one that will make the data parallel. Such a transformation was obtained, and its effect is shown in the right panel of Figure 1.11. This parallelism supports the bisection model.

Technical problems of monotone analysis are considered in Anderson (in press, Chapter 5), but one of them deserves mention here. A method that can make parallel the curves in the left panel of Figure 1.11 clearly has great potency. Perhaps it is too potent. Perhaps any set of curves that did not actually cross over could be made parallel. Fortunately, that is not the case and, indeed, the bisection model actually failed for the case of length (Anderson, 1977a). The use of factorial-type design can provide sufficient constraints to determine the best possible monotone transformation and still leave degrees of freedom for the necessary test of goodness of fit. Nevertheless, this problem of monotone analysis presents severe practical difficulties about which little is yet known.

Psychophysical Function. Success of the bisection model makes available a validated subjective scale of grayness. By the parallelism theorem, this scale is obtained from the marginal means of the factorial design. The plot of these psychological values against the physical reflectance values then defines the psychophysical law or function.

This function is shown in Figure 1.12. Separate functions for the six subjects were fairly similar though a little more variable. The curve is a power function fit and has an exponent of .14, somewhat more convex than the standard Munsell scale of gray, which yields an exponent of .33. This difference can also be seen in the unequal vertical spacing of the curves in the right panel of Figure 1.11. In terms of this bisection scale, equal steps on the Munsell scale are too close together at the white end of the scale.

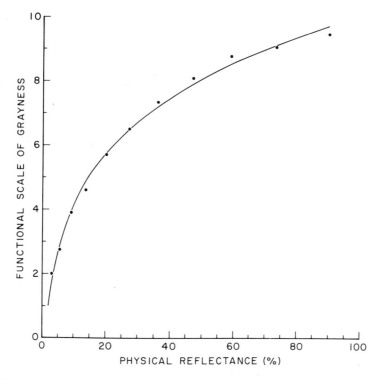

FIGURE 1.12. *Psychophysical law for grayness. Functional measurement scale of grayness obtained from bisection task is plotted against physical reflectance. (After Anderson, 1976c.)*

1.4 Multiplying Models

Multiplying models are hardly less frequent than adding-type models in cognitive algebra, and a variety of experimental studies are shown in the following section. Here are presented the elements of their analysis, which should suffice for understanding the empirical applications in the later chapers.

1.4.1 THE LINEAR FAN IDEA

The essential idea for analysis of multiplying models rests on a linear fan pattern. The model

$$R = bX$$

can be considered as a straight-line function of X with slope b. If the X values are on the horizontal, then each value of b defines a straight line through the origin. Several values of b would yield a radiating fan of straight lines. This linear fan pattern is the characteristic sign of a multiplying model.

Experimental application of this linear fan idea would begin with a two-way factorial design, just as in Figure 1.2. The subject would receive stimulus combinations (S_{Ai}, S_{Bj}) and make a numerical response (R_{ij}) to the combination. Suppose that this response can be represented as the product of subjective values of row and column stimuli:

$$R_{ij} = s_{Ai}s_{Bj}. \tag{11}$$

The two preceding equations, despite the different notation, are formally the same: b corresponds to s_{Ai}, and X corresponds to s_{Bj}. Hence R_{ij} is a straight-line function of s_{Bj} with slope s_{Ai}.

This last observation has a simple yet vital consequence. If the column stimuli S_{Bj} are spaced on the horizontal at their subjective values s_{Bj}, then each row of data from the design will form a straight line. No more would be needed, therefore, than to plot the factorial graph of the response data and look to see whether it did exhibit a linear fan pattern.

One seemingly absolute barrier blocks this simple linear fan analysis. The subjective values s_{Bj} are in general unknown; without them, the way is blocked. It is this barrier that has held back the study of multiplying models in psychology.

But there is a way to the goal. Under the given assumptions, the column means of the factorial design will provide the needed subjective values, s_{Bj}. If the column stimuli are spaced on the horizontal at the locations specified by the column means, then the several rows of response data will form a radiating fan of straight lines. If the assumptions are not met, of course, then the linear fan pattern will not in general hold. This procedure thus provides a straightforward test for multiplying models.

1.4.2 LINEAR FAN ANALYSIS

The linear fan analysis is fairly similar to the parallelism analysis of Section 1.2.2. The same notation will be used so that the implicit response in cell ij of the design is

$$r_{ij} = s_{Ai}s_{Bj}. \tag{12}$$

Linear Fan Theorem. *If the multiplying model of Eq. (12) is true, and if the observable response is a linear scale, then (a) the appropriate factorial plot of the data will form a fan of straight lines, and (b) the row means of the factorial design will be estimates of the subjective values of the row stimuli on linear scales, and similarly for the column means.*

Proof of Linear Fan Property. The assumption that the observable response is a linear scale means that $R_{ij} = C_0 + C_1 r_{ij}$. From Eq. (12), the overt response can therefore be written as

$$R_{ij} = C_0 + C_1 s_{Ai} s_{Bj}. \tag{13}$$

The entries in row i are thus a linear function of the s_{Bj}. If the column stimuli are spaced on the horizontal according to their s_{Bj} values, then the entries in row i will form a straight line with slope equal to $C_1 s_{Ai}$. Hence the several rows will form a linear fan.

All that is needed, therefore, is to obtain the s_{Bj}. But a linear function of these values will suffice, and that is available by virtue of the second conclusion of the theorem as will be proved next.

Derivation of Stimulus Scales. Let I be the number of rows so that the mean of the I entries in column j, $\bar{R}_{.j}$, has the expression

$$\bar{R}_{.j} = (1/I) \sum_{i=1}^{I} R_{ij}$$

$$= (1/I) \sum_{i} [C_0 + C_1 r_{ij}]$$

$$= (1/I) \sum_{i} [C_0 + C_1 s_{Ai} s_{Bj}]$$

$$= (1/I) \sum_{i} C_0 + C_1 s_{Bj} (1/I) \sum_{i} s_{Ai}$$

$$= C_0 + C_1 s_{Bj} \bar{s}_A, \tag{14}$$

where $\bar{s}_A$ is the mean of the s_{Ai}.

But $C_1 \bar{s}_A$ is a constant, say, C_1'. Hence Eq. (14) reduces to the linear form

$$\bar{R}_{.j} = C_0 + C_1' s_{Bj}. \tag{15}$$

Thus, the column means are a linear function of the column stimulus values, and conversely. By symmetry, the same holds for the row means, which completes the proof.

Comment. The linear fan theorem provides a remarkably simple and precise way to test the multiplying model. Hardly more is necessary than to run the experiment and plot the data. An observed linear fan supports the basic model. It also supports the auxiliary assumption that the response scale is linear. If either assumption is incorrect, then the linear fan pattern will not in general be obtained. Just as with parallelism, therefore, an observed linear fan accomplishes three simultaneous goals:

1. It supports the multiplying model.
2. It supports the linearity of the response scale.
3. It provides linear scales of the stimulus variables.

1.4.3 PROPERTIES OF THE LINEAR FAN ANALYSIS

Stimulus Measurement. The linear fan analysis employs the column (or row) means of the data as stimulus values. In practice, of course, these means are only estimates and, indeed, only provisional estimates because their validity depends on the model. If the model fails the statistical test of goodness of fit, that vitiates these provisional estimates. On the other hand, success of the model confers validity on these provisional estimates. In this way, the model provides a validational base for stimulus measurement.

Response Measurement. The linear fan theorem includes the assumption that the observable response is a linear scale. The discussion of response scaling for the adding-type models (Section 1.2.4) applies almost without change to the multiplying model. In particular, the success of the linear fan analysis in the experimental applications of Figures 1.13–1.18 provides strong validational support for the rating method as a true linear scale. Conversely, the development of experimental procedures to obtain linear rating scales was the key to a successful attack on the multiplying models.

Independence Assumption. Implicit in Eq. (13) is the independence assumption that the value of each stimulus is constant, independent of what other stimulus it is combined with. This assumption serves an equivalent role to the independence assumption for the adding model (Section 1.2.5). If independence did not hold, if the value of s_{Bj} in Eq. (13) was not constant but varied across rows, then Eq. (13) would not in general define a set of straight lines, and the linear fan property would fail. Conversely, of course, an observed linear fan pattern supports the independence assumption.

Testing Goodness of Fit. The graphical form of the linear fan test is straightforward, as already shown. In effect, this method of plotting constrains the column means to lie on a straight line. The model then requires that each separate row curve also be a straight line. This is a stringent test, for even one deviant point will in general cause all the separate row curves to be crooked.

This graphical test will usually need to be supplemented by a proper statistical test. The analysis of variance provides a special test that can be used for just this purpose. The regular Row × Column interaction term is split into two components, the linear × linear and the residual. The linear × linear component represents the linear fan pattern; the residual represents deviations from the linear fan. Thus, a complete test requires a significant linear × linear component and a nonsignificant residual. Details and complications are given in Anderson (in press).

1.4.4 COMMENTS AND QUALIFICATIONS

Most of the comments and qualifications on the adding model of Section 1.2 transpose directly to other algebraic models, including the multiplying model. Accordingly, only a few brief comments will be given here.

Generalized Adding–Multiplying Model. It is important to recognize that the linear fan analysis does not discriminate between the simple multiplying model of Eq. (12) and the following, more general, model with additive factors and weights.

$$r_{ij} = w_0 + w_A s_{Ai} + w_B s_{Bj} + w_{AB} s_{Ai} s_{Bj}. \tag{16}$$

This is a considerable advantage for it enables the linear fan test to proceed without knowledge of the zeros of the stimulus scales. However, it leaves some indefiniteness about the exact form of the model.

There are situations in which the distinction between Eqs. (12) and (16) is psychologically important. In general, therefore, it should be kept in mind that the success of the linear fan test, as in the examples of the next section, does not of itself preclude separate additive components. To resolve this question requires collateral information about the zeros of the stimulus scales.

Dividing Model. The dividing model

$$R_{ij} = C_0 + s_{Ai}/s_{Bj}$$
$$= C_0 + s_{Ai} s_{Bj}^{-1},$$

also obeys the linear fan analysis. The row and column means provide linear scales of s_{Ai} and $1/s_{Bj}$, respectively. Of course, the reciprocal column means provide a linear scale of s_{Bj} only if $C_0 = 0$.

Averaging Model. Under certain conditions, the averaging model and the multiplying model produce similar factorial graphs. With unequal weighting within the row (or column) stimuli, the averaging model produces nonparallel curves (Section 1.6). If the weights are a monotone function of scale value, then the curves can form a diverging fan that is very nearly linear. In practice, therefore, it may be difficult to discriminate between averaging and multiplying models.

One instance of this arose in the experiment on Performance = Motivation × Ability of Figure 1.18. A conjunctive averaging model has perhaps as much plausibility as the multiplying model and would fit the data about as well (Anderson & Butzin, 1974, p. 602). The operative rule may depend on culture, or task instructions, or both. Singh, Gupta, and Dalal (1979) found an averaging rule in India, whereas unpublished work by Anderson (1978) found reasonably clear evidence for the multiplying rule.

Model–Scale Tradeoff. As with the linear model, there is a model–scale tradeoff for the multiplying model. For example, a log-type transformation on the response will transform a multiplying model into an adding model as long as both variables are positive. This very problem arose in the interpretation of the Height + Width rule in 5-year-old's judgments of area. The parallelism in Figure 1.9 was sufficiently surprising that the possibility of a log-response transformation had to be taken seriously. This possibility was ruled out in part on grounds of response generality (Section 1.8.2), because this age group followed linear or averaging models in other judgment tasks (e.g., Figure 1.8) for which a multiplying integration rule was not reasonable. More direct evidence for linearity of the rating response was obtained by using a two-operation model (Section 1.7.2).

1.5 Experimental Studies of Multiplying Models

The seven studies of this section illustrate the potential of multiplying-type models in psychological theory. All but one of these models has been considered by previous workers but, as with the adding-type models, satisfactory tests were not previously available owing to lack of a capability for measuring subjective values. One happy outcome of the present research program has been the substantial empirical support for the multiplying model. This work did not merely affirm the conjectures of previous workers, however, for in every case it led to significant conceptual changes. The first two examples illustrate this point in the contrast between normative theory and descriptive theory in decision making.

1.5.1 SUBJECTIVE EXPECTED VALUE

Life is a succession of chances. Some people enjoy fool's luck; others see good risks turn sour. But although chance limits our control of our lives, even chance can be handled wisely. The happiness we may expect along life's path depends on how wisely we place our bets.

Subjective Expected Value Model. The probabilistic nature of our world is recognized in the law of expected value from mathematical decision theory. In terms of objective probabilities and values

$$\text{Expected Value} = \text{Probability} \times \text{Value}, \qquad (17a)$$
$$\text{EV} = \text{P} \times \text{V}.$$

It is a natural speculation that the mind obeys a similar law of expected value but with objective probabilities and values replaced by their subjective counterparts. By analogy, this law of subjective expected value would read

$$\text{Subjective Expected Value} = \text{Subjective Probability} \times \text{Subjective Value},$$
$$\text{SEV} = \text{SP} \times \text{SV}. \qquad (17b)$$

The validity of this SEV law was a central question in modern decision theory but little progress was made, owing to lack of methods for measuring subjective probability and value. This problem was resolved by the linear fan theorem, and more important, by the associated experiments that demonstrated the empirical validity of the SEV law.

Experimental Test. Subjects made intuitive judgments of personal value of lottery tickets of the form

You have _____ chance to win _____.

Chance to win was defined in terms of the roll of a die; the amount to be won was defined by coins. These stimuli were varied in factorial design, as shown in Figure 1.13, in which the probability stimuli are spaced on the horizontal axis at values equal to the corresponding column means. The data curves clearly exhibit the linear fan pattern, in support of the multiplying model. Fundamental work by Shanteau (1970b, 1974) has amply verified this result.

The unequal spacing of the objective probability stimuli on the horizontal axis of Figure 1.13 deserves attention. This spacing is according to subjective value, as prescribed by the linear fan theorem, and means that subjective and objective probabilities are nonlinearly related. The test of the SEV model was only feasible, therefore, because functional measurement could allow for subjective values.

Descriptive Theory and Normative Theory. The similarity between the EV and SEV models of Eqs. (17a,b) is only analogical. Epistemologically,

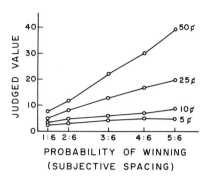

FIGURE 1.13. *Linear fan pattern supports multiplying rule for subjective expected value. Subjects judge personal value of lottery tickets, given specified chance (horizontal axis) to win specified money (curve parameter). (After Anderson & Shanteau, 1970.)*

the two are entirely different. The EV model is a normative model from statistical theory that prescribes a rational mode of behavior. The SEV model is a descriptive model that is embedded in empirical theory. Although the SEV model was suggested by the analogy, it is inadequate to think that it is the same as the EV model except for the use of subjective values. The ultimate basis for the SEV model lies in psychological integration processes that have no relation to normative theory (Sections 1.3.3, 1.8.1).

This epistemological difference is easy to demonstrate. It is only necessary to exhibit situations in which the subject continues to use SEV-type models when normative theory prescribes otherwise. One such instance was shown in Shanteau and Anderson (1972, Figure 2; see Anderson, 1974d, Figure 13). Another striking instance is noted in the next section.

1.5.2 STUD POKER: MULTIPLICATION OF PROBABILITIES

Mathematical theory yields simple addition and multiplication rules for probabilities of compound events, and many writers have speculated that human judgment obeys subjective versions of these same models. Lopes (1974, 1976b) applied integration theory to obtain perhaps the first rigorous test of such a probability model.

Experimental Test. Long-term subjects played in a simplified, computerized version of five-card stud. On each trial, the computer showed the subject the four upcards and the bet of each of two opponents: A, who maintained a conservative playing style; and B, who maintained an average playing style. The subject rated his probability of winning, or made an even-money side-bet with the computer, between 1¢ and 30¢, that he could beat both opponents, or made both responses. The computer would then turn over the hole card of each opponent, determine the winner, and pay off or collect from the subject.

The theoretical hypothesis was the subjective analog of the multiplication rule for independent probabilities:

$$\begin{aligned}
&\text{Subjective Probability (beating both A and B)}\\
&= \text{Subjective Probability (beating A)}\\
&\quad \times \text{Subjective Probability (beating B).}
\end{aligned} \tag{18}$$

To test this model, the two terms on the right were varied by seminaturalistic control of the strengths of A's and B's hands as given by their upcards and bets.

Figure 1.14 plots the rated subjective probabilities of winning the side bet as a function of hand strength of the two opponents. The three curves represent the three levels of A's hand strength, and the values on the horizontal represent the levels of B's hand strength. The essential feature of these data is the linear fan pattern. That pattern supports the probability multiplication rule and was further verified in the analyses for individual subjects.

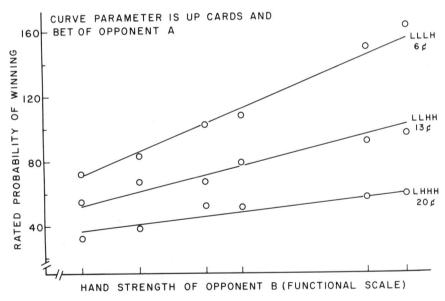

FIGURE 1.14. *Linear fan pattern supports multiplying rule for independent probabilities in stud poker. Subjects rate probability that their hand will beat two opponents, A and B, whose hands are varied in strength as implied by up cards and amount bet. (After Lopes, 1976b.)*

Irrational Betting. A striking outcome in Lopes's further work was that the size of the side bet was proportional to the rated subjective probability of winning. Thus, the amount bet also showed the linear fan pattern. To bet in proportion to confidence seems intuitively reasonable to many people, but a little reflection shows that it is irrational. If subjective probability of winning is less than .5, then the even-money side-bet is subjectively unfair and the subject should bet the minimum. Similarly, if subjective probability is greater than .5, then the subject should bet the maximum (of 30$^¢$). By betting in proportion to confidence, a player can lose as much as 12% for a uniform distribution of bets, even if the subjective probabilities are veridical.

This result of Lopes has fundamental ramifications. It illustrates that behavior may follow verbal judgment even though it is suboptimal. It suggests the pervasive importance of algebraic models in the cognitive economy. And it demonstrates the essential inadequacy of the normative approach used in classical utility theory.

1.5.3 FUZZY LOGIC

A central problem in semantic theory is that of representing truth value as a continuous quantity. In traditional, two-valued propositional logic, class membership is all or none. In common thought, however, class membership is often a matter of degree. Animal categories, for example, do not have sharp boundaries. It is somewhat true that a bat is a bird, yet not totally true that a penguin is a bird.

Fundamental work by Oden (1974, 1977a,b, 1978b) has shown how the conceptual structure of integration theory is ideally suited to represent fuzzy semantics and logic. The theory provides a natural representation for degree of truth together with a cogent methodology for experimental analysis.

Experimental Test. In one of Oden's studies, subjects responded to questions of the form, "How true is it that a sparrow is a bird, or a penguin is a bird?" The truth value of the first component is high, that of the second medium. The question is how these two truth values are integrated to determine the truth value of the compound statement. One suggested rule has a multiplying form that can be most simply stated as follows: The falsity value of the compound equals the product of the falsity values of the components.

This multiplying rule is supported by the results shown in Figure 1.15. Various winged creatures were used to vary the truth values of the two component statements according to the indicated 3 × 3 factorial design. The predicted linear fan pattern is clear. Other work by Oden

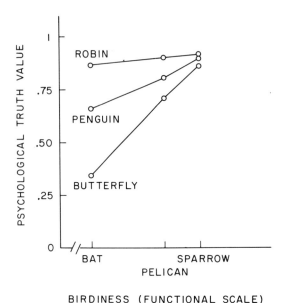

FIGURE 1.15. *Linear fan pattern supports multiplying rule for fuzzy logic. (After Oden, 1977b.)*

has provided further support for a cognitive algebra of fuzzy logic. Fuzzy logic appears to follow the same laws that are found in other areas of judgment theory.

Two Implications for Psycholinguistics. Two features of this integration-theoretical approach are important for psycholinguistics. First, semantic constraints need to be treated as continuous variables. Linguistic semantics typically employs the convenient fiction that semantic variables obey a two-valued logic. Similar all-or-none conceptions are found in most current theories of semantic memory that are primarily propositional. It is generally recognized that an adequate theory must allow for continuous semantic variables, but little has been done in this direction. Integration theory provides a natural representation for continuous semantic variables, and Oden's work points to cognitive algebra as a cornerstone for semantic theory.

Second, psycholinguistics must be able to analyze at the individual level, both for the unit proposition and for the single subject. Averages over groups of subjects or over groups of propositions can be useful, but they are often inadequate and misleading for experimental or theoretical analysis. As Oden's work has demonstrated, information integration theory provides a useful approach to individual linguistic analysis.

1.5.4 ADVERB × ADJECTIVE

Quantifier words have a natural representation as scalar multipliers. Previous workers have accordingly suggested a multiplying rule for the integration of quantifier adverbs with adjectives.

Functional Measurement Test. Subjects judged likableness of persons described by phrases such as *somewhat honest, very honest,* and *extremely dishonest.* These judgments are shown in Figure 1.16 with the adverbs spaced on the horizontal axis as prescribed by the linear fan theorem. The linear fan shape provides seemingly good support for the hypothesized model. The clarity and precision of these data illustrate the advantage of functional measurement analysis compared with other measurement theories that have been applied to this problem (Section 5.3).

"As-if" Multiplication. Closer conceptual analysis of the Adverb × Adjective task suggests that subjects do not actually multiply, but that

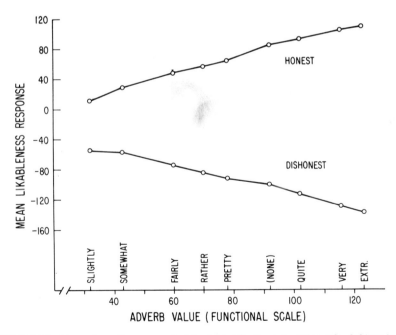

FIGURE 1.16. *Linear fan pattern supports Adverb × Adjective integration rule. Subjects judge likableness of person described by adverb-adjective phrases from the factorial design indicated in the figure. (After Anderson, 1974b. [Copyright © 1974 by Academic Press. Reprinted by permission.])*

the model form reflects different processing operations. In this view, the task is properly a valuation operation. Each adjective is assumed to define its own dimension of judgment, and the quantifier adverbs are assumed to lie at the same proportionate locations along different adjective dimensions.

However, a further step in valuation is required to get from the adjective dimension to the dimension of judgment, in this case, personal likableness. This relation is assumed to be linear so that a constant increment on the adjective dimension will yield a constant increment on the judgment dimension. These two assumptions, of proportionate location and of linearity, imply a linear fan form (Anderson, 1974b,d).

In this view, therefore, there is no actual multiplication. The linear fan results from quite different operations. Aside from its substantive interest, this example illustrates the distinction between form and process in cognitive algebra.

1.5.5 MOTIVATION × EXPECTANCY × VALENCE

Multiplying-type models have long been popular with writers on learning and motivation. This mirrors a dichotomy between concepts such as motivation and habit, which are considered characteristics of the organism, and concepts such as reward and punishment, which are associated with environmental goals. In this view, motivation is seen as an energizing factor in goal-directed behavior. To allow for probabilistic outcome, an expectancy factor is often introduced. Thus, the three-factor MEV model becomes

$$\text{Action} = \text{Motivation} \times \text{Expectancy} \times \text{Valence}, \tag{19}$$

where Valence represents the subjective incentive value of the goal.

Almost nothing is known about the validity of this model. The occasional attempts to provide a serious test have not been satisfactory owing to lack of a capability for psychological measurement (Anderson, 1978a). However, a partial test, of the Expectancy × Valence hypothesis, was obtained in Klitzner's study of snake phobics.

Snake Avoidance. Subjects were selected to have moderate to strong fear of snakes on the basis of a standard test. On each trial of the experiment, they could purchase chances to avoid a snake stimulus by inserting coins in a marble dispenser. Winning a special gold marble would avoid the snake stimulus. Probability of winning the gold marble and aversiveness of the stimulus were varied in factorial design (Klitzner, 1977).

Figure 1.17 gives the factorial graph of the number of chances pur-

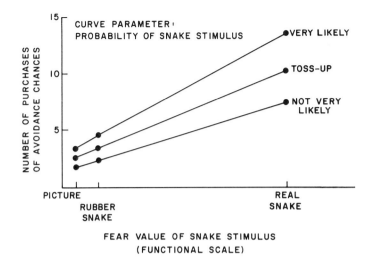

FIGURE 1.17. *Linear fan pattern supports Motivation × Expectancy integration rule for snake-phobic subjects who pay money (vertical axis) to avoid exposure to a motivational fear stimulus (horizontal axis) with specified probability (curve parameter). (After Klitzner, 1977.)*

chased under the various stimulus conditions. The linear fan shape is clear. These data thus provide good support for the Expectancy × Valence hypothesis.

An interesting suggestion from this integration-theoretical approach is that it may be possible to treat the motivation term in Eq. (19) as an individual difference variable. Subjects may be considered as a third factor in the design and, formally, the Subject × Expectancy graph, for example, should also exhibit a linear fan form. In this case, the two factors are not independent, of course, since the expectancies are themselves in the subjects' heads. However, the linear fan prediction will hold as long as expectancy values are proportionate across subjects. Although this analysis had mixed success, it does suggest an interesting line of inquiry.

Behavioral Algebra. Serious evidence on the MEV models has been largely nonexistent. The continued popularity of these models lies in their intuitive appeal rather than in experimental data. The concepts and structure of the models originate from and are still very close to common language concepts. Although ostensibly models of behavior, they seem rather to be judgmental conjectures of the kind considered in the next section (see also Klitzner & Anderson, 1977).

An important property of the snake avoidance experiment is that the response measure, number of plays on the marble dispenser, is behavioral. These results, like those of Lopes in Figure 1.14, support the hypothesis of an algebra of behavior.

1.5.6 PERFORMANCE = MOTIVATION × ABILITY

Two major determinants of achievement, at least in the conceptualization of common language, are motivation and ability. Attempts to predict some person's performance, as when admitting a graduate student or hiring a research assistant, will reflect some integration of these two kinds of information. A multiplying rule follows naturally from the longstanding conception that motivation is an energizer of ability.

Predictions about behavior are of course quite distinct from the behavior itself. The concepts of motivation and ability may be valid in one realm but not in the other. If they are valid, then the multiplying rule may hold in one realm but not in the other. The following experiment provides the first quantitative test in the judgmental realm.

Experimental Test. Subjects received information about the motivation and ability of persons trying out for college track and judged their expected performance. The two stimulus variables were varied in a 4 × 4 design, as indicated in Figure 1.18. The linear fan shape supports the multiplying model

$$\text{Performance} = \text{Motivation} \times \text{Ability}.$$

Although the amount of fanning may appear small, it is actually substantial. The easiest assessment is made by measuring the vertical spread between top and bottom curves; that increases about 60% over

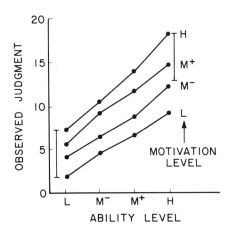

FIGURE 1.18. *Linear fan pattern supports prediction rule: Performance = Motivation × Ability. Subjects judge expected performance of tryouts for college track, given information about levels of motivation and ability. (After Anderson & Butzin, 1974.)*

the given range of ability, as indicated by the equal-length vertical bars. Theoretically, the curves will intersect at zero ability, which would be about six steps leftward, by rough extrapolation. The given range of ability thus constitutes a third of the total range, a proportion that seems not unreasonable in this college track scenario.

Cognitive Algebra Is Not Mathematical Algebra. This experiment had another part, in which the subject received information about ability and performance and made judgments about motivation. This is a common judgment in university life, where disproportion between ability and performance is not infrequent. From the above equation, it follows mathematically that Motivation = Performance ÷ Ability.

The actual judgments, however, exhibited parallelism and so demonstrated the operation of a subtracting rule. One interpretation is that a dividing rule is difficult and subjects tend to slip into an easier, subtracting, mode. In any case, the complete set of judgmental equations was

$$\text{Performance} = \text{Motivation} \times \text{Ability}$$
$$\text{Motivation} = \text{Performance} - \text{Ability}$$
$$\text{Ability} = \text{Performance} - \text{Motivation}.$$

Essentially this same cognitive algebra was obtained in a total of four experiments (Anderson & Butzin, 1974; Graesser & Anderson, 1974). These judgments do seem to obey a simple cognitive algebra, but this is not a mirror of mathematical algebra.

Construct Validity. Cognitive algebra can provide construct validity for the terms of the model. Concepts such as "motivation" and "ability" have only the face validity conferred by the phenomenology of common language. It is an attractive conjecture that such concepts constitute proper scientific constructs, but it seems more likely that they will require purification and redefinition. In any case, some validational base beyond phenomenology is needed to assess the cognitive reality of such concepts. Algebraic models can be helpful for this purpose because their success implicitly confers construct validity on the terms of the model. Unless these terms are veridical, they are not likely to obey an exact model.

Thus, the preceding results support the hypothesis that motivation and ability are not merely words, but represent operative cognitive entities. Qualitative analysis would seem generally unable to go beyond statements that the use of such words is correlated with the operative entities. Although model analysis is not conclusive, the exactness of these tests provides cogent support that might not otherwise be obtainable (see, similarly, Section 1.3.6). In this way, integration theory can utilize the knowledge embedded in common language as a starting point for the development of rigorous theory.

1.5.7 MEMORY SEARCH

The following experiment shows how a novel search task, together with functional measurement methods, provided a tool for studying organization of memory search in problem solving.

Response Time Models. In one part of this experiment (Shanteau, 1976), subjects thoroughly memorized a list of eight foods (*lime, tuna, beef, fish, pear, cake, veal, bean*), and then searched the list, in serial order, for answers to questions of varied difficulty. Illustrative easy, medium, and difficult questions were: "Which do not have the letter H?" "Which have exactly two vowels?" and "Which have exactly two consonants?" Seven of the eight list items, *fish* excepted, are correct answers to each question. Subjects responded overtly only for correct answers.

In the left panel of Figure 1.19, the foods are given in their list order on the horizontal axis. The vertical axis represents the cumulative time for successive responses, measured from the beginning of the list. These curves form a linear fan, and this fan shape reflects the processing structure.

In the model proposed by Shanteau and McClelland (1972, 1975), processing time for any list item is the sum of a retrieval time, which depends on the list item, plus a decision time, which depends on question difficulty. Because processing time is additive for each item, cumulative time obeys a multiplying model. This model structure becomes manifest in the linear fan pattern. This ingenious analysis automatically takes into account the processing time for the nonresponse items in the list.

In this graph, the slope of each curve represents question difficulty. This varies markedly, even for the two cited questions about consonants and vowels, which are lexically equivalent for this list. The functional spacing on the horizontal represents retrieval time for successive list items, except for *pear*, which follows the incorrect item, *fish*. Retrieval times are roughly constant at .5 seconds. This graph illustrates how the structure of the search task allows a straightforward separation of retrieval time and decision time. Related data and analyses provided estimates of input and output times.

Practice Effects. A striking practice effect can be seen by comparing the first and second replications in the left and right panels of Figure 1.19.

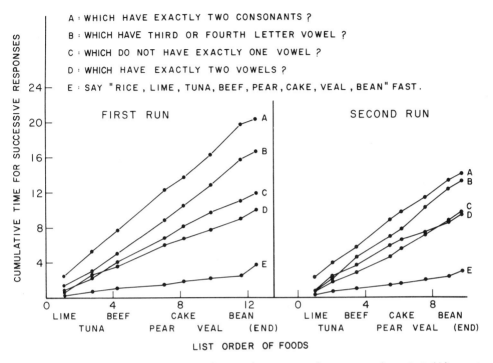

FIGURE 1.19. *Linear fan pattern reveals processing structure of memory search; see text. (After Shanteau, 1976.)*

Response times are generally shorter when the questions are given a second time. One locus of practice is in retrieval time, which shows a modest improvement of about 15% as measured by the range on the horizontal axis. However, the major locus of practice is in decision-processing time, which decreases an average of 1 second per item for the hardest question.

In Shanteau's (1976) interpretation, the processing steps for a given question are basically the same the first and second times. Questions of various types were intermixed so that subjects could not simply recall previous answers. Instead, they had to reconstruct the answers. "The effect of repetition was to facilitate this reconstruction through faster memory-retrieval and decision-making times [Shanteau, 1976, p. 3]."

The cause of such marked facilitation in the decision-making time is not clear. A generalized practice effect that would apply equally to new questions seems unlikely in this experiment, and two speculative suggestions may help bring the problem into focus. One possibility is that assembled fragments of the original decision processing are still intact and can be utilized without reconstruction. Alternatively, the processing steps may be identical in the second replication but have a lower threshold from activation remaining from the first replication and so occur more quickly. Varying the time between replications and increasing the number of replications might shed light on the Shanteau effect.

Multiple-Answer Search Task. An important characteristic of this search task is that it allows multiple answers to a given question. The interanswer times provide analytical leverage not available with standard search tasks that have only one answer. Although considerable model analysis had been done on standard search tasks, the most popular approach rested on assumptions about distributional properties for reaction times and other underlying processes that had tenuous relations to the data. The multiple-answer search task was developed specifically to bypass these uncertain assumptions and to make the structure of the underlying processing immediately visible in the pattern of the raw data (Anderson, 1966b, 1969b). This philosophy and approach agree with that of Sternberg (1966, 1969).

The method and model developed by Shanteau illustrate the analytical power of this search task and suggest many opportunities for more fine-grained analyses. Search modes other than strict serial search can also be considered. Most important, the task is not limited to simple recognition, but is especially suitable for decisions that require more or less complex problem solving. Problem-solving tasks would be useful for testing the hypothesis that the construction that yielded the first response remains in partially assembled form to be utilized for the second response. The position still seems valid that "a potentially good experimental task—one that would hold the mirror up to nature—deserves discussion as much as any theory [Anderson, 1969b, p. 163]."

1.6 The Averaging Model

Adding and other linear models have been widely considered in psychology and have dominated research in some areas of decision theory and other fields. The intuitive plausibility and mathematical simplicity of these models have been major reasons for their popularity. When appropriate experimental tests have been made, however, the linear models have seldom succeeded. Simple critical tests have regularly ruled out linear models in many of the areas in which they have been most popular.

Many of these tests have also supported the operation of an alternative averaging model. Averaging and adding models have basically different natures. The implications of the averaging phenomenon are thus pertinent to almost every aspect of theory and method.

1.6.1 EVIDENCE FOR AVERAGING

This section presents three of the many experiments that have provided evidence for the averaging hypothesis. All three follow the same logic. One part of the experiment is designed to produce parallelism: that pattern supports both the adding and the averaging hypotheses. The other part of the experiment provides a critical test between adding and averaging.

Critical Test. The experiment in Figure 1.20 was discussed in Section 1.3.1 and is repeated here to illustrate the logic of the critical test between adding and averaging. Subjects judged likableness of a man described by a personality trait (curve parameter) and a characteristic behavior toward other people (horizontal axis). As noted previously, the near-parallelism of the four solid curves supports a linear-type model, either adding or averaging.

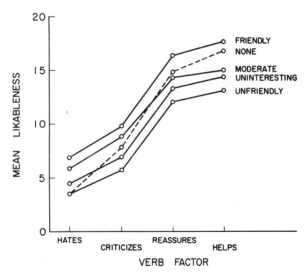

FIGURE 1.20. *Crossover interaction eliminates adding model and supports averaging model. Subjects judge likableness of person described by trait-adjective (curve parameter) and behavior toward others (verb factor on horizontal axis). Near-parallelism of solid curves supports adding-type model; crossover of dashed curve is critical evidence against adding model but supports averaging model. (From Anderson, 1974i.)*

The critical test between adding and averaging is obtained by comparing the dashed curve labeled "none" with the solid curve labeled "moderate." For both curves, the judgment is based on the behavior information listed on the horizontal. The "none" curve contains no more information; the "moderate" curve contains the added information that the man is moderate. The test revolves on the effect of this added information.

If an adding process was operative, then adding the mildly positive information that the man is moderate would cause the "moderate" curve to lie above the "none" curve at every point. But the curves cross over, thereby eliminating the adding model. This conclusion holds not only for the exact adding model, but also for any qualitative, directional adding model. Thus, the crossover eliminates an entire family of adding-type models.

The averaging model provides a simple explanation of the crossover. If moderate information is averaged in, that will lower the response to the very positive behavior, "helps people." Similarly, it will raise the response to the very negative behavior, "hates people." The data thus agree quite well with the averaging hypothesis.

Meals. Judgments of meals might be expected to obey an adding rule. A meal is a commodity bundle, and each additional food should add to the value of the meal. An exact adding rule might fail because of diminishing returns, for example, but that would only complicate a basic adding operation.

In fact, meals seem to elicit averaging processes. The left panel of Figure 1.21 represents likableness of meals with a main course varied in value, as indicated on the horizontal, and a vegetable varied in value, as indicated by the curve parameters. The parallelism of these three curves supports both adding and averaging models.

The critical test between adding and averaging is given in the right panel of Figure 1.21. The N curve represents a main course with one neutral vegetable; the NN curve represents the same meal with a second neutral vegetable.

The crossover of the N and NN curves eliminates the adding formulation. If the added neutral vegetable is positive (negative, or zero), then the adding formulation requires the NN curve to lie above (below, or on) the N curve at every point. The adding formulation cannot be saved by an assumption of diminishing returns, for example, because that still makes the same qualitative prediction.

The crossover supports the averaging formulation. Since the added vegetable is nearly neutral, it should average up the low main course,

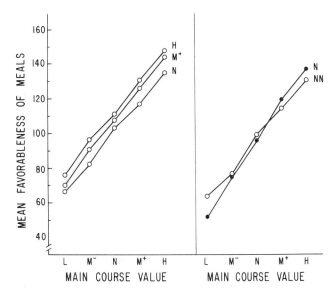

FIGURE 1.21. *Parallelism in left panel supports adding-type model; crossover in right panel eliminates adding rule and supports averaging rule. Subjects judge desirability of meals with specified main course (horizontal axis) and vegetable (curve parameter). (After Oden & Anderson, 1971.)*

average down the high main course. Taken together, the parallelism in the left panel and the crossover in the right panel provide strong support for the averaging model.

One feature of this experiment has special significance. In Figure 1.20, the crossover was between curves with one or no trait adjectives. These two might not be informationally equivalent. The qualitative difference between *some* and *no* adjective information might itself introduce some change in valuation or integration. In the meals experiment, however, the crossover is between curves with two or one vegetables. These are informationally comparable, varying only in the amount of information about vegetables.

Averaging by Children. Although only a few experiments are yet available, they suggest that children average even more readily than adults. The left panel of Figure 1.22 represents children's judgments of how much they would like to play with combinations of two toys constructed from a 3 × 3 factorial design. The near-parallelism of the three curves supports an adding or averaging model.

The right panel of Figure 1.22 replots the "medium" curve from the left panel together with the dashed curve that represents judgments

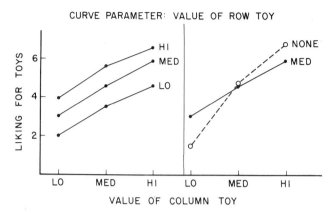

FIGURE 1.22. *Parallelism in left panel supports adding-type model; crossover in right panel eliminates adding rule and supports averaging rule. Children judge liking for one or two toys in factorial type design. (After Butzin & Anderson, 1973.)*

based only on the one toy listed on the horizontal axis. The crossover eliminates the adding model and supports the averaging model by the logic indicated previously.

It had been hoped that developmental studies would shed light on the origin of the averaging rule. To date, this hope has been disappointed by the fact that children seem to average as young as they have been tested, at present almost down to 3 years of age. However, this early appearance of the averaging rule and its ubiquity suggest that it reflects a basic property of cognitive integration.

1.6.2 THE MODEL EQUATION

It is a peculiarity of the averaging model that each stimulus has a two-parameter representation. Besides the scale value, s, a weight parameter, w, is also required. The weight can be considered as the importance of the stimulus with respect to the given dimension of judgment (see also Section 4.4).

For a given set of stimuli, $\{S_0, S_1, S_2, \ldots\}$ the model is written

$$
\begin{aligned}
r &= (w_0 s_0 + w_1 s_1 + w_2 s_2 + \cdots)/(w_0 + w_1 + w_2 + \cdots) \\
&= \Sigma\, w_i s_i / \Sigma\, w_i.
\end{aligned}
\tag{20}
$$

The numerator is a weighted sum of the stimulus values, and the denominator is the sum of the weights.

Absolute and Relative Weights. A distinction is made between the absolute weight, w_i, and the relative weight, $w_i/\Sigma\, w_i$. The independence assumption is applied to the absolute weights: these are assumed to be constant across different sets of stimuli. However, the relative weight of any stimulus depends on the other stimuli in the set. In each set, the denominator term forces the relative weights to sum to 1, the condition for an average.

This averaging property reflects a configural quality of the model. The effect of any one stimulus depends on what other stimuli it is combined with. A stimulus of medium weight will dominate the response when paired with a lightweight stimulus but will play a subordinate role when paired with a heavyweight stimulus. This configural property of the averaging model has an important theoretical role. For example, it provides a natural interpretation of the fact that attitudes based on more information are harder to change. Similar applications are noted later in the discussion of implicit additivity.

Initial State. The S_0 stimulus ordinarily refers to an internal state, not an external stimulus. In attitude theory or decision theory, for example, S_0 is used to represent the prior attitude or belief of the subject before exposure to the experimental stimuli. Accordingly, it may be called the *initial attitude* or the *prior belief.* One function of S_0 is to represent the mass of previous information that led to the present attitude or belief. It has basic theoretical importance. Interesting empirical support for the S_0 concept has been obtained by Kaplan in studies of predispositions (see Section 4.3.1).

Three properties of the concept of initial state deserve mention. The first property is that it enables the averaging model to account for the set-size effect in which added information of equal value can produce a more extreme response. To illustrate the essential idea, suppose that the prior attitude is neutral and that all experimental stimuli have the same positive value. The attitude produced by a given number, k, of such stimuli is the average of the k positive values and the neutral value. As k increases, so does this average. Averaging in more information of equal value can thus make the response more extreme.

This point deserves emphasis to avoid the incorrect assumption that the averaging model must imply that the response to k stimuli of equal value is constant, independent of k. The ability of the model to give an exact, quantitative account of the set-size effect is an important component of the theoretical structure (Section 2.4).

An awkward consequence of the initial state is that the response to a single stimulus is not in general a linear function of its scale value. From Eq. (20) the response to a single stimulus S_i is the average of that stimulus and the initial state:

$$r_i = (w_0 s_0 + w_i s_i)/(w_0 + w_i).$$

This equation implies that r_i is not in general a linear function of s_i. Exceptions occur when $w_0 = 0$, as in psychophysical averaging, or when the experimental stimuli, S_i, all have equal weight. In the former case, $r_i = s_i$. In the latter case, the denominator in the equation is constant, so r_i is a linear scale of s_i. With differential weighting, however, parameter estimation is more difficult.

Finally, S_0 need not be considered as a unitary entity. Rather, it may be some complex field of cognitive elements. Regardless of the structure of this internal field, it may be treated as a molar unit by virtue of the principle of cognitive unitization (Section 1.6.5). The w_0 and s_0 parameters may therefore represent the resultant of some integration operation over internal stimulus field. For analysis of the given judgment, fortunately, nothing need be known about the structure of S_0 or about the integration operation; these are exactly and completely represented in the parameters.

1.6.3 MODEL ANALYSIS

The analysis of the averaging model will be illustrated under the same conditions used for adding-type models in Section 1.2. The same assumptions about response linearity and stimulus independence are made here as there. For the averaging model, of course, the independence assumption applies to both absolute weight and to scale value, and both are assumed to be constant for a given stimulus across different sets. It is also assumed that the stimulus sets are constructed from a two-factor design, as in Figure 1.2 (see Section 1.2).

The averaging model can be written in a form analogous to the adding model of Eq. (8) as

$$r_{ij} = (w_0 s_0 + w_{Ai} s_{Ai} + w_{Bj} s_{Bj})/(w_0 + w_{Ai} + w_{Bj}). \tag{21}$$

The cases of equal and unequal weighting require rather different analyses and are considered separately.

Equal-Weight Case. Equal weighting means that all row stimuli have the same weight so that $w_{Ai} = w_A$ and, similarly, all column stimuli have equal weight so that $w_{Bj} = w_B$. The denominator of Eq. (21) then has the same constant value $(w_A + w_B + w_0)$ for every pair of stimuli. The averaging model is then a linear function of the two scale values with constant coefficients. It has the same form as Eqs. (2) and (8), and so the entire development of Section 1.2 applies directly. With equal weighting, therefore, the averaging model implies parallelism, and the marginal means provide linear scales of the subjective stimulus values.

Differential-Weight Case. If the row (or column) stimuli do not all have equal weight, then the denominator of Eq. (21) is not, in general, constant. The averaging model is then nonlinear, and the parallelism theorem does not apply. Standard methods of analysis are still available but exact tests and parameter estimation can be markedly more difficult.

Because of the greater difficulty of the differential-weight case, it can be worthwhile to avoid it by selecting stimuli to have approximately equal weight. In the attitude study of Figure 1.5, for example, the paragraphs were written to carry approximately equal amounts of information.

Qualitative Tests. Qualitative tests may be adequate for some differential-weight analyses, as in the critical test between adding and averaging. The crossovers of the dashed and solid curves in Figures 1.20–1.22 reflect unequal weighting of the corresponding row information. The crossover also shows, incidentally, that the averaging model is inherently nonlinear.

A useful principle for qualitative analysis can be illustrated for the case in which stimuli for, say, rows 1 and 2 have equal weight, but the column stimuli have unequal weight. From Eq. (21), the difference between rows 1 and 2 is readily shown to be

$$r_{1j} - r_{2j} = [w_A/(w_0 + w_A + w_{Bj})](s_{A1} - s_{A2}). \qquad (22)$$

The vertical distance between the two row curves is thus proportional to the *relative* weight of the row stimulus as given by the quantity in brackets. The greater the absolute weight of the column stimulus, w_{Bj} in the denominator of the bracketed term, the less the relative weight of the row stimulus and the closer the two row curves. When two row stimuli have equal weight, therefore, the vertical separation between the two row curves at any point is an inverse index of the weight of the corresponding column stimulus.

Three applications of this principle may be mentioned. First, the weight of a given column stimulus might be experimentally manipulated, as by instructions about its importance or reliability. That should produce predictable effects on the separation of the two curves (Section 4.4.3).

The second application deals with the not infrequent case of extremity weighting, in which stimuli more extreme in scale value also receive greater weight. Extremity weighting on one or both factors will cause convergence of the row curves toward the extremes (Section 4.4.2). The factorial graph will thus form a converging fan if only positive stimuli are used, or a slanted barrel shape if both positive and negative stimuli are used.

The last application requires independent evidence on the rank order of the stimulus weights. For example, the subject may be asked to rate or rank the column stimuli on importance. These self-estimated weights should then relate directly to the separation of the row curves. The method of self-estimated weights has been little studied, and its prospects are not clear (Section 4.4.4). It deserves more attention, especially as a tool for analysis of interactions in which the independence assumption for weight may not hold.

1.6.4 A CONSTANT-RATIO RULE DERIVED FROM AVERAGING THEORY

Certain situations involve two or more competing responses. For example, a judgment may be required about the relative likelihood of two possible causes of an observed action or event. Each possible cause is assumed to correspond to a covert response tendency, and the overt response is some compromise between these two covert tendencies. Under the averaging hypothesis, with S_0 ignored, this compromise becomes

$$R = (w_1 s_1 + w_2 s_2)/(w_1 + w_2).$$

The weight parameters in this expression represent the strengths of the covert responses. With only two response alternatives, the scale values may arbitrarily be set equal to 0 and 1; that is, to the endpoints of the likelihood scale. This yields the simple ratio

$$R = w_1/(w_1 + w_2). \tag{23}$$

Experimental applications to equity theory and decision theory are cited in Section 1.7.4.

This ratio model has the same form as various ratio models that have been used in choice theory, but there are fundamental differences. In particular, the choice models apply only to threshold differences because imperfect, probabilistic choice is necessary to obtain usable choice proportions. The integration ratio model is more general for it also allows numerical responses based on suprathreshold, nonprobabilistic differences. It can thus cover a substantial stimulus range using single-subject design and analysis.

1.6.5 COGNITIVE UNITIZATION IN AVERAGING THEORY

Informational stimuli generally have a complex structure. In the study of attitudes toward United States presidents (Figure 1.5), the unit physical stimulus was a set of two paragraphs, each containing diverse arguments, both independent and interrelated. Even the single adjectives in the person perception task of Figure 1.2 may produce multiple meanings and implications.

A complete representation of the stimulus must allow for such complex structure. That can be difficult, for these molecular components are not generally observable. Fortunately, averaging theory justifies application of the principle of cognitive unitization (Section 1.1.5). That enables a complex stimulus to be treated as a molar unit, thereby allowing substantially simpler design and analysis than when cognitive unitization does not hold (e.g., Farkas & Anderson, 1979).

Let w_i and s_i be the weight and scale value of the molecular components of some stimulus. Theoretically, the aggregated effect of these molecular components can be treated as a molar unit with parameters w and s defined by

$$w = \Sigma w_i, \tag{24a}$$

$$s = \Sigma w_i s_i / \Sigma w_i. \tag{24b}$$

where the sum is taken over all the molecular components. Thus, the weight of the molar stimulus is the sum of the component weights; its scale value is their mean scale value (Anderson, 1971a).

This unitization property has fundamental importance because it makes possible valid and exact analyses at the molar level. The parallelism theorem, for example, can be tested at some molar level that is experimentally convenient. The stimuli may be single words, extended prose passages, handshakes, laughter, etc., that require complex valuation operations. The parallelism test takes cognizance of all the molecular processing for it is all represented in the molar parameters. Exact, quantitative analyses are thus possible without concern for molecular detail. Furthermore, such molar results provide a useful tool for molecular analysis because they constitute boundary conditions that molecular analysis must obey.

1.6.6 COMMENTS ON THE AVERAGING MODEL

*The **w–s** Representation.* The concept of averaging entails the existence of a separate weight parameter for each stimulus. Two parameters are necessary to provide an adequate theoretical representation of the stimulus. The success of the averaging model provides construct validity for this *w–s* representation.

There is, of course, no more necessity for different stimuli to have equal weight than to have equal scale value. The equal-weight case is not a separate model, merely a special condition that may not be fulfilled experimentally. It will often not be fulfilled, which, unfortunately, can make analysis markedly harder than if some simpler model had been true.

In compensation for its greater difficulty, the averaging model does have some notable advantages. Foremost is that it accounts for the data, which have a moderately complex pattern (Chapters 2–4).

Stimulus Comparisons. A second advantage of the averaging model is its remarkable capability for comparing variables that are qualitatively different. With appropriate design and analysis, the weights of the several stimulus variables are estimable on ratio scales with common unit. Similarly, the scale values are estimable on linear scales with common zero and common unit. In this way, valid comparisons in both importance and value become possible among stimulus variables that may be quite different in nature. This comparison capability, which is not generally available with linear models, is one compensation for the difficulties of working with the averaging model (see e.g., Anderson, 1976b, in press; Norman, 1976a; Zalinski & Anderson, in press).

Measurement Theory. The averaging model points up a need for change in the traditional outlook of measurement theory. As its name indicates, scaling has traditionally been a one-dimensional quest for scale values. Weight parameters were ignored or considered as extraneous constants, not an integral part of measurement theory.

The importance of the weight parameter is reflected in the fact that the response need not be a monotone function of scale value. An increase in scale value of some stimulus can decrease the response by virtue of a corresponding decrease in its weight. Such disordinality does not seem readily amenable to ordinal or nonmetric approaches to measurement.

The empirical evidence for the averaging formulation makes clear the inadequacy of an outlook that limits measurement theory to scale values. Weight must be placed on equivalent footing, and both parameters are required for an adequate theoretical representation. Functional measurement seems to be the only approach so far developed that is conceptually adequate to handle this situation.

Implicit Additivity. Much theory and research rests on implicit assumptions of additivity. Deviations from additivity accordingly tend to be interpreted as signs of stimulus interaction or as the result of a nonlinear response scale. In the former case, substantive processes are assumed in order to account for the interaction. In the latter case, it is claimed that the response scale should be transformed so as to force the data to be additive. Both alternatives may be misguided if the averaging hypothesis applies.

Deviations from additivity may reflect the nonadditive nature of the averaging model. For example, one of the earliest instances of nonparallelism in this research program was later found to reflect a negativity weighting effect rather than a nonlinearity in the response scale (Section 4.4.2). It would not be appropriate to eliminate such nonadditivity either by monotone transformation of the response or by treating the response as an ordinal scale.

In other instances, deviations from additivity have been taken to imply the operation of cognitive processes to "explain" the deviation. For example, the second of two otherwise equivalent stimuli will generally produce less change in response than the first. Under an additivity assumption, this reduced effect seems to require some explanatory process in terms of a redundancy interaction, for example, or an expectancy–impact hypothesis. The averaging model, in contrast, predicts the diminished effect, so there is no discrepancy to "explain" (see, e.g., T. Anderson & Birnbaum, 1976; Bogartz, 1976; Harris, 1976; Himmelfarb, 1973, 1974).

This point also illustrates that qualitative model analysis can be very useful without concern for quantitative fits. Indeed, it is the conceptual structure of the averaging model that is basic. The tests of goodness of fit are a means to establish this conceptual structure.

Social Significance. Aside from its theoretical importance, averaging theory has practical social significance. Adding favorable information can actually produce less favorable judgments. This result was first obtained in person perception (Anderson, 1965a) and it has since been observed in many other situations.

One application of this result appears in the preparation of vitae for seeking jobs, promotions, or grants. Eager to make the best possible impression, many persons add everything possible to their vita; they implicitly assume an adding or summation process.

However, a technical report, for example, that would have a favorable impact by itself may average down the favorable impact of published articles. Similarly, an organization that seeks to improve its image by presenting favorable information may incur the opposite effect (Zalinski & Anderson, in press).

The empirical situation is not simple; adding mildly favorable information can act both ways, either to increase or to decrease the response. Theoretically, this reflects the action of the initial impression. Prediction in any situation thus requires the more detailed theoretical analysis of Sections 2.3 and 2.4.

1.7 Other Algebraic Models

A few other algebraic models have appeared fairly often in the experimental studies and will be discussed briefly here. For the most part, these models are straightforward generalizations of models already considered and involve the three basic operations of adding, multiplying, and averaging.

1.7.1 MULTILINEAR MODELS

The adding and multiplying models of Sections 1.2 and 1.4 are easily generalized to allow for more than two stimulus variables and also to allow for both adding and multiplying operations within one model. These models may be studied experimentally by treating each stimulus variable as a design factor.

Multifactor Adding Models. The parallelism theorem of Section 1.2 may be applied directly to each and every pair of factors in a multifactor adding or linear model: Each two-way factorial graph should be a set of parallel curves. Analysis of variance again provides exact tests of goodness of fit. If all variables are integrated by adding-type operations, then all interactions are zero in principle and are expected to be nonsignificant in practice.

Multifactor adding-type models have frequently been applied in studies of serial integration that involve many stimuli. Informational learning, for example, may require accumulation of information over many trials. By treating each serial position as a design factor, the model analysis can dissect the response into its separate serial components (Section 2.5). Knowing only the response at the end of the sequence, therefore, the model analysis can reveal its step-by-step buildup.

Multifactor Multiplying Models. Linear fan analysis can also be generalized to handle more than two factors. Indeed, each and every pair of factors in a multifactor multiplying model should exhibit the linear fan pattern. The analysis of variance tests have corresponding generalizations.

Three-factor multiplying models occasionally arise, as in the MEV formulation of Section 1.5.5, but no four-factor multiplying model is known. Indeed, there is some suggestion that subjects may simplify even a three-factor model by adding rather than multiplying (Shanteau & Anderson, 1972).

Adding–Multiplying Models. Many integration tasks, especially in decision theory, entail both adding and multiplying operations. Analysis of such compound models is straightforward. Linear fan analysis applies to any two or more factors separated only by multiplication signs; parallelism analysis applies to any two or more factors separated by a plus sign. The corresponding interaction tests from analysis of variance are again applicable and so provide a useful tool for diagnosing the underlying integration operations.

In the experiment on subjective expected value of Section 1.5.1, for example, subjects also judged combined worth of two independent chances to win two distinct objects. The natural hypothesis is that this judgment will obey the compound model

$$SEV = SP_1 \times SV_1 + SP_2 \times SV_2, \tag{25}$$

where SP and SV denote the subjective probability and subjective value of the two events, respectively.

In this compound SEV model, two of the two-factor graphs should exhibit the linear fan form, namely, $SP_1 \times SV_1$ and $SP_2 \times SV_2$. These linear fan predictions were supported in both group and individual analyses. The multiplying operation was thus verified for these duplex bets as well as for the single bets of Figure 1.13.

All other interactions in Eq. (25) should exhibit parallelism. That was verified for the two-way interactions, but the higher-order interactions were significant. This unexpected discrepancy has been verified in fundamental work by Shanteau (1974, 1975a), who discovered a general subadditivity effect.

Compound Probability Model. Another experimental illustration of compound models is given by the probability model of Section 1.3.3

$$Prob(White) = Prob(Urn\ A)\ Prob(White|Urn\ A)$$
$$+ Prob(Urn\ B)\ Prob(White|Urn\ B). \tag{26}$$

Besides the adding operation between the two conditional probability terms on the right side of this equation, Prob(Urn A) and Prob(Urn B) = $1 - $ Prob(Urn A) have a multiplying role. In the actual experiment, all three factors were varied independently. The three two-way factorial graphs are shown in Figure 1.23.

The left panel of Figure 1.23 has already been seen in Figure 1.6. As

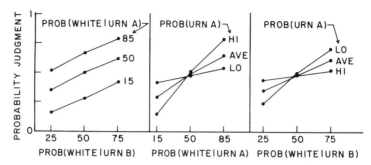

FIGURE 1.23. *Data patterns support three-factor integration rule for subjective probability. Urn A has 15, 50, or 85 white beads in 100; Urn B has 25, 50, or 75 white beads in 100. Choice of Urn A over Urn B is to be made with Hi, Ave, or Lo probability, and one bead is to be selected at random from the chosen urn. Plotted response is the judged probability that a bead chosen in this way will be white. (After Anderson, 1975b.)*

noted there, the parallelism supports the adding operation for the two conditional probability terms on the right side of Eq. (26).

The center and right panels of Figure 1.23 both show linear fans. The center fan reflects the multiplication, Prob(Urn A) Prob(White|Urn A). The right fan reflects the multiplication, [1 − Prob(Urn A)] Prob (White|Urn B). Taken together, these three panels provide excellent support for this probability model, confirming previous work by Wyer (1975a,b).

1.7.2 TWO-OPERATION MODELS AND RESPONSE SCALE VALIDITY

Tasks with two integration operations can be important for resolving certain questions about linearity of the response scale. This capability is sufficiently important that it deserves detailed consideration.

A striking instance of two-operation logic appears in the equity experiment of Figure 1.25. The parallelism in the four right panels provides strong support for response linearity. That, in turn, supports the interpretation of the deviations from parallelism in the two left panels as genuine, due not to nonlinear response, but rather to the comparative ratio process. Overall, the data patterns in the six panels provide interlocking support for the compound integration model and for the response scale.

In many tasks, primary concern is with a two-variable integration that is expected to be interactive or configural and hence to produce deviations from, say, the parallelism pattern predicted by a nonconfigural integration rule. But observed nonparallelism could result merely from nonlinearity in the response output function and so does not warrant a configural interpretation. However, if the task includes a second integration operation that does obey a simple model, that model can establish response linearity. Response linearity, in turn, would allow a definite conclusion about configurality or interaction at the integration stage.

One instance of this two-operation logic arose in the experiments on the SEV model of Eq. (25). The multiplying operation worked well, but the adding operation did not. The success of the multiplying operation supported response linearity; it thereby also supported localization of the nonadditivity at the integration stage.

This two-operation technique can also be employed specifically for response scaling. A second integration operation, perhaps of no interest in itself, may be included to assess response linearity. In the Height + Width rule for 5-year-olds' judgments of rectangle area (Section 1.3.6), for example, a natural concern was that the response function might be logarithmic. Because the log of a product is the sum of the logs, a true multiplicative Height × Width integration would then appear additive in the observable response scale.

To test this interpretation, children were asked to judge combined area of pairs of rectangles, the natural expectation being that this judgment should be a sum of the two subjective areas. All factorial graphs showed parallelism, supporting a (Height$_1$ + Width$_1$) + (Height$_2$ + Width$_2$) rule and, thereby, also a Rectangle 1 + Rectangle 2 rule. An antilog response transformation would change the Height + Width rule for each separate rectangle into a Height × Width rule, but it would simultaneously imply a multiplying rule, Rectangle 1 × Rectangle 2 for the sum of the areas, a result that seems strange and improbable. The observed success of the adding rule for the two separate rectangles makes perceptual sense, supports the assumption of response linearity, and hence affirms the Height + Width interpretation.

Two-operation logic may be employed more generally to use one operation as the scaling frame. The observed response would be transformed monotonically to fit one operation, and goodness of fit could be tested for the other operation, a two-phase approach that has important statistical advantages. Successful application requires one valid integration operation, of course, and so places a high premium on establishing such operations. This response scaling technique could be especially useful with behavioral or physiological measures that are not linear in raw form (Anderson, 1978a). Possession of even one such integration operation would provide a reference standard for quantification with unique advantages for theoretical analysis.

1.7.3 MULTIATTRIBUTE MODELS

Everyday life demands judgments about a variety of complex objects or events, from a movie or picnic to an article review or an academic promotion. Such events are often represented as sets, or bundles, of features or attributes, and judgments about them are represented as weighted sums or averages of attribute values. Such models are widely used in applied decision theory and are generically termed *multiattribute models.*

Since multiattribute models typically have a multilinear form, the parallelism and linear fan analyses are applicable in principle. However, these analyses ordinarily assume experimental manipulation of the attributes, and that is often infeasible and sometimes impossible in practical applications. For this and other reasons, multiattribute models have rarely received adequate tests of goodness of fit. Despite their undoubted usefulness in practical decision making and prediction, therefore, multiattribute models have lacked a foundation as descriptive models in psychological theory.

Theoretical Foundation for Multiattribute Models. Information integration theory may provide a theoretical foundation for multiattribute models. Two somewhat different approaches are possible and will be discussed briefly here.

The first approach views multiattribute models as part of cognitive algebra. It seems reasonable to assume that the same integration processes are operative whether or not the attributes are under experimental control. If this assumption is correct, then the results on cognitive algebra from the present research program provide a foundation for multiattribute models that has substantial empirical solidity.

This assumption is open to test. Multiattribute models are often applied to objects or events for which at least some attributes are subject to experimental manipulation. Items can be added to or subtracted from a package deal, for example, or new information bearing on probabilities or values may be obtainable. Such manipulation can provide model tests in the manner illustrated in previous sections. An important practical prediction from averaging theory is that adding a favorable item to a package may actually decrease its value (Section 1.6.1).

The second approach takes up problems of measurement, both of stimulus parameters of the individual attributes and of overall response to the multiattribute object or event. With respect to response scaling, the present research program has placed the common rating response, which is widely used in multiattribute studies, on a solid theoretical footing. Practicable techniques for monotone analysis of rank order data have also been developed.

With respect to stimulus scaling, it is necessary to develop methods for self-estimation of attribute parameters. In typical multiattribute studies, parameters are obtained on some convenient basis such as by asking directly for ratings or rankings of value and importance. Such self-estimation procedures seem reasonable, but an adequate criterion for their validity has been lacking.

The need for a validity criterion may be illustrated by the problem of obtaining self-estimates of importance or weight. A number of studies have compared self-estimated weights with weights obtained from regression models. Finding little or no relation, they concluded that people have little insight into their judgment processes. But such comparisons are not valid. Among other reasons, standard methods in multiattribute theory confound the weight parameter with the scale unit. The fault may lie in the methods used to assess people, not in the people themselves.

Integration theory provides a validational base for developing self-estimation methodology. The simplest approach is to obtain self-estimates in a task in which functional measurement scales can also be obtained. These scales then constitute a validational criterion against which to judge the self-estimates—and thereby the method of obtaining them. Experimentally controlled tasks may thus provide a basis for development of a self-estimation methodology that could then be employed where experimental control is not convenient or not possible. The first published study of this kind (Shanteau & Anderson, 1969) found reasonable but not excellent agreement between functional scales and self-estimates. Other evidence is noted in Anderson (in press) and Zalinski and Anderson (in press).

Experimental Study with Self-Estimated Parameters. An alternative approach is to use the self-estimated parameters directly in the model analysis on the principle that a successful test provides joint support for both model and parameters. A striking illustration of this approach is given by the date-rating study of Shanteau and Nagy in Figure 1.24. The stimuli were photographs of certain males; female subjects rated them on desirability as dates. Two attributes were assumed for each photograph: physical attractiveness and probability that the male would go out with the subject. Each subject made self-estimates of both attributes for each photograph. On the basis of an SEV model established in a

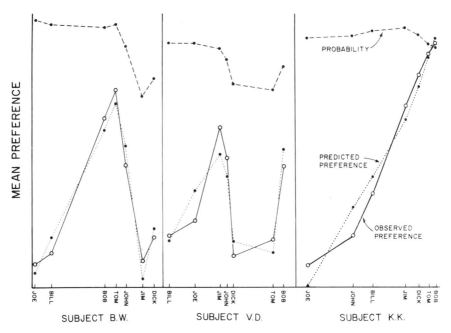

FIGURE 1.24. *Functional measurement test of multiattribute model for date ratings. Female subjects rate photographs of males on (a) physical attractiveness (horizontal axis), (b) probability of acceptance (upper dashed curve), and (c) desirability as a date (solid curve). Dotted curve gives theoretical predictions from multiattribute model, Attractiveness × Probability. Note that the three females exhibit different rank orders for physical attractiveness of the seven males on the horizontal axis of each panel. (After Shanteau & Nagy, 1976, 1979.)*

preliminary factorial experiment, these two attributes were hypothesized to multiply

Date Preference = Probability × Physical Attractiveness.

Observed date preferences are given by the solid curves in Figure 1.24 for three representative females. Theoretical predictions obtained with the self-estimated parameters are given by the dotted curves. Data and theory agree quite well, even at the individual level. As a collateral result, the theoretical preference values correctly predicted about 19 of 21 paired comparisons choices among the photographs.

Of special interest in Figure 1.24 is B. W.'s low preference for the two most attractive males, shown in the downturn of the solid curve at the highest values of physical attractiveness. This paradoxical downturn reflects the lower value of the probability attribute, which is given by the dashed curve at the top of the panel. Thus, the best-looking males are less-preferred dates, presumably because of a fear of rejection.

This impressive study accomplishes a threefold goal. It provides validational support for the integration model as well as for the method of self-estimated parameters. No less important, it provides construct validity for the specified two-attribute representation of the photographs.

1.7.4 RATIO MODELS

One other model has been sufficiently frequent to deserve separate mention. This is the ratio model

$$R_{ij} = s_{Ai}/(s_{Ai} + s_{Bj}). \tag{27}$$

This formal model has appeared in decision theory (Leon & Anderson, 1974), equity theory (see Figure 1.25), speech perception (Massaro & Cohen, 1977; Oden, 1979a; Oden & Massaro, 1978), and psycholinguistics (Oden, 1974, 1977a, 1978b). One conceptualization of the model is as an averaging compromise between competing responses (Section 1.6.4).

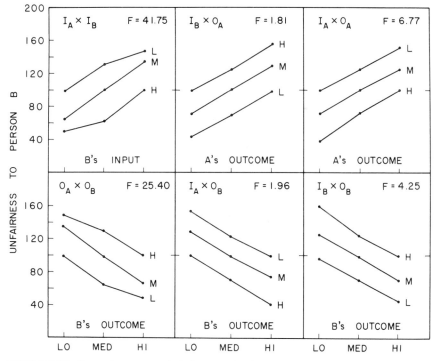

FIGURE 1.25. *Processing structure in judgments of unfairness revealed in factorial data patterns; see text. (From Anderson & Farkas, 1975. [Copyright © 1975 by the American Psychological Association. Reprinted by permission.])*

This conceptualization seems reasonable for some of the cited applications, but whether they all reflect a common integration process is uncertain.

The data pattern predicted by this ratio model depends on the values of s_{Ai} and s_{Bj}. If both variables have the same levels, and these form a geometric progression of at least three terms, then a slanted barrel pattern like that in the left panels of Figure 1.25 is obtained (see also Farkas & Anderson, 1979). If the values of one variable are generally greater than those of the other, then the pattern tends toward a fan shape.

Analysis of Comparison Structure. How model analysis can diagnose the structure of cognitive processing appears in the following experiment from equity theory. The essential idea of fairness, that one's outcome, O, should be proportional to one's input, I, was first formalized by Aristotle. For two persons, A and B, Aristotle defined the state of fairness by the equality

$$O_A/O_B = I_A/I_B. \quad \text{(Aristotle's model)}$$

In words, the ratio of rewards for the two persons should equal the ratio of their contributions.

In modern times, the following alternative was proposed by J. S. Adams, who defined fairness by the equality

$$O_A/I_A = O_B/I_B. \quad \text{(Adams's model)}$$

Nearly all current literature on equity models has used this formulation.

These two models have quite different comparison structures. The two ratios in Adams's model represent within-person comparisons of outcome to input separately for persons A and B. These are followed by the between-person comparison of the two ratios. Since O/I may be viewed as a "piece rate" of payment, this model befits a capitalist society.

The opposite order of comparison appears in Aristotle's model. The two ratios represent initial between-person comparisons, separately on the outcome and input dimensions. These two comparisons are then themselves compared. This model places primary emphasis on interpersonal comparison.

Integration theory has led to still a third model. This stems from the ratio model of Eq. (27) and leads to the following definition of fairness:

$$O_A/(O_A + O_B) = I_A/(I_A + I_B). \quad \text{(Averaging model)}$$

This model has the same basic comparison structure as Aristotle's model but differs in process detail and in mathematical form.

But these three models are mathematically equivalent. Each can be transformed into the other by simple algebra. Judgments about fair division are inadequate to diagnose the underlying comparison structure.

This problem can be resolved by asking for judgments of degree of unfairness. Each of the above fairness models can be transformed into an unfairness model by taking the difference between the two ratios. These three unfairness models may be written in terms of unfairness to person A as

$$U_A = O_A/O_B - I_A/I_B. \qquad \text{(Aristotle)}$$
$$U_A = O_A/I_A - O_B/I_B. \qquad \text{(Adams)}$$
$$U_A = O_A/(O_A + O_B) - I_A/(I_A + I_B). \qquad \text{(Averaging)}$$

These three unfairness models are not mathematically equivalent and can be readily discriminated, as illustrated in the following experiment.

Models of Inequity. Subjects received information about two persons, A and B, who had worked together on a common task. The input information specified how much each person had accomplished; the outcome information specified how much each had been paid. Subjects judged fairness on a graphic scale defined by *unfair to A* at one end, *unfair to B* at the other.

Each panel of Figure 1.25 plots one two-way factorial graph from this experiment. These graphs provide strong, simple tests of the three models. To illustrate, consider the $O_A \times O_B$ interaction. In Adams's unfairness model, O_A and O_B are separated by a minus sign; hence their factorial graph should exhibit parallelism. In Aristotle's model, however, O_A and O_B are separated by a division sign; hence their factorial graph should exhibit nonparallelism. The actual graph, in the lower left panel of Figure 1.25, shows a highly significant nonparallelism, thereby infirming Adams's model and supporting Aristotle's model. Each of the other five factorial graphs may be analyzed in similar manner. The net outcome is readily summarized as follows.

Adams's model is seriously incorrect. It implies that the two left panels will exhibit parallelism and that the two right panels will form a linear fan diverging to the right. These four predictions are all incorrect.

Aristotle's model does quite well. It predicts parallelism in the four panels that exhibit parallelism and it predicts nonparallelism in the two panels that exhibit nonparallelism.

These same predictions are obtained from the integration theory model. This averaging model is a little better than Aristotle's model because it predicts the observed barrel shapes in the two left panels, whereas Aristotle's model predicts linear fans. The importance of this analysis is not the quantitative accuracy, however, but the ability to delineate the comparison structure of the underlying processing (see also Anderson, 1976a; Farkas, 1977).

Cognitive Algebra and Cognitive Structure.   As this equity experiment indicates, cognitive algebra can lay bare the underlying structure of information processing. Comparison processes have basic importance in this task, and their structure is revealed by the data. This approach has also been used to study a serial–parallel processing question in social judgment (Farkas & Anderson, 1979) and to evaluate two models of comparative judgment for geometrical illusions (Clavadetscher & Anderson, 1977). As these applications illustrate, cognitive algebra provides a powerful new methodology for cognitive analysis.

1.8 Research Philosophy

The nature of any theoretical approach is perhaps best defined in its empirical applications. It may be useful, however, to comment on some general characteristics of information integration theory. The organization of this section is topical rather than systematic, and takes up a number of different aspects of research orientation and conceptual structure. These comments, partly personal, are intended to help establish a frame of reference for the direction and goal of the experimental studies.

1.8.1 INFORMATION INTEGRATION THEORY

The goal of this research program is to develop a unified theory of judgment and behavior. Some measure of success may be claimed, for experimental applications have done reasonably well in diverse areas (Sections 1.3, 1.5, and 1.7). As these applications show, the theory provides a unified conceptual framework that can be applied across many different substantive areas. It has helped resolve a variety of methodological and theoretical problems and it has unified strands of thought that were previously isolated. This section comments on several characteristics of integration theory.

Multiple Causation. The distinctive character of information integration theory flows from its central concern with multiple causation. Virtually all thought and behavior has multiple causes, being integrated resultants of multiple sources of stimulus information. Accordingly, the processes that are involved in integrating stimulus information constitute fundamental problems for psychological investigation. Although problems of multiple causation have long been of interest in every area of psychology, the present approach provides a new capability for analysis. In many respects, indeed, integration theory embodies a new way of thinking, a way that is out of phase with many prevailing orientations. As a consequence, it has been generally difficult for it merely to gain consideration.

This difficulty stems in considerable part from the approach to multiple causation. Previous approaches did not possess very good methods for analysis of stimulus integration. In the main, they had to bypass multiple causation and develop issues and problems that could be studied with available tools. Naturally enough, the nature of theory was accommodated to these issues and these methods. Equally naturally, such approaches attained a conceptual existence and inertia that carried beyond their domain of usefulness.

Conceptual orientations that lack effective methods for analysis of multiple causation are generally too narrow to permit growth of adequate theory. The need for theory that can handle multiple causation has appeared repeatedly in every empirical area that has been studied in this research program. Illustrative examples include continuous representations in semantic theory, additivity in decision theory, the psychophysical law, integrational capacity in children, and meaning-constancy in person perception. In all these areas, an integration-theoretical approach has led to restructuring of basic issues.

The present formulation is deeply indebted to the work of previous investigators. Problems of stimulus integration have been of concern throughout the field of psychology, and the present development may be viewed as part of the Zeitgeist (Section 1.1.1). The debt to previous workers is difficult to express because previous work is intimately interwoven with the present formulation. Virtually every idea in integration theory has evolved from ideas of previous writers. Every experimental application has benefited greatly from previous work. Every procedure of design and data analysis is an application or direct outgrowth of previous methods. Neither the theoretical structure nor the wide range of applications would have been possible without the background of ideas and results of innumerable previous workers.

Nevertheless, the present approach does require significant shifts from previous habits of thought. This shift in thinking touches every aspect of inquiry, from mundane details of procedure to the epistemology of investigation. Some aspects of this difference in ways of thinking are highlighted in the following topical discussions.

Inductive Theory. Integration theory has operated in a primarily inductive mode in which generalizations are sought as emergents from experimental analysis. The theory rests on a general conception of important problems and a framework for thinking about and studying such problems.

An alternative view sees science from an axiomatic or hypothetico-deductive perspective. A few fundamental postulates are sought as the basis for chains of deductive consequences. This axiomatic approach has produced impressive achievements in physics. In psychology it has had little success.

Of course, no one would deny the value of deductive inference in everyday research. Some local element of deductive thought is pervasive in the study of specific problems. However, that is quite different from deductive theory. An example from the present research program appears in the critical test between averaging and adding as integration rules. This test required a directed, deductive search, and the conclusion would otherwise have been slow to emerge. But that does not give the averaging hypothesis the status of an a priori axiom. It is, obviously, a theoretical proposition, not a simple generalization from data. However, its nature is that of an empirically based generalization that has been incorporated into the theory. The averaging rule can be used as a premise for deductive inference once it has been established, but that should not obscure its inductive origins.

Many psychological theories claim to operate in the deductive mode. It is often considered to be the ideal, sometimes the only, truly scientific way of thinking. Workers in the deductive mode often have difficulty comprehending inductive theory; to them, it appears formless and uncertain. To workers in the inductive mode, however, the deductive mode appears simplistic, not to say specious, beyond the local level. Among other reasons, what passes for deductive theory in psychology is typically an awkward form of inductive theory.

For the plain fact is that deductive theories are rarely abandoned when their deductions fail. Instead, they are modified, first in their auxiliary simplifying assumptions, later in their basic conceptual assumptions. Deductive theories in psychology typically exhibit a short, initial period of deductive flourish, followed by slow, grudging assimilation of inductive change. Open acceptance of a more inductive approach as a basic research orientation would seem developmentally more truthful, not to say more efficient.

Inductive theory views science not as formalized knowledge, but as living inquiry. It recognizes and incorporates background thinking and experimental lore, including pesky problems of apparatus and organism, that are obscured or lost in deductive formalizations. And it is more open to nature, which continually reveals new riches to surprise and delight her students.

Normative and Descriptive Theory. Normative theories, which are concerned with how people "should" behave, have several attractions. Typically, they have genuine logical or rational appeal. Typically, they have some axiomatic basis and appear as deductive theories. But, typically, they have little relevance to the study of cognition; thought and behavior appear to embody quite different processes.

That might seem a fatal defect; in the long run no doubt it is. In the short run, however, this defect is often turned to presumptive advantage by treating the normative prediction as a standard. Any deviation from this standard can then be viewed as a clue to understanding and can even be treated as a phenomenon to be explained.

Descriptive theories are concerned with actual behavior. From the descriptive standpoint, normative approaches seem typically irrelevant, typically, indeed, a hindrance to understanding. In decision theory, for example, deviations from normative prediction may be suboptimal or irrational in the sense that they reduce the payoff. In a practical way, therefore, they are to be avoided. That does not make them true phenomena, however, for they exist only by reference to a conceptual framework that lacks psychological relevance.

This conclusion has appeared many times in this research program. Even when behavior does exhibit the same form as the normative models, understanding must still be sought elsewhere. Decision theory provides the most experimental illustrations because of the prominence of normative models (e.g., Sections 1.3.3, 1.5.1, 1.5.2, 1.7.1). Thus, extension of the concept of subjective probability to a more general concept of weight (Anderson & Shanteau, 1970, p. 450; Payne, 1973, p. 449; Kahneman & Tversky, 1979; Rapoport & Wallsten, 1972, p. 147) was slow of acceptance because it did not fit the prevailing normative framework.

A theory need not be correct to be useful. Some of the normative theories have provided an organizing framework that has called attention to important problems and led to useful research. At the same time, however, they have frequently hindered the growth of psychological theory.

Consciousness and Psychophysics. Since Fechner's original investigations, psychophysics has been dominated by two ideas. First, that sensation is conscious; second, that the link between the physical stimulus and the conscious sensation constitutes a unitary law of simple form. The first idea appears in Fechner's doctrine of panconsciousness and in the modern use of verbal reports of sensation magnitude (Section 5.4). The second idea began with Fechner's law that subjective sensation is proportional to the logarithm of the physical stimulus, and continues in Stevens's attempt to replace the log law with a power law. Perhaps no other idea in the history of psychophysics has been as attractive. The very term, *the psychophysical law*, both reflected and reinforced this idea of finding a simple natural law.

Neither idea seems appropriate from an integration-theoretical view. In the weight–size illusion of Section 1.3.7, the sensory effect of the physical weight is preconscious, not accessible to introspection. The same holds for many physical cues that govern judgments of size and distance or of taste. What does attain consciousness is often, perhaps always, a result integrated across different sense modalities at preconscious stages. Whatever does attain consciousness certainly deserves study, but is not itself a sufficient field for psychophysical analysis.

Moreover, recognition of preconscious sensory integration also emphasizes that there are multiple stages between the physical stimulus field and conscious sensation. The psychophysical law represents the concatenation of these stages. To lump their action is hardly likely to produce a meaningful unitary law. A more fruitful approach would shift focus to problems of stimulus integration.

This integration-theoretical view is consistent with a long tradition in perception. To this tradition it contributes a new capability to analyze multiple causation, a capability that can dissect conscious sensation into preconscious components. This capability for analysis can also shed new light on the processing chain between the physical stimulus and the conscious sensation. Conscious sensation is no longer seen as an ultimate, but as a working tool for perceptual analysis. The one-variable approach of traditional psychophysics is thus replaced by the many-variable approach of a new psychophysics that has more in common with the larger area of perception (Section 5.4; Anderson, 1975a, p. 480).

Phenomenology and Explanation. Psychology may seem to be a privileged science in that consciousness and phenomenology promise direct access to knowledge. Certainly it is reasonable to expect that phenomenological concepts embodied in the common language will have some degree of validity. At the same time, as the failure of the introspectionist school indicates, conscious phenomenology is not enough.

This last point was brought home sharply in the research on meaning constancy and the halo hypothesis (Chapter 3). It seemed transparently clear that personality adjectives changed in meaning when combined with other adjectives in a person description. But this phenomenological truth turned out to be an illusion. Conscious knowledge is not merely insufficient, but is sometimes seriously false.

Nevertheless, many theories in psychology have the practice of treating terms from the common language as though they were established scientific constructs with established explanatory power. Such theories can seem attractive and impressive. When considered more carefully, however, they often become unsatisfactory and empty, trading on circular explanations and on surplus meaning in common language terms. This problem of the role of phenomenology in explanation has arisen repeatedly in this research program. The issue of cognitive consistency provides one illustration.

An attractive explanatory postulate is that the mind strives for consistency. Diverse writers have referred to needs to find cognitive clarity, to predict and control the environment, to reduce uncertainty, to defend belief systems, to achieve balance, etc. Such postulates have a fine ring and have often been adopted as primary theoretical assumptions with secondary concern for empirical foundations.

Unfortunately, the idea that the mind abhors inconsistency has about the same explanatory power as the idea that nature abhors a vacuum. Both are true to some degree, but require explanation rather than provide it. In physics, this became clear when it was discovered that nature abhorred vacuums less at higher altitudes. In psychology, similarly, experimental analysis showed that the mind often has substantial tolerance for inconsistency (e.g., Anderson & Jacobson, 1965). For the most part, therefore, the consistency theories have not provided genuine explanations and have not been very fruitful.

The problem is how to get past the phenomenological starting point. To ignore common language concepts seems unwise because they may contain invaluable clues not otherwise obtainable. A way is needed to develop the true knowledge contained in these concepts without becoming disoriented by the false knowledge.

One source of help may be found in cognitive algebra. In the cited problem of cognitive consistency, for example, model analysis helped clarify the nature of the problem and also contributed to more effective methods of analysis. More generally, it seems reasonable to expect that only concepts that are psychologically real will obey a simple cognitive algebra. Studies of cognitive algebra can thus provide one important source of construct validity with which to develop and purify the constructs of intuition.

1.8.2 THE ROLE OF MODELS

Algebraic models have a much broader role in research than is recognized in the common stereotype that centers on their quantitative properties. Qualitative, conceptual properties are not less important (see also Anderson, 1969b). The comments of this section are intended to bring out something of the conceptual role of algebraic models in the present research program.

Cognitive Algebra. Some problems cannot be solved without exact model analysis. Thus, definition and validity of the concepts of weight and scale value depend ultimately on model analysis. Psychological measurement is essential for testing the law of subjective expected value (Section 1.5.1) and the laws of subjective probability (Sections 1.5.2, 1.7.1). The conclusion that subjects do not use density as a cue in lifted weights rests on exact model analysis (Section 1.3.7), as does the conclusion that 5-year-olds make quantitative judgments in a qualitatively different way than adults (Section 1.3.6). Models also play a vital role in delineating comparison structure in areas as far apart as social fairness and geometrical illusions (Section 1.7.4). The models need be exact only under certain conditions to provide a penetration not otherwise possible.

For conceptual analysis, models can be useful because they help focus on problems of cognitive structure. This conceptual emphasis can be helpful regardless of the exact form of the model. In psychophysics, the model approach has made clear the inherent inadequacy of the one-variable experiment and the necessity of a many-variable, integration-theoretical approach (Section 5.4). In psycholinguistics, Oden's work shows how the model framework has provided a natural and effective representation of continuous truth value. In decision theory, work by Shanteau has been basic to the development of a descriptive cognitive algebra to replace the traditional normative framework (Section 5.6.8). In attribution theory, causal schemes have been reconceptualized as integration models (Anderson, 1978b). Studies in developmental psychology have led to sharp upward revision of previous beliefs about children's integrational capacities (see Anderson, 1980). And in attitude theory, the model approach has contributed to conceptualizations about attitude as a constructive process, about the relations between verbal memory and attitude, between attitude and behavior, and the important distinction between attitude and attitudinal judgment.

It may seem surprising that the operation of cognitive algebra was not recognized much earlier, for the idea seems natural. Aristotle prescribed a model for equitable division:

A's reward : B's reward :: A's input : B's input (Section 1.7.4).

William James (1890/1950, p. 310) suggested the half-serious model:

$$\text{Self-Esteem} = \text{Success} \div \text{Pretensions}.$$

Algebraic models have been ubiquitous in modern decision theory and have been considered by numerous workers in nearly every area of psychology. Until recently, however, these models have been mainly schematic verbalisms, not true equations.

Among the reasons for the slow development of cognitive algebra was that the complete picture is by no means simple. The behavior encompassed by the averaging model, for example, had considerable surface complexity (Chapters 2 and 4). Natural preference for the adding model, coupled with doubts about the rating response, amplified the uncertainty. At the same time, the fact that much of the initial work was done in person perception put off many who did not recognize the unusual cognitive potential of this task as well as many who found it difficult to accept that so complex a process might embody so simple a model. It is an interesting twist that the major initial support for cognitive algebra emerged in an area where it was least expected.

General-Purpose Integration Processes. Ability to integrate stimulus information is a basic characteristic of behavior; one of the most useful, it enables the organism to utilize multiple cues. Depth perception, for example, is better with two eyes than with one because two eyes bring in additional cues. Again, integration in the form of informational learning serves as memory through which past experience can guide future behavior.

Ability to accumulate partial information across diverse sources obviously has high survival value, and it may be conjectured that integration processes have evolved for such purposes. Many will be sense-specific, of course, but general processes are also needed for cross-sensory integration and, especially, for integration of conceptual information. Indications of general-purpose integration processes can be seen in various kinds of context effects (e.g., Figure 1.10; Section 4.1.8). Such context effects may be conjectured to be manifestations of a general tendency to integrate whatever is present in the stimulus field, a tendency that may be related to a primitive capacity of forming cognitive units.

Algebraic integration rules may themselves have a general-purpose nature. Evidence for averaging processes, in particular, has been found in so many situations that it seems reasonable to consider averaging as a basic capacity of adaptive behavior. Indeed, averaging may be a form of the general integration process already mentioned.

Some indication that multiplying rules reflect a general-purpose operation can be seen in the appearance of such rules even when they are inappropriate from a normative view of optimal behavior (e.g., Sections 1.5.1, 1.5.2). In addition, considerable evidence points to the operation of a general-purpose adding-type rule in children's judgments (Section 1.3.6; Anderson, 1980). Algebraic rules may thus represent a general mode of cognitive functioning.

Cognitive Units. Search for explanation is often a search for primitive concepts. In physics, the concept of velocity was still unclear, still embedded in the common language, at the time of Galileo. Much the same situation holds for psychological concepts today. Loudness, area, and similarity, to take three different examples, are generally taken for granted as unitary cognitive entities. The same holds for more social notions such as likableness, deservingness, unfairness, and motivation. These concepts have undoubted face validity, but, except for motivation, the issue of their construct validity has received little explicit attention. Perhaps these words are little more than convenience labels for ill-assorted aggregates, as indeed seems to be the case with the various typologies of motivation.

Integration theory provides two lines of attack on the problem of cognitive units. One is with the principle of molar unitization (Sections 1.1.5, 1.6.5), by virtue of which heterogeneous stimulus aggregates can be treated as functional units. That can markedly simplify analysis, but, unfortunately, it does not always hold. An illustration comes from the concept of input, or deservingness, in equity theory. Multiple determinants of input may not be integrated into a unitary value of input; instead, each may act separately to determine a partial value. Such lack of unitization can complicate theoretical analysis (Farkas & Anderson, 1979, p. 893). Similar results have been obtained in other problems that involve comparative judgment.

The other line of attack utilizes cognitive algebra as a basis for construct validity of the stimulus and response terms. To obtain a functional scale is to obtain a fortiori evidence for the reality of the concept being scaled. Thus, the parallelism theorem includes an implicit assumption that the design factors correspond to operative cognitive units. If this implicit assumption did not hold, then parallelism would not in general be obtainable. Similarly, the success of the Height + Width rule in children's judgments of rectangle area (Section 1.3.6) implies that height and width are cognitive units and, no less important, that area and shape are not. Other instances of the general-purpose adding rule in children's judgments of quantity also illustrate the useful precision of model analysis in making definite the operative cognitive units (Anderson, 1980, esp., pp. 14–15).

Cognitive algebra is not a routine procedure for establishing construct validity, because success of the model does not guarantee the reality of the terms of the model. On the stimulus side, at least, the cognitive units are not necessarily isomorphic to the terms of the model. Nevertheless, algebraic models can be helpful, and because of their precision, they provide a novel and cogent line of evidence for establishing construct validity.

Continuous Representations. Integration theory has been primarily concerned with continuous representations. On the response side, this is manifest in the reliance on numerical response measures as well as in the concept of general metric sense (Section 1.1.6). Although the stimuli may themselves be purely categorical, such as the personality trait-adjectives, the requirements of judgment–decision tasks tend generally toward continuous stimulus representations.

Most other cognitive theories have centered on discrete, propositional representations. These are not sufficient, however, as has been increasingly recognized. Thus, Lakoff (1972) speaks of all-or-none truth value as a "convenient fiction," and Zadeh *et al.* (1975) have attempted to develop a continuous-valued "fuzzy" logic. In similar vein, Rosch (e.g., 1973) and Smith (e.g., Rips, Shoben, & Smith, 1973) have stressed the need to treat class belongingness as a continuous variable. Such continuous concepts are awkward to handle within the conventional propositional formulations and accordingly have received little attention. Affective information, which also has a natural continuous representation, has been similarly neglected.

Fundamental work by Oden has extended information integration theory to the study of continuous semantic representations. In Oden's "fuzzy propositional" approach, the logical propositions of conventional semantic network models are adopted. However, the semantic primitives are either fuzzy or continuous predicates, and the logical connectives are operators that follow rules of cognitive algebra. This approach requires a minimum of additional conceptual apparatus beyond that of conventional semantic network models. Once the primitives are taken as continuous rather than discrete, complex concepts that are composed from these primitives also become continuous.

In this approach, class belongingness, for example, is naturally treated as a continuous dimension of judgment. Moreover, an effective theoretical and experimental analysis becomes available through the study of tasks that involve information integration, as illustrated in Figure 1.15. In Oden's hands, the integration-theoretical approach has had considerable success in modeling various psycholinguistic phenomena such as ambiguity resolution, quantifier understanding, implicit negation on continua, as well as letter and phoneme identification (Oden, 1974, 1977a,b, 1978a,b, 1979a,b; Oden & Anderson, 1974; Oden & Massaro, 1978).

A central characteristic of this approach is its reliance on functional measurement and cognitive algebra. As Oden emphasizes and as his work demonstrates, this allows rigorous analysis of single propositions within the value system of each individual. Because of its focus on the study of multiple causation, information integration theory provides a new capability for the study of continuous semantic representations.

The Nature of Information. Integration theory embodies a general conception of information as signal, without other a priori restrictions. This usage differs from the standard cognitive view that centers on semantic or symbolic information and seems to have difficulty with information of a more biological kind. Perhaps the simplest illustration appears in the continuous stimulus dimensions of psychophysics. Sensory stimuli obviously have a numerical, rather than a discrete, propositional representation. Moreover, they are presumably perceived without semantic or symbolic mediation. Yet they certainly function as signal information, as in motor skills and sports.

Affective information also has a natural place in the information-integration view. Under the two-memory hypothesis of Section 1.8.3, affective information may be extracted from semantic information but thereafter become independent of its semantic origin. More generally, the original information need not be in semantic form; it may act directly, analogous to sensory information. Mood, emotion, and drive state may constitute information per se, without necessity for semantic way-stations. Although such internal states may not have information channels as specific as the sense organs, they do not require cognitive labels for their operation.

These considerations also illustrate that integration theory embodies a functional or adaptive conception of information. Information is not a constant of the stimulus, but is always relative to the immediate needs and goals of the organism. These needs and intentions are embodied in the momentary valuation operation that extracts relevant information from the prevailing stimulus field. More generally, needs and intentions are themselves signal information that control the behavior of the organism. The organism is not "lost in thought" at the choice point, therefore, because "thought" is only part of the total information flow.

Process Generality and Outcome Generality. Two quite different goals can be pursued in psychological research: the goal of predicting behavior and the goal of understanding behavior. The goal of prediction usually leads to a search for *outcome generality;* that is, direct generalization of results at a more or less empirical level. The goal of understanding usually leads to a search for *process generality;* that is, generalization about cognitive processes at a theoretical level.

The distinction between these two goals is well illustrated by the case of linear models. The use of linear models originated in questions about practical judgment and decision, for which purposes they were often very useful. When attention shifted to theoretical process questions, however, the linear models could not make sense of the data. For example, the addition of a favorable piece of information could make the response more favorable or less favorable, a nonlinear outcome that is contrary to the linear models. At the surface level of outcomes there was no generality because almost any outcome could be obtained.

At a process level, however, a conceptually simple answer was obtained. This simplicity lay in the averaging model and in the w–s representation. This process model made understandable a complex array of results as shown in subsequent chapters.

The process–outcome distinction has particular relevance to current research in decision theory and in social judgment. In both these areas, attempts to obtain process generality are often encumbered by an ill-fitting inheritance of ideas and experimental procedures that have only partially evolved from an origin in outcome goals. Procedures and methods that have worked well in outcome studies have been carried over to process studies where they are not only ineffectual, but actively misleading. This intellectual inertia is well illustrated by the perseverance of weak inference methods (Section 1.2.8).

Although one may wish it to be otherwise, the goals of prediction and understanding are often incompatible. Each goal imposes its own constraints on design and procedure. It is a misleading half-truth to say that prediction is the touchstone of understanding and to ignore the genuine differences, both practical and theoretical, between the two classes of investigations. Attempts to pursue both goals within one study will usually require compromises in procedure that compromise the results, rendering them unsatisfying for either goal.

Although the outcome and process orientations are typically incompatible within any single study, they can be mutually beneficial. Outcome studies frequently raise new and important questions for process analysis. Such questions can ameliorate the tendencies toward dead-ending and trivialization that beset the process orientation. In return, process studies can be indispensable for testing and developing ideas suggested by outcome studies. The emphasis on process–outcome incompatibility is an argument against good-intentioned but ineffectual compromise in design and procedure. It should, however, clarify the need and capacity for mutual interaction.

Nomothetic–Ideographic Unity. The nomothetic approach seeks for regularities across individuals, whereas the ideographic approach seeks for meaningfulness within individuals. Nomothetic studies are typically oriented toward analysis of stimulus effects and often take average response across individuals as primary data. Ideographic studies are typically oriented toward individual dynamics. Those who employ the nomothetic approach tend to view it as the proper scientific way; those who advocate the ideographic approach tend to view it as the only meaningful psychological way.

Integration theory combines both nomothetic and ideographic approaches. The valuation operation allows for individual differences, including the multitude of hereditary, environmental, and chance factors that make one individual different from another. Furthermore, it emphasizes person–environment interaction, especially as a determinant of the response dimension and the valuation operation (Sections 1.1.3, 1.1.4). Nevertheless, regularities across individuals may still be expected, particularly in the form of the integration operation. Individual differences then appear in the parameters of the integration rule.

This joint nomothetic–ideographic approach was illustrated in the experiment on person perception shown in Figure 1.2. Individual analyses showed that both subjects obeyed the same integration rule—a nomothetic result. However, the personal values of the two subjects were different—an ideographic result. Individual differences in value were harmoniously accommodated within a search for processing regularities across individuals.

The basic unit of psychological analysis is the individual. Group averages are often useful and sometimes necessary, but the ideal situation allows for analysis at the level of the individual. Only individual analysis can provide the precision needed for cognitive theory.

Attitudes and Attitudinal Judgment. In attitude theory it has been useful to distinguish between attitude and attitudinal judgment (Anderson, 1976d, 1981b). This distinction has general relevance, just as does the concept of attitude itself. Although attitude theory has traditionally been sequestered within social psychology, similar concepts arise in many other areas. Currently popular ideas about schemas, for example, have much in common with the concept of attitude. Accordingly, the distinction between attitude and attitudinal judgment is relevant to many later chapters. This distinction also helps bring out the pervasiveness of constructive processes in judgment and decision.

The concept of attitude is commonly defined as an organized mental state that manifests itself as a readiness or predisposition to respond in certain characteristic ways (McGuire, 1969). Thus, a person might have a general attitude that manifests itself in diverse behaviors concerning another person. This definition of attitude is obviously vague, and many investigators, in a quest for precision, have defined attitude in terms of one-dimensional, pro–con response scales. This practice gains justification from its experimental usefulness but does injustice to the full concept of attitude.

Integration theory leads naturally to a distinction between attitudes, which are conceptualized in terms of cognitive structure, and attitudinal judgments, which include response along specific dimensions. One may have a general attitude toward another investigator but make quite different judgments about his conceptual understanding, his experimental abilities, and his substantive accomplishments. Diverse attitudinal judgments may thus derive from the same attitude. Attitudes subserve the valuation operation and play an important role in the cognitive economy.

The concept of attitudinal judgment itself has explicit definition, both conceptual and operational. It allows the precision and penetration of dimensional analysis without forcing the concept of attitude into the same mold. In particular, attitudinal judgments appear to follow a cognitive algebra much like other judgments.

Moreover, the attitude–judgment distinction re-emphasizes the basic role of constructive processes. Integration is a constructive process in an obvious sense: To integrate is to construct a more or less unitary response from multiple, often heterogeneous, stimulus cues. The valuation operation is also a constructive process. This is clear from the dependence of values on the dimension of judgment:

> In certain judgment tasks, the weight and value parameters can seemingly be considered as givens. For example, in judging the likableness of a person described by common trait adjectives, no thought process seems necessary to get the values of the adjectives. They have already been learned and need only be retrieved from storage.
>
> However, many, if not most, judgments would seem to require a chain or net of thought to get the stimulus values. In stimulus interactions, for example, the effective values of each stimulus will depend specifically on the other stimuli that are present. Since the number of such stimulus combinations is virtually unlimited, it seems unlikely that the appropriate response or the stimulus values are located in a passive memory store. In

general, they cannot have been learned beforehand, but instead must be computed on the spot.

In a very real sense, therefore, people do not know their own minds. Instead, they are continuously making them up [Anderson, 1974b, pp. 88–89].

In this constructivist view, the valuation operation has special importance because of its sensitive dependence on the cognitive background, or knowledge structures, of the individual. This background constitutes the interpretive framework through which the situational context and the present stimulus field acquire meaning. This view is not limited to attitudinal judgments, of course, but holds generally for ongoing activity of the individual.

The difficulties of analyzing cognitive organization need no emphasis. By the unitization principle, however, cognitive algebra can make complete and precise allowance for the influence of the background. Progress is thus possible without having to await analysis of the individual's cognitive organization because functional measurement can provide the values in which are encapsulated the outcome of the valuation operation. Moreover, these values constitute boundary conditions that any theory of cognitive organization must obey. In this way, cognitive algebra provides a cogent method for the analysis of cognitive organization.

1.8.3 COGNITIVE ORGANIZATION

This section comments on certain experimental results and on their implications for the study of cognitive organization. Most prominent is the group of studies that showed that verbal reports about interactive processes in person perception were not valid, but instead represented halo effects. Also important is an early finding that person memory or, more generally, idea memory, is distinct from verbal memory. These and a few other results are noted briefly here because they were important determinants of the direction of investigation and of the theoretical structure. These remarks therefore provide a background and context for the more detailed presentations of later chapters.

Meaning-Constancy Hypothesis. A perennially attractive hypothesis about the personality–adjective task is that the adjectives in a person description interact with one another in various ways. In this change-of-meaning view, the meanings of the adjectives are not fixed, but each has a complex dependence on the others. Subjects and experimenters alike give confident reports about such adjective interactions whose existence is phenomenologically clear, almost beyond disbelief.

But this confidence is misplaced, for the effect is only a cognitive illusion. The initial study with this task implied that the adjectives had fixed, constant meanings (Sections 1.2.1, 1.2.5), and later work has corroborated this hypothesis of meaning constancy. There is, indeed, a positive context effect, but that was shown to be a halo effect, not true change-of-meaning (Chapter 3).

A basic problem in the change-of-meaning issue is that of method and evidence. No one could doubt that person perception involves complex cognitive organization. The problem is to develop effective methods of cognitive analysis, methods that can get behind the sometimes illusory face validity of introspective report. Cognitive algebra has provided such a method. The model tests could have failed, thereby supporting the phenomenological view. The actual success of the model goes beyond disproof of the phenomenological view, however, to provide a useful tool for cognitive analysis.

Verbal Memory and Judgment. That judgment depends on memory hardly needs emphasis. Many judgmental studies conveniently and usefully take memory for granted, without concern for its structure or mode of action. However, the earlier discussion of constructive processes indicates that understanding valuation and integration depend on understanding memory—and vice versa. Some relevant results are noted in Section 4.2, but these relations still await systematic study.

There is, however, one experimental result that has exerted considerable influence on the present theoretical development. Because this influence has been largely implicit, and because it disagrees with the more common view, it deserves explicit mention here.

It appears that different memory systems operate in judgment and in verbal recall. This hypothesis emerged from an early experiment on person perception in which subjects had two tasks: to judge likableness of a person described by a short sequence of adjectives and to recall the adjectives themselves (Anderson & Hubert, 1963).

According to the verbal memory hypothesis that prevailed at that time, the person judgment is based on the contents of the verbal memory for the stimulus materials at the time of judgment. The order effects in the two tasks should therefore be similar. That was not verified, however, for recall showed very strong recency, whereas the judgment showed very weak recency. In a critical test with unwarned recall, opposite order effects were predicted and obtained in the two tasks: recency in the recall, primacy in the judgment.

To account for this difference in order effects, it was suggested that the judgment was based on a memory system different from the recall.

Work by other investigators has generally supported this hypothesis of different memory systems (Section 4.2).

This hypothesis about memory suggested the following hypothesis about semantic processing. As each adjective was received, the valuation operation extracted its implications for the task at hand. Further processing, especially the integration, was performed on these implications. The verbal material itself, no longer necessary, was transferred to a verbal memory or forgotten.

As just stated, this view is too simple to hold uniformly. Novel judgments about familiar persons or objects, in particular, will require a constructive process based on retrieval of relevant information from semantic memory (Anderson, 1981b). Nevertheless, the clear distinction between verbal memory and judgment memory in even a limited range of tasks has basic implications for the analysis of cognitive organization. From this standpoint, semantic memory itself plays a relatively passive and subsidiary role. The main cognitive apparatus consists of judgmental operations.

Few areas of psychology have received as intensive study as verbal learning and memory. From the viewpoint of judgment theory, however, this body of work seems strangely incomplete, largely untouched by concern with the use and function of memory. Verbal learning is seldom an end in itself; memory is typically in the service of judgment and decision. Memory retrieval, for example, can hardly be understood without reference to the immediate goals of the person, as reflected in the valuation operation.

The present conceptions of constructive process and of separate memory systems for words and meanings have certain similarities to the views of Bartlett (1932), which have recently begun to attract attention in verbal learning theory (e.g., Cofer, 1973). The present view goes farther, however, to emphasize the functional nature of memory in action. Not only is memory important in judgment, but judgment is important in memory.

Independence of Valuation and Integration. The extreme sensitivity of the valuation operation to contextual factors contrasts sharply with the marked insensitivity of the integration operation. The sensitivity of the valuation operation is reflected in the swift facility of semantic inferences; the insensitivity of the integration operation is reflected in the success of the independence assumption for integration, illustrated, in particular, by the meaning-constancy hypothesis. This sharp contrast indicates that valuation and integration are independent operations.

Independence of valuation and integration imposes structural constraints on any theory of semantic processing. For person perception with trait adjectives, in particular, the following picture is suggested. Task and context act together to define a response dimension and to set up a valuation operation for that dimension. This valuation operation represents inferences, based in part on similarity judgments, in part perhaps on nonce schemas. Each adjective has a fixed existence in semantic memory. Presentation of the stimulus adjective activates the memory representation for processing by the valuation operation. This operation extracts the implications of the given adjective for the dimension of judgment. Once obtained, these implications are transferred to an integration center and there combined to yield the overall judgment.

In this simple picture, the valuation process operates independently on each stimulus adjective and there is no interaction among them. That will not always be true, and interactions can certainly be induced by task and instructions (Section 3.4). The essential point still remains, however, that valuation and integration reflect distinct processes that act independently in some important instances.

The existence of this independence between valuation and integration is extremely fortunate. On it rest the success of the simple integration models and their capability for obtaining functional measures of semantic interrelations. These functional measures open a window into cognitive organization.

Cognitive Organization. Stimulus integration offers many advantages for studying cognitive organization, as may be illustrated by the valuation operation in person perception. Given some trait adjectives in a person description, people readily infer other traits and behavior. This network of inferential relations among the vast array of traits and behaviors that might go with any one person has been called *implicit personality theory.* Aside from its importance for person perception, this inference network has general interest as a realm of cognitive organization. The readiness with which people can make varied inferences from diverse bits of information testifies to a facile, intricate cognitive organization.

Mapping out this network of inferential relations thus becomes a central problem. However, progress on this problem has been halting. The state of knowledge is illustrated by the fact that the most popular index of inferential relations has the most flaws and, indeed, comes near to being self-contradictory (Anderson, 1977c, p. 155).

Information integration theory provides a theoretical and practicable foundation for analysis of implicit personality theory. Any aggregate of stimulus cues may be used as given information; any trait adjective may be included among the myriad of possible response dimensions. The inferences from the stimulus cues to the response dimension are the province of the valuation operation in the functional measurement diagram. These inferences yield the w–s representation, in which w represents stimulus–response relevance and s represents the quantitative projection of the stimulus on the response dimension. Embedding the valuation operation within an integration task makes it possible to obtain validated measures of these inferences.

This approach is not limited to person perception. In principle it applies to any realm of cognitive inference. The sole requirement is an algebraic integration model. The success of cognitive algebra across diverse areas (Sections 1.3, 1.5, 1.7) indicates the promise of this approach. Where it is applicable, it provides a novel tool for analysis of cognitive organization.

Information Integration and Information Processing. The potential of cognitive algebra can be emphasized by comparing it with the process tracing methods that are basic to many current theories of information processing. It seems fair to say that the process tracing methods are incapable of revealing the algebraic integration rules. Furthermore, the algebraic rules provide a capability for cognitive analysis that is not possible with process tracing methods.

This contrast is clearly illustrated by the problem of psychological measurement. Measurement, which is central to the study of cognitive algebra, is virtually ignored in other theories of information processing. Without a measurement capability, it is not possible to proceed very far in the study of continuous- and fuzzy-propositional representations in particular or in the analysis of multiple causation in general.

But the main importance of cognitive algebra is not as a basis for psychological measurement. Measurement is typically no more than a necessary incidental to the study of certain processes. The algebraic rules are important because their structure reflects cognitive processes and because they provide a method for cognitive analysis. In addition to the work on meaning-constancy already discussed, a few experiments from previous sections will be cited here to illustrate these uses.

The Height + Width rule in children's judgment of area (Figure 1.9) provides a simple example of cognitive analysis. This rule appears to reflect a general-purpose adding operation whose study has revealed something of the structure and development of quantitative concepts. Standard process tracing methods would be unlikely to recognize the Height + Width rule and incapable of demonstrating its validity.

A more complex example of cognitive analysis comes from the experiment on inequity (Figure 1.25). These data provided a clear delineation of the underlying comparison structure. Process tracing might perhaps yield hints about such comparison processes, but it would be unable to provide validational evidence. Cognitive algebra simultaneously reveals the processing structure and provides a validational base.

Several experiments have provided examples from semantic analysis leading toward a union of semantic theory and judgment theory (Figures 1.2, 1.4, 1.15, 1.16). In particular, cognitive algebra is well equipped to handle continuous semantic variables, thereby avoiding the artificial discreteness that has been typical of propositional representations (Section 1.8.2).

Much of the preceding discussion applies also to reaction time analysis, which constitutes a principal tool in several theoretical approaches. Although algebraic models of processing time have been studied (e.g., Section 1.5.7), processing time has only indirect relation to the informational content being processed. It seems doubtful, therefore, that reaction time approaches would be able to make much headway on the issues just cited or on those considered in previous sections of this chapter.

These remarks are not intended to criticize the narrowness of current information processing theories, but to bring out the potential of cooperation with cognitive algebra and general judgment theory. There are many problems that can arise in choice tasks, problem solving, and language processes for which cognitive algebra is insufficient. Process tracing and reaction time methods can be helpful for such problems. In return, cognitive algebra can help in process analysis. It can measure functional values, and these values provide boundary conditions that more detailed theories of processing must obey. It can also reveal certain forms of processing structure, and these algebraic structures provide sharp results that any deeper theory of processing must explain.

1.8.4 PERSON PERCEPTION AS AN EXPERIMENTAL TASK

The first phase of this research program was largely devoted to one experimental situation, namely, the personality adjective task illustrated in Section 1.2.1. This may seem an unlikely beginning, especially for a quantitatively oriented approach, but person perception has many characteristics that make it attractive for experimental study. It is basic to human cognition, socially important, and experimentally flexible. These three task characteristics deserve comment.

Social Cognition. Person perception has interesting similarities to object perception, but has more complex cognitive structure (Schneider,

Hastorf, & Ellsworth, 1979). Even face perception, which has been a focus of many perceptual studies, involves concepts such as emotion and intention that do not ordinarily arise in object perception, except perhaps as personification. More generally, perception of another person blends into attitudes toward that person and into the perceiver's value system, and all have forms of organization not found in perception of objects. Not only is the informational base far more variegated, but a complex network of consistency and similarity relations among the information is active in perception of the person.

Moreover, person perception is heavily involved with language processes, for the person lives in a sea of words from birth to death. Foremost among these linguistic processes are the valuation operations that extract relevant meanings and implications from the given words. No less important are associated nonverbal stimuli, for evaluation of another's words will depend on many context factors, including tone of voice and facial expression, not to mention past behavior. Indeed, person perception has special interest because of the joint importance and equivalent status of verbal and nonverbal cues. For both kinds of cues, moreover, the valuation processes are intimately related to the person's cognitive structures of knowledge, attitudes, and values.

Theoretical interest in person perception also arises from its functional or instrumental nature. Interpersonal interactions are generally motivated toward various goals, and perceptions of attributes and intentions of others are instrumental to achieving these goals. This functional aspect endows person perception with a broader outlook than is commonly found in cognitive theories, most of which seem overly involved with structure of semantic memory. Because person perception is centrally concerned with ongoing behavior, it places central emphasis on judgments and values. From this standpoint, the main cognitive apparatus is not in semantic memory, but in a superstructure of dynamic operations of judgment and decision that functions in a continually changing environment.

Social Relevance. The social importance of person perception is clear. The development of the child is overwhelmingly dependent on person perception within family, playground, and school. Adult life is dominated by interpersonal actions from requests for directions to seekings for affection and praise—actions that are intimately keyed to perceptions about traits and motivations of other persons. The primary role of marriage and family life, which are intensely interpersonal, emphasize the social importance of person perception. The unique status of self-perception also deserves consideration, in part for its intrinsic interest, in part for its dependence on perception of others.

Person perception has been called the naive, commonsense, or everyman's theory of personality. People do not have direct knowledge of one another, only theories. That these commonsense theories are somewhat lacking is illustrated in the frequent divorce complaint that "he/she doesn't understand me." The study of the cognitive processes that underlie these commonsense theories is important as a means to improve their functioning.

Experimental Advantages. Person perception has many advantages for experimental analysis. Foremost is that it is natural and meaningful to the subjects. That is not only a convenience but, more important, an indication that the behavior will embody important cognitive skills. At the same time, it is possible to embed experimental studies within the ongoing context of daily life (Section 4.5.2). In addition, person perception provides opportunities to apply developments from other areas of psychology, ranging from object perception to psycholinguistics and decision theory.

Experimentally, the task has great flexibility. The stimulus materials may range from single adjectives to letters of reference, case histories, and personal interviews. Moreover, a variety of different judgments may be obtained that provide contrasting perspectives on the cognitive representation of the given stimulus field. Results from any one experiment may be peculiar to procedural details and may be explainable in a variety of ways. Experimental flexibility is important, therefore, to allow the interlocking constraints that are necessary for theoretical analysis.

Such considerations of research strategy were one reason for concentrating on the personality adjective task in the initial phase of this research program. Although the experiment of Figure 1.2 showed a simple pattern of parallelism, the overall picture was moderately complex. Set-size effects, primacy–recency effects, and positive context effects had to be considered. Popular hypotheses about stimulus interaction, such as contrast and change-of-meaning, had to be assessed. Deviations from the parallelism property presented considerable difficulty because they could result from several different causes. In the early experiments, especially, the possibility that the rating response was invalid created ever-present uncertainty. Because of these complexities, initial effort was largely concentrated on this one task.

As it turned out, this complex of data could be accounted for by a simple theory with a minimum of assumptions. The basic results, presented in Chapters 2-4, show that the theory has survived a number of demanding tests, both qualitative and quantitative. It rests on a substantial interlocking network of evidence. It has required very little emendation and stands in essentially its original form.

The theory developed in this initial work with person perception has proved useful in other areas. Later work has been concerned with generalizations and extensions to many areas that will be covered in subsequent volumes: decision theory, psychophysics, learning and motivation, attitude theory, attribution theory, moral judgment, psycholinguistics, and developmental psychology. Almost every such attempt has led to new outlooks on old problems and to new results. This work adds confidence to the integration-theoretical approach and affirms the belief that it is leading to a unified, general theory.

2

Basic Experiments on Information Integration

2.1 Person Perception as Information Integration

2.1.1 THE PERSONALITY ADJECTIVE TASK

Most of the experiments in these next three chapters are built around a single task of person perception. A hypothetical person is described by a few personality traits such as *humorless, level-headed,* and *tactful.* The subject forms an impression of this person and rates the person on likableness. Among the advantages of this task are its experimental flexibility, its naturalness to the subjects, its ecological relevance, and its scientific interest (Section 1.8.4).[a]

From an integration-theoretical perspective, the operations of valuation and integration are central in this personality adjective task. The adjectives are discrete verbal signs that have to be interpreted and made meaningful within the cognitive system of the individual subject. Furthermore, they have to be integrated into some more or less unitary impression of the person. Valuation and integration are thus primary problems for investigation.

A popular, attractive view is that valuation and integration involve dynamic interaction among the adjectives. As they combine into a unified impression, the adjectives are hypothesized to influence one another's meanings in complex ways. If the adjectives were inconsistent, for example, some reinterpretation of their meanings would occur to help resolve the inconsistency. More pervasive and subtle changes of meaning have been claimed by Asch (1946), who made an extended study of responses to such adjective descriptions. This interactive view has strong introspective support. Indeed, Asch's main evidence consisted of subjects' verbal protocols. Because of this commonsense appeal, Asch's view has remained popular to this day.

Under experimental analysis, however, a quite different view has emerged. The valuation and integration operations, which were not distinguished in the popular view, turn out to be largely independent. Furthermore, the integration itself obeys algebraic rules. This work has led to basic reorientation for theory and experiment. The main evidence for this theoretical formulation is given in the present and following two chapters. [b]

2.1.2 FOUR BASIC PROBLEMS

This chapter discusses four basic problems that were taken up in the initial stage of research on information integration theory. The first problem was to test an adding-type model. It worked surprisingly well, as shown by the parallelism in the data. Contrary to general expectation, little sign of interaction among the stimulus adjectives was seen. Subjects appeared to form impressions of another's personality by a simple algebraic rule.

The second problem was additivity. There were two integration rules that could account for parallelism, namely, adding and averaging. A qualitative test was found that provided strong discrimination between these two rules. This critical test showed that the integration process was fundamentally nonadditive, thereby eliminating a general class of adding-type models. The averaging model itself, however, still faced several problems.

The most obvious problem for the averaging model was the set-size effect. If one piece of information yields a favorable response, two pieces of equal value will yield an even more favorable response. More generally, the response to a set of stimuli can be more extreme than the response to any single stimulus, which is clearly contrary to a simple averaging rule. The averaging model employed in integration theory was able to provide a good account of the set-size effect, however, by making use of the concept of initial impression.

Closely related to the set-size problem is the problem of serial integration. Information is typically received and integrated one piece at a time, in some particular order. This order makes a difference. The same information can produce substantially different judgments when given in different orders. Order effects seem to imply stimulus interaction, and that was how they initially were interpreted. The last section of this chapter shows how the averaging model was extended to provide an exact treatment of order effects without invoking stimulus interaction.

2.1.3 THE ADJECTIVE LIST

Since 1964 a master list of 555 personality trait adjectives has been used in this research program. These adjectives were rated on a 0–6 scale of likableness by 100 college students, and the mean ratings, together with auxiliary information, are given in Anderson (1968d).

This list is useful as a reference compilation. The normative ratings are not a validated functional scale, of course, but they are a guide to selecting adjectives for specific purposes. In particular, they are useful for selecting categories of adjectives that have graded normative value.

Four such graded categories have been used extensively and are denoted by H, M^+, M^-, and L. In most of the experiments, these labels refer to the following sublists of 32 adjectives, which are tabulated in Appendix B.

H : From *reasonable* (5.00) to *truthful* (5.45)
M^+: From *painstaking* (3.45) to *persuasive* (3.74)
M^-: From *unpopular* (2.22) to *dependent* (2.54)
L : From *spiteful* (.72) to *abusive* (1.00)

When fewer than 32 adjectives are needed, a further selection is sometimes made on the auxiliary indexes of meaningfulness and variability. The extreme words on this list (e.g., *sincere, honest, understanding, loyal, truthful;* and *malicious, dishonest, mean, phony, liar*) are usually reserved for use as end anchors.

2.2 The Parallelism Test

The parallelism test of Section 1.2 has been a key tool of integration theory. If the integration model is of the adding type, and if factorial design is used to construct the stimulus combinations, then the data should plot as parallel curves. This test is widely useful and quite simple. All that is necessary is to plot the raw data and look.

2.2.1 FIRST TEST OF PARALLELISM

Group Test. The first test of the integration model for the personality adjective task is summarized in Figure 2.1. Each stimulus person was described by three trait adjectives, and the subjects rated likableness of these stimulus persons on a 1–20 numerical scale. A three-factor design was used, with Low, Medium, and High adjectives as the three levels of each factor. Thus, there were 27 person descriptions, corresponding to the 27 data points in the figure. Each panel in the graph represents a 3 × 3 design for one value of the third adjective.[a]

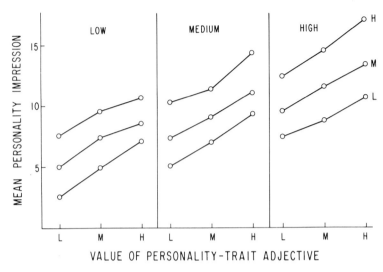

FIGURE 2.1. *Parallelism test in person perception. Mean likableness of 27 persons, each described by three personality trait adjectives. L, M, and H represent trait adjectives of low, medium, and high value. Adjectives were combined in a 3^3 design, as portrayed in the graph. Experimental detail in Note 2.2.1a. (After Anderson, 1962a.)*

The essential feature of Figure 2.1 is the parallelism. Each separate panel of data exhibits parallelism; moreover, the three panels are essentially congruent. There are some scattered deviations from parallelism, but no more than is to be expected from response variability. These data, therefore, give clear support to the integration model.

This group test automatically takes account of the different adjective values for each individual subject. Each subject served in all conditions, and so is represented at all 27 data points. If each subject obeys the model, then the corresponding plot will be parallel for each subject, and hence also for the group average. The analysis is effectively on an individual basis, therefore, although the test is only sensitive to deviations from parallelism that are systematic across subjects.

Individual-Subject Analysis. The group test of Figure 2.1 is incomplete because deviations from parallelism present in the individual data could cancel in the group average. This experiment, however, was specifically designed to allow individual-subject analysis. Each subject was run on 5 consecutive days and judged all 27 descriptions on each day. The data of the last 3 days were used for the individual analyses that are summarized in Table 2.1.

The test of the model is given by the F ratios for nonadditivity in

Table 2.1. They represent the pooled interactions and give an overall test of deviations from parallelism. F's are near the chance level of unity for the first nine subjects, though they are significant for the last three. Since two of these subjects, M. G. and N. B., had the same set of adjectives, the deviations could reflect interactions among these particular adjectives. However, the two-way data plots for these two subjects showed a divergence interaction of the form that would be expected if adjectives of lower value had greater weight (Section 2.2.4).

Graphs for the first two subjects have already been presented in Chapter 1; Figure 1.2 illustrates parallelism at the individual level. Overall, therefore, this experiment provided promising support for the model.[b]

Implications for Substantive Theory. The most important outcome of this experiment is, of course, the support for the integration model. The observed parallelism indicates that subjects integrate the adjective information with a simple adding or averaging rule.

This was a surprising conclusion. Stimulus interactions were generally expected to be prominent in person perception, with each trait adjective influencing the meanings of the others in diverse ways. Yet the

TABLE 2.1
Summary of Single-Subject Analyses of Integration Model for Person Perception

Subject	F for nonadditivity	F for additivity	MS_{Error}	Correlation: predicted and observed
FF-1	.84	54.50	3.42	.98
RH-1	.48	50.18	1.67	.98
AR-2	1.65	223.94	.57	.99
AT-2	.94	52.33	1.75	.97
JW-3	.94	126.87	.80	.99
DB-3	1.10	36.53	5.56	.95
LL-4	.86	66.23	4.02	.98
JZ-4	1.37	54.75	2.02	.96
BM-5	.67	31.41	6.11	.97
FM-5	2.53	74.65	1.47	.95
MG-6	2.84	105.97	2.90	.94
NB-6	2.45	74.88	2.79	.95

SOURCE: After Anderson (1962a). (Copyright © 1962 by American Association for Advancement of Science. Reprinted by permission.)

NOTE: Suffix on subject initials indicates group of stimulus adjectives. The F-ratios are pooled over all three design factors since all had L, M, and H adjectives as their three levels. Thus, the F's for Nonadditivity and Additivity have 20 and 6 df, respectively. Error term has 54 df. With 6/54 df, an F ratio of 2.28 is significant at the .05 level.

parallelism test rests on the independence assumption that the adjective values are fixed and constant. Interaction among the adjectives would in general produce deviations from parallelism (Section 1.2.5). Hence the observed results argue that the adjectives keep a fixed meaning across different combinations.

Even granted that the adjectives do not interact, it is still striking that the integration should follow a simple algebraic rule. The task instructions do not suggest any such algebraic rule. Furthermore, subjects' reports give little sign of such a rule, for they speak of lack of conscious awareness about the integration process or of complicated interactions among the adjectives.

Moreover, there was little theoretical basis for expecting a noninteractive rule. It would be easy to postulate an additive rule, of course, but not because it followed from any theory of person perception. Quite the opposite: the observed results were contrary to nearly all prevailing views. The outcome of this experiment, therefore, had far-reaching consequences for the study of person perception in particular, and for the study of information integration in general.

Implications for Measurement Theory. Beyond its substantive interest, the integration model also solves the two problems of measurement—stimulus and response—shown in the functional measurement diagram of Figure 1.1. By virtue of the parallelism theorem, the observed parallelism has two measurement implications:

It validates the rating response as a linear scale.
It yields linear scales of the functional stimulus values.

This measurement logic is not peculiar to person perception, but can be applied generally. Subsequent applications have shown promise in many areas, as already illustrated in Section 1.3.

2.2.2 SCALING SOCIAL DESIRABILITY

Functional Scales of the Stimulus. Measurement theory in psychology has traditionally been tied to stimulus scaling. This traditional emphasis is apparent in the popular Thurstonian methods, which view stimulus scaling as a methodological preliminary to substantive inquiry. Functional measurement takes an opposite orientation in which scaling is derivative from substantive laws. As a consequence, primary concern shifts to response scaling.

Because of this orientation, stimulus scaling loses much of its explicit interest. Many models are tested with only two levels in each design factor. In such designs, stimulus scaling may not be possible because the zeros and the units of the scales are arbitrary. Even three levels of a factor allow only one nonarbitrary stimulus scale value. Stimulus scaling thus involves design considerations somewhat different from those used in testing integration models.

A Scaling Experiment. The present experiment was designed to illustrate how integration theory can be used for scaling social stimuli. The judgment dimension was changed from likableness, which has been customary in this research program, to social desirability. This response dimension was used to emphasize the relevance of functional scaling to traditional scaling concerns. Twelve subjects made graphic ratings of person descriptions containing two adjectives in a 3 × 10 design. This long, thin design is desirable for scaling purposes since it provides a linear scale of the more numerous column stimuli with near-minimal cost.

The data are shown in Figure 2.2. The parallelism of the three curves validates the integration model. According to the parallelism theorem, therefore, the marginal means constitute a linear scale of the adjectives on the dimension of social desirability. These functional scale values are listed for the 10 column adjectives along the horizontal axis in each panel. These scale values correspond graphically to the mean of the three curves in the figure.

Affective Bimodality. An interesting incidental result from Figure 2.2 is the affective bimodality. The adjectives seem to break into two clusters, one negative, one positive. That is shown by the sharp jump in each curve, between *dependent* and *inoffensive* in the lower panel, and between *preoccupied* and *hopeful* in the upper panel. Within each cluster the trend is relatively mild.

This bipolar clustering is similar to the bimodality in the master list of 555 traits. Neutral traits are infrequent. Subjects appear to give a definite value, positive or negative, to each personality trait. Similar bipolarity has been observed in constructing a collection of president paragraphs for use in attitude change experiments (Anderson, Sawyers, & Farkas, 1972). Bipolarity in the valuation operation may thus be a fairly general phenomenon, at least for affective dimensions.

It is worth noting how the integration design demonstrates the reality of this bipolar clustering. If only the middle curve had been obtained, it could be argued that the clustering was merely bias in the response scale, that the subjects tended to avoid neutral responses. On this interpretation, the observed clustering would reflect bipolarity or nonlinearity in response usage, not true bipolarity in the stimulus values.

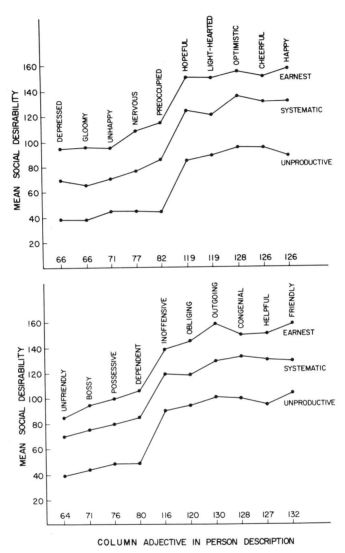

FIGURE 2.2. *Functional scaling of social desirability. Mean social desirability of persons described by two personality trait adjectives. Each point is the mean of one judgment by each of 12 subjects. Data in each panel were averaged over two different replications of the row adjectives for each of six subjects. (After Anderson, 1973b.)*

In the three curves from the integration design, however, the jumps occur at different places in the response scale, and all three curves maintain their parallelism. That rules out the response bias interpretation; it also illustrates the validational power of the parallelism property.[a]

2.2.3 PARALLELISM AS NATURAL PROCESSING

A not infrequent objection to the parallelism results is that they may represent superficial processing. These experiments ordinarily require a considerable number of judgments from each subject. In the experiment of Figure 2.1, for example, the same 27 three-adjective descriptions were judged in each of five sessions. In the experiment of Figure 2.2, 60 two-adjective descriptions were judged in the single session. Faced with this many judgments, it is argued, subjects may abandon their normal complex mode of processing and adopt some easy rule to get through the experiment. In this view, the simple integration model might apply to snap judgments but not to more considered judgments of other persons.

Evidence against this objection was obtained in the following comparison between Paragraph and No-paragraph conditions. The No-paragraph condition used essentially the standard procedure: Subjects read the three adjectives in the description and then judged the person on likableness. The Paragraph condition required subjects to spend 2 minutes writing a description of the person in their own words before judging likableness. Paragraph subjects should therefore have a much more articulated and developed impression. So if parallelism results from superficial processing, then the Paragraph condition should yield nonparallelism.

The data are shown in Figure 2.3. The main result is the parallelism of the filled-circle curves for the Paragraph condition. This result shows no support for the argument that parallelism results from superficial processing.[a]

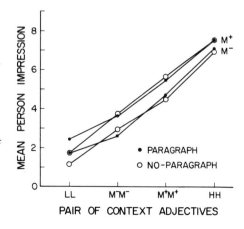

FIGURE 2.3. *Test of superficial processing objection. Mean likableness of persons described by three personality trait adjectives. Paragraph and No-paragraph conditions each contained 48 subjects, each of whom judged the eight descriptions from the 2 × 4 design portrayed in the figure. Similarity of Paragraph and No-paragraph conditions rules out superficial processing objection. (From Anderson, 1971c, Experiment 2. [Copyright © 1971 by Academic Press. Reprinted by permission.])*

Also important is the near-identity of data from Paragraph and No-paragraph conditions. Subjects who work over the stimulus material in detail arrive at the same final response as subjects who use more casual procedure. This comparison provides a second source of support for the generality of results obtained with the standard procedure.[b,c]

2.2.4 PARALLELISM AND NONPARALLELISM

The simplicity and parsimony of parallelism analysis make it attractive for the study of stimulus integration. Of course, the parallelism theorem has value only if it is empirically applicable. The initial studies in person perception were important because they indicated an empirical validity for parallelism analysis.

However, many cases of nonparallelism have also been found, and they present a serious theoretical problem. Whereas parallelism points to a simple theoretical interpretation, nonparallelism is theoretically ambiguous and uncertain (Section 1.2.6). Nonparallelism might result from nonlinear bias in the response measure or from a nonlinear integration model—or from both. If scale and model were wrong together, it would be difficult to determine what was what.

Moreover, prevalent nonparallelism raises a concern that the parallelism obtained in the cited experiments was accidental or, at best, limited to special conditions. Indeed, there were two good grounds for such concern. One was the plausible belief that person perception could not follow a simple algebraic model. The other was the prevailing belief that rating scales were generally biased and nonlinear. Until nonparallelism was understood, therefore, it caused concern that what parallelism had been obtained was fortuitous.

As it happened, there was a straightforward explanation for much of the nonparallelism that has been obtained. In this interpretation, stimulus integration is seldom truly additive or linear. Instead, averaging processes are predominant. The averaging model appears additive under special conditions of equal weighting, and that accounts for the observed parallelism. Inherently, however, the averaging model is nonadditive and predicts systematic deviations from parallelism under specifiable conditions.

The averaging rule provides a theoretical unification of most of the results that have been obtained. Furthermore, it indicates the experimental conditions that are needed to apply parallelism analysis. It is essential, therefore, to demonstrate the empirical validity of the averaging rule. Logically enough, the critical evidence for averaging is to be found in the nonparallelism itself. This special manifestation of nonparallelism is the theme of the next section.

2.3 The Averaging Hypothesis

Two quite different integration rules can account for the parallelism results of Section 2.2. These are the adding rule and the averaging rule discussed in Section 1.6. This issue of adding versus averaging has basic importance for person perception as well as for general theory. This section gives the main, direct evidence on this issue and shows that the adding hypothesis is untenable. Serious problems for the averaging hypothesis are also found that need to be resolved in later sections and chapters.

2.3.1 QUALITATIVE TEST BETWEEN AVERAGING AND ADDING

The averaging and adding hypotheses can be distinguished by a simple critical test. The idea is to add mildly positive information to very positive information. That should increase the response if the adding hypothesis is true; it may decrease the response if the averaging hypothesis is true. An observed decrease would therefore argue against an adding model and support an averaging model.[a]

This test is qualitative, or scale-free. It depends on a difference in direction only, not in amount. This test is robust, therefore, in that only a monotone response scale is required.

This qualitative test is robust in an additional sense. It rules out variant forms of the adding hypothesis that do not assume exact addition. An adding rule with diminishing returns, for instance, would make the same directional prediction as the strict adding model. Similarly, the test rules out some, though not all (Section 4.1.1), models that assume stimulus interaction overlaid on a basic adding process. Thus, the qualitative test has the great advantage that it eliminates a large class of models all at once.

2.3.2 FIRST EXPERIMENTAL TEST BETWEEN AVERAGING AND ADDING

Disproof of Adding. The results of the first test between adding and averaging in person perception are shown in Table 2.2. Stimulus persons were described by sets of two or four trait adjectives of selected value. Their likableness was rated on an open-ended numerical scale with 50 specified as neutral.

The four critical entries are in boldface type. In the top row, the comparison is between HH and HHM$^+$M$^+$. The mean likableness of a person described by two very favorable adjectives, HH, is 72.8. The addition of two mildly favorable adjectives, M$^+$M$^+$, causes the response to decrease to 71.1, contrary to the adding hypothesis.

TABLE 2.2
Test of Averaging and Adding Hypotheses in Person Perception

Set type	Response	Set type	Response	Set type	Response
HH	**72.8**	HHM⁺M⁺	**71.1**	HHHH	79.4
M⁺M⁺	57.6			M⁺M⁺M⁺M⁺	63.2
M⁻M⁻	42.2			M⁻M⁻M⁻M⁻	39.5
LL	**23.7**	LLM⁻M⁻	**25.7**	LLLL	17.6

SOURCE: After Anderson (1965a).
NOTE: Entries are mean judgments of likableness on open-ended rating scale with neutral point at 50. H, M⁺, M⁻, and L stand for personality traits of very high, moderately high, moderately low, and very low value.

The bottom row tells the same story. The LL person, described by two very unfavorable adjectives, is rated at 23.7. Addition of mildly unfavorable, M⁻M⁻ information makes the person more likable.ᵃ This result also contradicts the adding hypothesis. As stated in the original report (Anderson, 1965a), "Since the critical result is qualitative, resting on a direction of difference, it cannot reasonably be attributed to a shortcoming in the scale of measurement [p. 397]."

Trouble for Averaging. Qualitative tests are much better at disproof than at proof. Disproof of the general adding hypothesis does not prove averaging, and in fact Table 2.2 contains three difficulties for the averaging model. The first of these is a small deviation from parallelism among the negative adjectives; that will be taken up in Section 4.4.

The second problem for averaging is that the response to HHM⁺M⁺ is not the simple mean of the responses to HH and M⁺M⁺. Indeed, the two values are quite far apart: 71.1 for HHM⁺M⁺ compared to 65.2 for the mean response to HH and M⁺M⁺. A similar discrepancy can be seen for the negative adjectives. A simple averaging model cannot account for these results. The third problem for averaging is the set-size effect. Comparison of the first and last columns shows that four adjectives of equal value produce a considerably more extreme response than two adjectives of the same value. For example, descriptions with two and four H adjectives yield responses of 72.8 and 79.4, respectively. This set-size effect appears at all four value levels, H, M⁺, M⁻, and L. A simple averaging model would predict the same response for four adjectives as for two adjectives.

Both of these last two results were actually predicted from the averaging hypothesis. Theoretically, both results reflect the action of an initial or general impression, I_0, an internal state variable that is averaged in with the external stimuli. This aspect of averaging theory will be considered in the discussion of the set-size effect in Section 2.4.

2.3.3 FURTHER TESTS BETWEEN AVERAGING AND ADDING

Hendrick (1967, 1968a). In his doctoral dissertation, Clyde Hendrick gave a careful, extensive study of the averaging–adding issue. This work had special value since it was the first independent replication of the original study.

Hendrick's data, in Table 2.3, rule out the adding hypothesis by virtue of the same qualitative test discussed above. In each experiment, the addition of a mildly positive M trait reduces the likableness of a person described by a very positive H trait. Hendrick's corroboration of the averaging effect is especially valuable because each of his three experiments was done under different conditions in order to assess the generality of the effect.

An incidental result in Table 2.3 is that likableness for the redundant pairs averages out a bit more extreme than for the nonredundant pairs. A real redundancy effect would go in the opposite direction. This result is surprising at face value but is not atypical of redundancy effects in person perception. Redundancy effects can be obtained, but they are commonly weak and often irregular.

A Test with Serial Presentation. Nearly all tests of averaging–adding have used simultaneous presentation in which all the information is present at one time. In practice, however, information is usually received a piece at a time and is integrated in a serial manner. Accordingly, it is important to assess the averaging hypothesis using serial presenta-

TABLE 2.3
Hendrick's Test of Adding versus Averaging

Experiment	N	H	Set type		M
			Redundant HM pairs	Nonredundant HM pairs	
1	80	16.2	15.4	15.0	11.4
2	64	16.1	15.2	14.8	11.5
3	84	15.6	14.0	14.3	10.7

SOURCE: After Hendrick (1967, 1968a).
NOTE: Entries are mean likableness judgments on a 20-point scale of persons described by one or two personality trait adjectives. H and M denote one-adjective descriptions of high and medium value, respectively. N denotes number of subjects in each experiment.

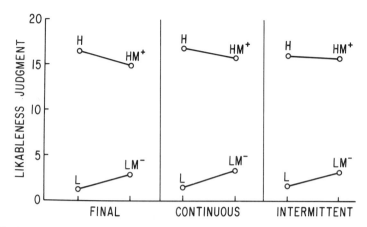

FIGURE 2.4. *Critical test between adding and averaging. Mean likableness of persons described by three or six trait adjectives presented in serial order. H and L denote sequences of three adjectives, HM⁺ and LM⁻ denote sequences of six adjectives. Subjects in final, continuous, and intermittent conditions responded only after all adjectives had been presented, after each adjective was presented, or after the third and sixth adjectives, respectively. (After Anderson, 1968a.)*

tion. Data on this question were obtained as part of a larger experiment and are summarized in Figure 2.4.

In Figure 2.4, H and L denote sequences of three High and three Low adjectives, respectively. Similarly, HM⁺ and LM⁻ denote sequences of three H followed by three M⁺ adjectives, and three L followed by three M⁻ adjectives. All subjects saw the adjectives one at a time in sequential order but responded under one of three conditions represented in the three panels of the figure. In the final condition, subjects responded only at the end of the sequence. In the continuous condition, subjects responded in a cumulative manner after each successive adjective. In the intermittent condition, subjects responded only after the third and sixth adjectives in the sequence. Only the response at the end of the sequence is plotted here.

The critical aspect of Figure 2.4 is the slope of each connected pair of points. Any kind of adding process would imply that the top curves should slope upward, from left to right, because M⁺ is added to H. Similarly, the bottom curves should slope downward, from left to right, because the negative M⁻ is added to L. The slopes are exactly opposite in every case. That rules out the adding hypothesis but is completely consistent with the averaging hypothesis.

All three conditions show roughly the same magnitude averaging effect. That rules out possible demand effects in the continuous responding condition, in which subjects revised their response to take account of each added adjective.

Gollob and Lugg (1973). Gollob and Lugg were concerned that the critical averaging result of Section 2.3.2 was a peculiarity of instructions and procedure. They conjectured that more naturalistic experimental conditions would yield adding rather than averaging. Accordingly, they included several conditions in which the adjectives were embedded in a sentence, or the "equal importance" instructions were omitted, or stimulus presentation was self-paced. Their work is notable for its meticulous attempts to maintain exact comparability, including an exact replication of the original experiment reported in Anderson (1965a).

Their results uniformly supported averaging. The overall effect of adding the mildly polarized information was to decrease response polarity; the differences among the five instruction conditions did not approach significance. As Gollob and Lugg concluded, the averaging effect appears to be experimentally robust.

A Choice Test. All the preceding tests of averaging–adding have used a numerical rating response. However, ratings are often suspected of biases and extraneous influences. A choice test removes objections that might be leveled against the rating experiments.

Such a choice experiment is summarized in Table 2.4. Subjects read aloud each of two person descriptions, then chose which of the two persons would be more likable. The upper left quadrant shows that HH was chosen 68 times against 52 choices of HHM$^+$M$^+$. The upper right quadrant shows that LL was chosen only 47 times compared to 73 choices of LLM$^-$M$^-$. Both sets of data contradict the adding hypothesis; both support the averaging hypothesis. Both comparisons in the two lower quadrants tell the same story.

This choice test served its purpose, but it was also a sharp reminder of the difficulties and inefficiency of choice experiments. The choice between HH and HHM$^+$M$^+$ naturally requires different H adjectives in the

TABLE 2.4
Choice Test of Adding–Averaging

HH	HHM$^+$M$^+$	LL	LLM$^-$M$^-$
68	52	47	73
HH	HHM$^+$M$^+$M$^+$M$^+$	LL	LLM$^-$M$^-$M$^-$M$^-$
64	56	37	83

SOURCE: After Anderson and Alexander (1971, Experiment 2).
NOTE: The pair of entries in each quadrant of the table is based on two choices by each of 60 subjects so that the chance level for each entry is 60. In each quadrant, the comparison favors the averaging hypothesis.

two descriptions. Using the same HH pair in both cases would be inappropriate because it would allow a possible cancellation strategy and so vitiate the interpretation. But since the effect is theoretically small, it was considered mandatory to equate the H adjectives across pairs for each separate subject before beginning the experiment. Considerable precautions were used in presenting the two descriptions in each pair in order to avoid biases from order of presentation and other attentional factors. These precautions were successful, but it would seem generally advisable in any future choice work to get direct ratings of degree of preference in addition to the choices themselves. Indeed, as was suggested in the article, the preference for choice data may be misplaced; rating responses to single descriptions may actually be less influenced by extraneous factors.

2.3.4 IMPROVED QUALITATIVE TEST BETWEEN ADDING AND AVERAGING

The test between the adding and averaging hypotheses in the experiments of the two previous sections is demanding in two respects. Theoretically the averaging effect is small (see Section 2.4). To maximize the effect requires taking the M^+ (or M^-) information as close to neutral as possible; otherwise, the set-size effect may override the effect expected from stimulus averaging alone. But coming close to neutral entails the risk of going too far; accordingly, it is necessary to provide separate evidence that the M^+ and M^- traits are indeed positive and negative, respectively. For the experiment of Table 2.2, the needed evidence is provided by the set-size effects for the M^+ and M^- traits (see Note 2.3.2a). Without due care, however, the averaging and adding models may make the same directional prediction.

A less demanding test can be obtained by adding the same medium information to both high and low information. That should change the response in the same direction according to an adding formulation, either up or down, depending on whether the medium information is positive or negative. In contrast, an averaging formulation implies that adding the medium information can make the response less extreme in both cases. This test allows a maximal averaging effect without requiring any assessment of the polarity of the medium information. However, it is still necessary to allow for set-size effects.

Lampel and Anderson (1968). The first application of the improved qualitative test is shown in Figure 2.5. Females made date-ratings of males described by a photo, plus zero or two personality traits.

The main design in this experiment is the 4 × 3, Adjective × Photo-

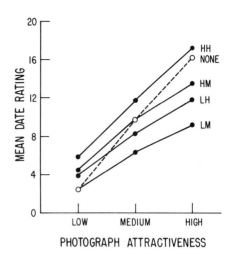

FIGURE 2.5. *Crossover interaction supports averaging, eliminates adding. Mean date rating of males described by photograph (horizontal axis) and zero or two trait adjectives (curve parameter). (After Lampel & Anderson, 1968. [Copyright © 1974 by Academic Press. Reprinted by permission.])*

graph factorial represented by the four filled-circle curves. These data show marked nonparallelism, with the four adjective curves diverging as the attractiveness of the photograph increases. This nonparallelism has a straightforward interpretation in the averaging model. The weight or importance of the photograph is assumed to vary inversely with attractiveness. That makes social sense. Owing to social pressures, females may not wish to be seen on a casual date with an unattractive male, regardless of his personality. In a more attractive male, personal qualities would begin to take on greater importance.[a]

In the averaging model the weights must sum to 1. When the photograph has a higher absolute weight, the adjectives must have a lower relative weight. This relative weight controls the vertical spread of the four adjective curves since the adjective scale values are fixed for each curve. Thus, the vertical spread of the curves is a direct function of the effective weight of the adjectives and an inverse function of the effective weight of the photograph. This pattern of nonparallelism occurs fairly frequently in integration theory. In the present experiment it was taken as a sign of differential weighted averaging (see Eq. [22], Section 1.6.3).

Another interpretation should also be considered. It has been argued that the nonparallelism in these data results from a rating response that is biased, not a true linear scale. This argument has two bases. The first is that the four filled-circle curves could be made parallel or additive by a monotone transformation of the response. The second is that data that can be made additive should be made additive. This latter statement seems to rest on an idealistic faith in additivity, a faith that is at least understandable in view of the difficulties of interpreting nonparallelism discussed elsewhere.[b]

However, there are three reasons to rule out the additive interpretation of these data. One is pragmatic, based on response generality. This experiment used the same methodology that had given parallelism in previous work. That background experience gives some measure of confidence that the ratings are a valid linear scale. This pragmatic reason is not conclusive, of course, but it does mean that the present nonadditivity should not be lightly scaled away.

A compelling argument against any additive interpretation comes from the no-adjective dashed curve of Figure 2.5. The critical feature is that the no-adjective curve crosses over the LH and HM curves. The LH curve represents the addition of the near-neutral LH combination to the photograph. It pulls up the response to the low photograph and pulls down the response to the high photograph, exactly in accord with the averaging hypothesis. The HM curve, which represents the addition of combined high and medium adjectives, supports the averaging hypothesis in the same way. This crossover interaction shows that the integration process is inherently nonadditive.

Auxiliary support for the averaging interpretation comes from the interplay of the two integration operations in this task and design. The four filled-circle curves represent a 2×2, Adjective $\times$ Adjective design that yielded a nonsignificant interaction. That agrees with the parallelism results for adjective integration cited in Section 2.2. It also implies that the unequal weighting must be in the photograph factor. And most important for present discussion, it supports the linearity of the response scale. That implies that the nonparallelism in Figure 2.5 represents real nonadditivity in the other integration operation, involving photograph and adjectives.

A different interpretation of Figure 2.5 is that the adjective acts as an intensifier or multiplier of the photograph. The adjective–photograph integration could then exhibit a linear fan pattern. Under the simplest multiplying model, however, the curves would intersect at the affective zero on the photograph scale, which would therefore have to be lower than the value of the low photograph. This interpretation was rejected on the presumption, unfortunately not testable in this design, that the affective zero should be in the middle of the set of photographs.

Actually, the linear fan pattern of Figure 2.5 agrees with the averaging model. With adjectives equally weighted and photographs differentially weighted, the semilinear form of the averaging model applies (Anderson, in press, Section 2.3). This model does have a multiplicative component and is consistent with the data pattern in Figure 2.5.

Himmelfarb (1973). An extended test between averaging and adding is portrayed in the data of Figure 2.6. The adjectives in each person description are listed by each data point. The topmost solid curve shows the standard set-size effect: Response becomes more extreme as the number of high H adjectives is increased from one to four. Each dashed descending branch shows that adding a near-neutral N adjective decreases the response. The lower panel of the graph shows the complementary pattern for low L adjectives. The bottommost solid curve shows the set-size effect, with response becoming more negative as the number of L adjectives is increased. Each dashed ascending branch shows that adding a near-neutral N adjective increases the response. Himmelfarb's results thus rule out the adding hypothesis and provide good support for the averaging hypothesis.

Oden and Anderson (1971). In one part of this experiment, subjects rated command effectiveness of hypothetical graduates of Annapolis Naval Academy who were described by their class standing on a global academy assessment, augmented by two or four personality trait adjectives. Table 2.5 shows the crossover pattern that signals an averaging process. The set of MMLL adjectives in the first row has the same scale value but greater number, and hence greater weight, than the set of ML adjectives in the second row. The slope of each row curve reflects the relative weight of the academy assessment, which theoretically will vary inversely with the number of adjectives. The two-adjective curve should

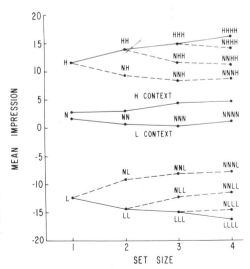

FIGURE 2.6. *Critical test supports averaging, rules out adding. Mean likableness of persons described by one, two, three, or four trait adjectives of L, N, or H value. (From Himmelfarb, 1973.* [*Copyright © 1973 by Academic Press. Reprinted by permission.*])

TABLE 2.5
Judgments of Command Effectiveness

Personality traits	General class standing on command ability				
	L	M⁻	N	M⁺	H
M M L L	44	52	84	108	111
M L	35	39	80	106	116
M M H H	74	86	117	142	148
M H	66	81	113	144	150

Source: After Oden and Anderson (1971). (Copyright © 1971 by the American Psychological Association. Reprinted by permission.)
Note: Ratings on a 0–200 mm graphic scale.

rise more rapidly, therefore, and cross over the four-adjective curve, as indeed it does. The same crossover pattern appears in the lower two rows of data.

The crossover is marginal in these data, a reflection of the heavy weight placed on the academy assessment in judgments of command effectiveness. Stronger crossovers were obtained for judgments of meals (Figure 1.21) and of criminal groups in this same experiment (see also Leon, Oden, & Anderson, 1973).

An interesting feature of this experiment is that the crossover is for curves based on different amounts of adjective information. The test of Figure 2.5 might be objected to on the ground that the critical curve represented judgments with no adjectives and so might be qualitatively different than the adjective curves. However, the crossovers of Table 2.5 are based on qualitatively similar conditions and so rule out this objection.[c]

2.3.5 OTHER TESTS BETWEEN ADDING AND AVERAGING

A number of other reports have also been concerned with the adding–averaging issue in person perception. Many have gone no further than demonstrating a set-size effect or its equivalent, failing to realize that a set-size effect is not sufficient either to prove adding or to disprove averaging (Section 2.4). Several reports have used more cogent tests, however, and these will be considered briefly here (see also Section 4.4.3).

Other Response Dimensions. The method of Section 2.3.2 was used by Hamilton and Huffman (1971) to test the averaging hypothesis on three dimensions simultaneously. Subjects rated on an evaluative scale of likableness ("good-bad") and also on scales of potency ("strong, domineering-weak, submissive") and activity ("active-passive"). The results clearly favored the averaging hypothesis in all four stimulus replications for evaluation and for potency, and for three of four stimulus replications for activity. These results help extend the generality of the averaging formulation.

In principle, of course, the averaging model should hold regardless of the response dimension, be it *sincerity, punctuality,* or *intelligence.* However, the report by Hamilton and Huffman is one of the few that has ventured beyond the generalized likableness dimension in the personality adjective task. Although differential weighting and nonparallelism may be expected to complicate exact analysis for other response dimensions (Section 4.4), that fortunately does not affect the qualitative test between adding and averaging.

Summation Theory. Fishbein's summation theory was originally developed around the personality adjective task. These experiments on personality impressions, which formed the experimental base for Fishbein's attitude theory, were mainly concerned to compare his summation model with the congruity model of Osgood and Tannenbaum (1955). The first of these reports (Triandis & Fishbein, 1963) is typical of the later ones. The essential results were the correlations between predicted and observed: .65 for the summation model; .53 for the congruity model. The later experiments (see Fishbein, 1967) yielded similar results and suggest that the congruity model is incorrect. The congruity model is a very specialized averaging model, however, and its failure does not raise any question about the general averaging hypothesis.[a]

None of Fishbein's experiments provide serious evidence on the summation model itself because they all rest on "weak inference" methods. The cited correlation of .65 not only seems low for this type of work, but also suffers from shortcomings discussed elsewhere (Section 1.2.8; Anderson, in press). It is quite possible that these data actually disagreed with the model.

The adding-averaging tests cited previously are critical for Fishbein's theory. Fishbein and Ajzen (1972, 1975) have argued that an adding or summation model could account for these results by assuming that the subject infers additional attributes from the given adjectives. This argument is theoretically incorrect; functional measurement treats the adjective as a molar unit and thereby automatically allows for such inferred attributes as part of the valuation operation (Section 1.1.5). Fishbein and Ajzen also argued that their model could account for the results by assuming stimulus interactions. This argument is empirically incorrect. Fishbein and Ajzen present hypothetical numerical examples but no actual evidence. The actual evidence indicates that interactions in these experiments are typically small and unable to account for the crossover outcomes either quantitatively or qualitatively.

A doctoral dissertation by one of Fishbein's students (K. J. Kaplan, 1969, 1972, Table 5) included a critical test between adding and averaging. Subjects received an initial set of four trait adjectives that described a Mr. Jones, rated their impressions of him, and wrote a series of recommendations about hiring and placing him. Then they received a final set of two more descriptive traits and rated their final impression of Mr. Jones. The initial and final sets of traits each had positive, neutral, and negative values in a 3 × 3 design.

The critical result concerns the change in impression when the final set of traits is neutral. These data provide a qualitative test of the type discussed in Section 2.3.4. Fishbein's model implies that adding the neutral traits should change the impression in the same direction regardless of the value of the initial set of traits. In contrast, an averaging model implies that the neutral traits should raise or lower the impression according to whether the initial set was negative or positive. The data agreed with the averaging hypothesis and disagreed with Fishbein's summation model.

Subtracting–Averaging Model. An interesting task was introduced by Schmidt and Levin (1972), who presented descriptions of two persons, A and B, and asked for judgments of difference in likableness. A suitable choice of adjectives in the two descriptions provided a novel qualitative test between averaging and adding.

Let A and B be descriptions based on single adjectives. The judged difference in likableness between A and B is assumed to equal the difference between the two separate likableness judgments:

$$R_{A-B} = C_0 + R_A - R_B.$$

Here R_{A-B} denotes the judged difference in likableness and R_A and R_B denote the separate likableness values. The term C_0 denotes the nominal value on the overt response scale that corresponds to the true zero point of no difference.

Now suppose that a second adjective X is added to each description, a different adjective but with the same value in each case. If an additive rule holds, then the same value is added to each description and the difference between them should be unchanged. If AX and BX denote the two-adjective descriptions, then the adding rule implies that

$$R_{AX-BX} = C_0 + R_{AX} - R_{BX} = C_0 + R_A - R_B = R_{A-B}.$$

However, if the information in each person description is integrated by an averaging model, then it is straightforward to show that the judged difference between the two-adjective descriptions is less than between the one-adjective descriptions:

$$R_{AX-BX} < R_{A-B}.$$

An easy way to see this is to consider an X adjective that lies between the A and B adjectives. Averaging in X pulls both descriptions toward the value of X, thereby decreasing the distance between them.

The results clearly supported the averaging model. Over six value ranges, the mean difference judgment between the two-adjective descriptions was 17.1; the mean difference judgment between the corresponding one-adjective descriptions was 25.7.

However, two complications appeared in these data. If the adjectives have equal weight, then the difference judgment, R_{AX-BX}, should be independent of the specific value of the added X adjective. In fact, this difference showed a steady decrease as the value of X became more negative. This result, as Schmidt and Levin pointed out, can be interpreted in terms of differential weighting, in line with results on negativity effects (Section 4.4.2).

The second complication was more serious, for it raised a question about the basic assumption that the difference judgments obeyed a subtracting model. If A, B, and C are any three person descriptions, listed in decreasing order of favorableness, then the subtracting model implies

$$R_{A-B} + R_{B-C} = C_0 + R_{A-C},$$

regardless of what rule governs the integration of information within each description. The C_0 term can presumably be set equal to zero, which was the nominal scale value for no preference. This theoretical equality thus provides a test of fit for the subtracting model. In fact, the sum on the left was generally larger than the term on the right. This may mean that the basic subtracting model is not correct. An alternative interpretation is that the response scale was nonlinear, with subjects tending to avoid the neutral point of no difference.

A subtracting–adding model was also employed by Birnbaum (1974a) in an attempt to resolve a question about nonlinear response. The initial experiments in this article obtained marked nonparallelism in judgments of likableness for two-adjective descriptions. This nonparallelism was due in part to nonlinear bias in the rating response, as Birnbaum noted (p. 558). Despite this bias problem, it was possible to get suggestive evidence that real nonadditivity was also present. The main conclusion of Birnbaum's paper, so far as it is warranted by the data, thus agrees completely with the preceding results on crossover interactions: Personality impressions are nonadditive.[b]

Unfortunately, the interpretation of Birnbaum's data is problematical. As Birnbaum emphasized, the nonparallelism in his first three experiments was ambiguous because it could result from nonadditivity in the integration rule or from nonlinearity in the rating response. Experiment IV, which was claimed to resolve this problem, is subject to the very same ambiguity.

In Experiment IV, subjects judged the difference in likableness between two persons, each described by a pair of adjectives. There is thus a covert integration to obtain the likableness of the two separate persons prior to the overt judgment of difference between them. Because the two person descriptions were combined in factorial design, the hypothesis that the difference judgments obeyed a subtracting model could be tested directly with the raw data.

Furthermore, if the subtracting model succeeded, then the design also allowed a test of the hypothesis that the implicit integration to form the person impression obeyed an adding model. Let A, B, X, and Y be arbitrary trait adjectives, let R_{AX-AY} be the judged difference between the person described by adjectives A and X, and the person described by adjectives A and Y, and similarly for R_{BX-BY}. If the adding and subtracting models both hold, then it is straightforward to show that

$$R_{AX-AY} = R_{BX-BY}.$$

The derivation of this equality follows essentially the same theoretical reasoning used by Schmidt and Levin.

Unfortunately, the subtracting model failed the test of goodness of fit. The difference judgments in Birnbaum's Table 1 show substantial, significant deviations from parallelism. This nonparallelism could result either from an invalid subtracting model or from a nonlinear response scale, but the data provided no way to distinguish between these two possibilities, even with monotone transformation.

The data did indicate that the adding and subtracting models could not both be correct. Birnbaum assumed that the subtracting model was valid and so concluded that the adding model was invalid. That may well be the case, and the pattern of results is consistent with the weighted averaging interpretation suggested by Schmidt and Levin (1972). However, Birnbaum's claim that his procedure resolved the model–scale ambiguity is not warranted by the given analyses.

Despite the difficulties with this particular experiment, Birnbaum's approach is interesting and represents an independent application of the essential logic of two-operation models (Section 1.7.2). In particular, as Birnbaum pointed out, a valid subtracting model allows a valid test of the adding model even with a monotone response scale. The easiest way to see this is to note that any equality, including the one just cited, remains true under any response transformation, including monotone transformation. On the other hand, Birnbaum's test is less general than that of Schmidt and Levin because it fails to distinguish between the strict adding model and the equal-weight averaging model. Birnbaum's test has the further limitation that it tests only the strict adding model; the crossover tests are more general because they rule out an entire class of adding-type models as discussed previously. Nevertheless, further work to determine what experimental conditions may yield valid subtracting rules is desirable (Section 1.3.4).

Two Choice Tests. In Hewitt's (1972) experiment, subjects chose which of two person descriptions would be more likable from among H, HM^+ pairs and L, LM^- pairs. After their choice, they classified the M^+ or M^- trait as desirable, neutral, or undesirable. Since this classification involves selection artifacts as well as positive context effects, only the overall choice data will be considered here. These data supported the averaging hypothesis for positive traits with 15.2 choices of H and only 10.6 of HM^+. For the negative traits, however, the choice difference was contrary to the averaging hypothesis.

Two difficulties with Hewitt's design should be noted. First, the set-size effect is largest between one and two adjectives and that could override the effect expected from stimulus averaging alone. More seriously, the same H or L trait was used in both descriptions of a given pair. That could induce a cancellation strategy, as noted in the choice test of Section 2.3.3, which would make the results moot.

A similar test was used by Lugg and Gollob (1973), except the subjects were instructed to consider that the two adjectives had been presented serially. They judged whether the second adjective raised, lowered, or left unchanged their liking of the person as described by the first adjective. Of those cases in which some change was produced, the response became more extreme in 83%, in line with predictions from an adding model. However, this design is open to the same two objections as Hewitt's design. In addition, as Lugg and Gollob note, there may have been a demand characteristic, with the subjects tending to respond directly in terms of the sign of the added adjective (see also Note 3.3.3b).

Cross-Cultural Tests. Japanese adjectives were used by Takahashi (1970, Tables 1 and 2) to make a critical test similar to that of Section 2.3.2. These data showed a higher response to HHM^+M^+ than to HH. This result is consistent with an adding rule and raises the disquieting possibility that the averaging model may be culture-specific. However, Takahashi's M^+ adjectives were apparently far above neutral. The obtained incremental effect of the added M^+ adjectives could thus reflect a set-size effect, as illustrated in Section 2.4.2.

No other tests of the averaging hypothesis in the personality adjective task seem to have been done in other cultures. However, two related studies by Singh and his students in India have supported averaging using the critical test of Section 2.3.4. In Singh, Sidana, and Saluja (1978), children judged attractiveness of playgroups described by the personal attractiveness of the members of the group and the number of toys that the group had to play with. In Singh, Sidana, and Srivastava (1978), children judged parental groups described by personal attractiveness of mother and father. Both studies obtained parallelism as well as an auxiliary crossover interaction, in complete agreement with the averaging hypothesis.

2.3.6 AVERAGING PARADOX

Outside the area of person perception, studies of the averaging hypothesis have led to a paradoxical result that has general importance for integration theory. Adding information that is clearly irrelevant or nondiagnostic can cause the response to become less extreme. This result is contrary to normative theory.

This paradox has its clearest form in Shanteau's work with the Bayesian two-urn task. This task employs two urns with complementary proportions of red and white beads, for example, 70:30 for urn A and 30:70 for urn B. One urn is secretly chosen at random. Samples drawn from the chosen urn are shown to the subject, who judges the probability that urn A is the chosen urn.

A sample of two red and two white beads is uninformative and nondiagnostic in this situation because it is equally likely to come from either urn. From a normative view, such balanced samples have zero diagnostic content. Subjects who receive such a sample should ignore it and maintain their prior opinions unchanged. But that does not happen. The nondiagnostic information can have substantial effects on the opinion.

Shanteau (1975b) used a procedure of serial presentation in which successive samples were drawn from the selected urn. If the first sample was nondiagnostic, the response naturally remained at the equal probability value of .50. If the first sample was diagnostic, however, and moved the response away from .50, then the nondiagnostic sample would move it partly back. Parallel results were obtained by Leon and Anderson for simultaneous presentation (1974, Figure 2). Related results have been reported by Nisbett and Ross (1980).

Even stronger demonstrations of the paradoxical effect were obtained by Troutman and Shanteau (1977), especially in the condition in which 50 blue beads were added to each urn. When both urns contain equally many blue beads, a sample containing only blue beads is clearly irrelevant and nondiagnostic. Nevertheless, these irrelevant samples had effects similar to the neutral samples in Shanteau (1975). This is called a *water down effect* (Troutman & Shanteau, 1977) or a *dilution effect* (Nisbett & Ross, 1980).

In attitude theory, a similar paradoxical effect has been reported by Himmelfarb (1974; Youngblood & Himmelfarb, 1972). In these experiments, subjects received fictitious messages about geographic, economic, and political aspects of the country of Mauritius and judged the Mauritian people on an evaluative scale of attitude. Nondiagnostic, neutral messages that contained no information relevant to the judgment made the attitude more resistant to change. This result has been confirmed and extended by Nisbett, Zukier, and Lemley (1981), who discuss a process model similar to one considered in information integration theory (Anderson, 1974a,d).[a]

These results were predicted by the averaging hypothesis and are readily explained within integration theory. Because the nondiagnostic information pertains to the objective situation that is to be judged, it is integrated by virtue of the general integration process discussed in Section 4.1.8. The nondiagnostic information has a scale value of zero and so would have no effect in the normative adding model. In the averaging model, however, this information has an effect because its weight is nonzero.

The exact value of the weight parameter will of course depend on many situational specifics. Weighting depends on the valuation operation, which, in the two-urn task, will involve relevance–similarity factors and amount of information (Section 4.4.1). In the Troutman and Shanteau task, for example, neutral samples of equally many red and white beads would have higher weight and more effect than the same number of irrelevant blue beads because the latter sample is less similar to the underlying dimension of judgment, which depends on the red:white proportions. Similarly, a larger nondiagnostic sample would have a greater dilution effect because it represents a greater amount of information and hence a greater weight.

This averaging paradox is one more illustration of the inadequacy of the normative approach to psychological decision theory. It seems irrational to integrate information that has no diagnostic content. It is not sensible to be swayed by clearly irrelevant facts. Yet such behavior occurs—as predicted by integration theory. Other results that bear on the descriptive–normative distinction are discussed elsewhere (e.g., Sections 1.5.1, 1.5.2, and 1.8.1).

2.4 The Set-Size Effect

2.4.1 PROBLEMS PRESENTED BY THE SET-SIZE EFFECT

If you like a woman and learn something new and good about her, you will like her even more. This illustrates the set-size effect, which in the simplest case may be stated thus: When all information has the same scale value, the response becomes more extreme as more information is added. Although the set-size effect is obvious to common sense, it presents a basic theoretical problem. Any theory must be able to handle set-size effects, both qualitatively and quantitatively. That is not so easy to do.

At the time the following set-size study was designed, the theoretical situation was as follows. Success of the parallelism prediction implied that the integration process followed some adding-type rule. That provided a seemingly secure basis for further analysis. The two obvious candidates were adding and averaging rules, but both had serious shortcomings, at least in their simplest forms.

The simplest averaging model predicts no set-size effect; adding information of equal value does not change the average value. The set-size effect thus presents a strong argument for an adding rule. However, adding mildly favorable information to highly favorable information can actually decrease the response (Section 2.3.2), in agreement with averaging but contrary to an adding rule. Both formulations faced difficulties, therefore, and neither could succeed without some additional explanatory concept. For the adding model, a contrast explanation was possible, an issue that is taken up in Section 4.1. This section and the next show how the averaging model used in integration theory fared with the set-size problem.

Serial and simultaneous presentation lead to somewhat different forms of the averaging model. Set-size effects for serial presentation had been observed in a jury trial experiment (Anderson, 1959b), and a serial averaging model with initial impression had given a reasonable account of most of those results. At that time, however, adequate tests of goodness of fit had not yet been developed, so the model remained uncer-

tain. Since the serial model is more complicated, it is deferred to Section
2.5. This section takes up the case of simultaneous presentation.[a]

2.4.2 AVERAGING-THEORETICAL ANALYSIS OF THE SET-SIZE EFFECT

Set-Size Equation. Consider a set of n stimuli that have equal value, s,
and equal weight, w. Let the value and weight of the initial impression
be s_0 and w_0. The response $R(n)$ is the weighted average

$$R(n) = C_0 + [nw/(nw + w_0)]s + [w_0/(nw + w_0)]s_0.$$

In this form the response is an average of the two scale values, s and s_0,
with relative weights given by the two expressions in brackets. For
simplicity, C_0 will be assumed to be zero.

This set-size equation is a growth-type curve. With no external
stimulus information, $n = 0$ and $R(0)$ equals the initial impression, s_0.
As n increases, the relative weight of s_0 decreases toward zero, and the
equation asymptotes at the value s. The averaging equation thus provides a qualitative account of the set-size effect.

The concept of initial impression plays an essential role in this
analysis. If there was no initial impression, that is, if $w_0 = 0$, then $R(n)$
would be constant regardless of n. Only by assuming the existence of an
internal state variable that is averaged in with the external stimuli does
the theory account for the set-size effect.

The concept of initial impression is clearly necessary because the subject may have prior knowledge of the person or object being described.
Such prior knowledge is assumed to be evaluated to yield weight and
value parameters, w_0 and s_0, in much the same way as the external
stimuli. In the personality adjective task, however, the description is
typically of a hypothetical person of whom there is no specific prior
knowledge. In this case, the initial impression is considered to represent
a background expectation. Its value would typically be near neutral,
therefore, and its weight parameter would reflect a lack of information
or an expectation based on previous experience within the experiment
(see also Section 4.3).

The set-size equation also clarifies a related difficulty for the averaging model. When both H and M^+ adjectives are used, an added M^+ may
raise or lower the response, depending on the exact parameter values.
Since this point is important in understanding the adding–averaging
tests of Section 2.3, an example using parameter values suggested by
Table 2.2 will be given to illustrate the point. The weight and value of the
H adjectives are taken as 1 and 100; the weight and value of the M^+
adjectives are taken as 1 and 70; and the weight and value of the initial

impression are taken as 2.5 and 50. The theoretical responses to H and HM+ descriptions are then 64.3 and 65.6, respectively. Here the set-size effect overrides the averaging effect, and the added M+ actually increases the response. However, the theoretical responses to HH and HHM+M+ descriptions are 72.2 and 71.5; these show a decremental effect of the added M+ information of roughly the size observed in Table 2.2. In general, the response must lie above the M+ asymptote before added M+ information will produce a decrement. That is more easily done with larger amounts of information, of course, as in this numerical example.

Goodness of Fit. A strong test of the set-size equation can be obtained with simplifying assumptions that would be reasonable in certain situations. Specifically, it is assumed that C_0 and s_0 can be set equal to zero or, more generally, equal to the midpoint of the response scale, and that s can be set equal to an endpoint of the response scale. This last assumption would not be reasonable for M+ adjectives, but it would be reasonable for H adjectives if end anchors were omitted.

It is no restriction to normalize by setting $w_0 + w = 1$ since that merely establishes the arbitrary unit of the weight scale. With this convention, $w_0 = 1 - w$ is the weight of the initial impression relative to a single piece of external stimulus information. In the set-size equation, therefore, the only remaining unknown is w, and it may be solved for as $w = R(n)/[ns - (n - 1)R(n)]$.

In this way, a value of w can be estimated from the response for each set size. The model requires that w be constant, independent of set size. The following tests are based on this constancy prediction.

Initial Test. A rough initial test of the set-size equation was made on the data of Table 2.2 (Anderson, 1965a). The w-parameter was estimated from the data for LL for each subject separately, and these w-estimates were used to predict the response to LLLL. Mean predicted and observed values were 16.6 and 17.6, a difference of 1.0 in a 100-point response range. Similarly, mean predicted and observed values for HHHH were 79.7 and 79.4, a difference of .3.

These results were promising, but the experiment had been designed for other purposes and included only two set sizes. The two following experiments were designed to give a more careful test over a greater range of set sizes.

2.4.3 QUANTITATIVE TEST OF THE SET-SIZE EQUATION

Design. In Experiment 1, 40 subjects rated likableness of stimulus persons described by sets of one, two, three, four, or six H or L adjectives chosen to be homogeneous in value and to avoid obvious redundancies. In Experiment 2, sets of nine H or L adjectives were added.

Because this is a basic experiment for averaging theory, some degree of procedural detail is given in Notes 2.4.3a and 2.4.3b. Two aspects of this procedure deserve explicit mention. First, two rating scales, −20 to +20, and −50 to +50, were used in Experiment 1. This was a routine check that different rating scales would give equivalent results. They did, as will be seen in Figure 2.7, and so only the −50–+50 rating scale was used in Experiment 2. Second, the usual end-anchor sets were omitted in order to justify the assumption that the scale values of the adjectives were equal to the endpoints of the rating scale. This aspect of procedure became important in the interpretation of the results.

Set-Size Curves. The data in Figure 2.7 follow the expected growth curve pattern. As set size increases, the stimulus person becomes more extreme in likableness or dislikableness. The curves for the positive and negative adjectives are fairly similar in each panel. A useful incidental outcome is the equivalence of groups 20 and 50 of Experiment 1. That means that subjects adapted their responses in exactly proportionate manner to the two rating scales.

Test of the Set-Size Equation. The w-estimates were made separately for each subject for each set size. The mean w-values from Experiment 1 are shown in the upper half of Table 2.6. According to theory, these values should be constant across each row. In each of the upper four

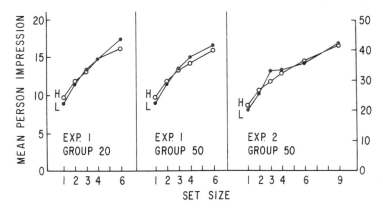

FIGURE 2.7. *Set-size curves. Mean likableness of persons described by sets of one, two, three, four, six, or nine adjectives, all of homogeneous high value (open circles) or low value (filled circles). (After Anderson, 1967b. [Copyright © 1967 by the American Psychological Association. Reprinted by permission.])*

TABLE 2.6
Weight Parameters as a Function of Set Size

Experiment	Rating scale	Adjective value	Set size					
			1	2	3	4	6	9
1	20	High	.48	.45	.43	.46	.53	—
		Low	.44	.43	.44	.47	.65	—
	50	High	.49	.44	.44	.44	.48	—
		Low	.45	.44	.47	.50	.61	—
		Mean	.46	.44	.44	.47	.57	—
2	50	High	.44	.37	.36	.36	.37	.49
		Low	.40	.38	.45	.39	.42	.58
		Mean	.42	.38	.40	.38	.40	.54

SOURCE: After Anderson (1967b). (Copyright © 1967 by the American Psychological Association. Reprinted by permission.)

rows the constancy prediction is essentially true for $n = 1, 2, 3$, and 4. But at $n = 6$, the w-estimates are markedly too large.

Two possible explanations of this discrepancy spring to mind. First, redundancy may begin to be important at larger set sizes. Second, information overload may cause the subject to partially neglect information in the larger sets. Both possibilities had been anticipated; the instructions and the procedure of having the subject read the traits aloud had been specifically chosen to avoid such effects. As it happens, neither explanation is viable because both imply that w should decrease as set size increases. Under either explanation, the effective values of n in the set-size equation would be less than the actual number of adjectives and so w would have to decrease in order to maintain the equality.

An ad hoc explanation of the discrepancy in Experiment 1 was given in terms of a response-scale end effect. When a response is near the end of the scale, there may be a tendency to use the scale endpoint as the response. That would affect primarily the response to the sets of six adjectives because they had the most extreme response in Experiment 1 (see Figure 2.7). Thus, the response to the sets of six adjectives would be biased toward the scale endpoints and that would cause the w-estimate to be too large.

This interpretation created an experimental difficulty. The hypothesized end-bias effect could presumably be eliminated by including the usual end-anchor sets composed of more extreme adjectives. But these anchors had been purposely omitted to simplify the parameter estimation. With such anchors present, the effective s-values of the regular adjectives would have been some unknown value less extreme than the scale endpoints and that would complicate the parameter estimation.

This difficulty was resolved in Experiment 2 by adding three adjectives of equal value to the sets of six to obtain sets of nine adjectives. The logic is straightforward. If the discrepancy for sets of six adjectives in Experiment 1 represented a true defect in the model, then it should reappear in Experiment 2. On the other hand, if the end-bias explanation is true, then the discrepancy should shift to the added sets of nine adjectives, and the sets of six adjectives should no longer be discrepant. This follows because the sets of nine adjectives give the most extreme response by virtue of the set-size effect, and hence the end-bias should concentrate on them.

This end-bias interpretation is well supported, as can be seen in the lower part of Table 2.6. The discrepancy has indeed shifted to the sets of nine adjectives, and the w-estimates for sets of six adjectives are no longer discrepant. It was concluded that the discrepancy reflected an experimental artifact, and that the averaging-theoretical analysis of the set-size effect was correct.[c]

Averaging-with-Adding. This was the third basic experiment in the initial stage of this research program. The outcome is theoretically important because it shows that the averaging formulation can provide an exact account of the set-size effect. In particular, the results rule out the hypothesis that the integration follows a mixture of adding and averaging processes.

This possibility that both adding and averaging processes may be operative deserves explicit consideration. Virtually all discussion of the integration question has taken the useful but treacherous either–or view. The qualitative tests of Section 2.3 provide clear support for an averaging process. They show that an adding process is inadequate; they do not show that no adding process is operative.

In fact, averaging-with-adding could provide a nice account of most observed results. Parallelism is straightforward: Either process can account for parallelism separately, and they will do the same if acting together. The averaging process can account for the qualitative tests of Section 2.3, while the adding process can account for the set-size effect. This two-process hypothesis has the special attraction that the concept of initial impression is not required. Thus, the second problem posed by the results of Table 2.2 would have a ready interpretation in terms of a joint averaging–adding rule.

However, the averaging model with initial impression was sufficient to provide an exact, quantitative account of the set-size effect. If the data resulted from joint operation of averaging and adding processes, then the w-estimates from the set-size averaging model would not be constant in Table 2.6. The observed constancy thus rules out even a partial adding process, an instructive illustration of the usefulness of exact analysis.

There is a real possibility that averaging and adding processes may operate together in certain tasks. Shanteau's (1974) subadditivity effect may be an example; judgment of the value of a bundle of commodities clearly calls for some kind of adding process. The subadditivity may not be a law of diminishing returns, however, but instead reflect a pervasive averaging process overlaid on the adding process. For the personality adjective task, it should be noted, no indication of need for a joint averaging–adding rule has appeared.

2.4.4 OTHER INTERPRETATIONS OF THE SET-SIZE EFFECT

Various other interpretations of the set-size effect have been suggested by other workers. Because of the importance of the problem, some of these interpretations will be summarized briefly here.

Artifact Interpretations. A puzzling aspect of the set-size effect is that it is easily obtainable in within-subject design, but not in between-subject design. Large effects are obtained when each subject judges sets of various sizes, as shown in Figure 2.7. But when each subject judges only one size of set, effects are typically small and nonsignificant (Rosenbaum & Schmidt, 1968; Sloan & Ostrom, 1974). One design or the other may therefore suffer from an artifact.

Within-subject design has been criticized as being susceptible to demand characteristics. HHHH would be rated more favorably than HH, for example, not because the subject really felt that way, but because the difference in amount of information caused the subject to think that a corresponding difference in response was expected. No actual evidence has ever been presented for this interpretation. On the contrary, it implies that HHM^+M^+ would also be rated more favorably than HH, and that is false (Table 2.2).

The between-subject design, on the other hand, suffers from two difficulties. It has higher variability, which could obscure any real set-size effect, especially at larger sizes where the set-size curve is nearing asymptote. More seriously, set-size effects would be expected to wash out in between-subject design owing jointly to the lack of a frame of reference and the relative nature of rating scales (Anderson, in press, Section 1.1.8).

Experimental reports by Sloan and Ostrom (1974) and Kaplan and Major (1973) have done much to clarify the nature of the set-size effect, though some uncertainties still remain. Sloan and Ostrom give a comprehensive compilation of results on the set-size effect and suggest that previous experiments had lacked adequate power to detect the effect in between-subject design. Their own careful between-subject experiment yielded a very strong set-size effect for likableness but not, oddly enough, for confidence.

The report by Kaplan and Major contains several interesting results, including a set-size effect in a between-subject design. However, this set-size effect appeared to require that subjects be explicitly instructed that more or less information might have been given. Kaplan and Major also suggest that confidence acts as a mediator for the set-size effect, an interpretation that is consistent with their Experiment 2, although not with the more substantial results of Sloan and Ostrom. Their interpretation appropriately relates confidence to weight and stresses that confidence judgments are also relative. Whether confidence has a true causal role is a delicate question.

Confidence Interpretation. An apparently attractive interpretation of the set-size effect is that it is caused by increased confidence. This interpretation has been suggested by a number of people, and it is consistent with the usual finding that confidence increases with set size, that is, with amount of information.

An alternative interpretation considers confidence to be a correlate of the weight parameter. Increase in set size would increase confidence by virtue of the corresponding increase in the total weight of the information.[a]

If confidence is merely a phenomenological reflection of the weight parameter, no conceptual problem arises. However, the confidence interpretation is sometimes used in a way that suggests that something more is intended, that confidence somehow has a direct causal effect. The idea seems to be that HH and HHHH induce equivalent underlying impressions but that the subject is less confident with less information and somehow holds back his response. A clear statement of this notion has not been given, however, and it remains questionable whether confidence is more than a phenomenological pseudoexplanation.

Thurstonian Adding Model. Several studies, beginning with Thurstone and Jones (1959), have attempted to adapt the method of paired comparisons to analyze an adding model for stimulus combinations (see Bock & Jones, 1968, Chapter 9). In a typical application, subjects choose between pairs whose elements may be either single items or two-item combinations. Application of paired comparisons scaling technique then yields scale values of the single items and their combinations. The hypothesis to be tested is that the scale value of a two-item combination equals the sum of the scale values of the two single items.

The test of the adding model is slightly complicated by the fact that paired comparisons scaling yields no more than a linear scale, so the zero point is unknown. If s_i, s_j, and s_{ij} denote the estimated scale values for two single items and their combination, respectively, then

$$(s_i + s_j) - s_{ij} = c_0,$$

where c_0 is the scale zero. Although the difference on the left is not in general zero, it should be constant for every pair of stimuli. This relation provides a test of the joint hypothesis that the adding model is correct and that the scale values are a valid linear scale (Thurstone & Jones, 1959).

For present purposes the most relevant experimental test is that of Hicks and Campbell (1965), who used three judgment tasks: disturbance of mental patients described by one or two behavior items; seriousness of offense of persons who committed one or two specific traffic violations; and desirability of one or two objects considered as a birthday gift. Scale values were estimated using Case V of Thurstone's method of paired comparisons for each of three groups of stimuli for each of the three judgment tasks. Hicks and Campbell hypothesized that an averaging model should apply to judgments of disturbance, and an adding model to the other two judgments. All three judgments yielded similar results and they concluded that the adding model held for all three.

Testing the Thurstonian adding model revolves around the prediction that the zero-point values, c_0, should be constant for every pair of stimuli. Hicks and Campbell relied primarily on an additivity index defined as "the ratio between the standard deviations of zero-point determinations and the standard deviations of corresponding composite stimuli [p. 797]." However, they also followed Thurstone and Jones in reporting scatter plots and correlations between s_{ij} and $(s_i + s_j)$. They concluded from these analyses that the Thurstonian adding model fit the data quite well, at least within each stimulus group. Although there was systematic variation of zero points across stimulus groups, that was interpreted as a psychological context effect from the range of values, which included mainly low, medium, or high values across the three respective stimulus groups.

A different conclusion was reached by Shanteau (1970b, 1975a), who reanalyzed these data using different estimation procedures. Marked deviations from additivity appeared, of the same form as a subadditivity effect that Shanteau had obtained in his own work on commodity bundles. Shanteau also found subadditivity in reanalyses of published reports by other workers, and he pointed out that the scatter plot analyses used in previous work could easily cover up real subadditivity (see Section 1.2.8). The additivity index of Hicks and Campbell does not seem useful. In particular it ignores any correlation between estimated zero points and scale values within a stimulus group. Visual inspection of the nine panels of Figures 1–3 of Hicks and Campbell shows clear, strong correlations in almost every case. This visual test shows that the zero points are not constant, contrary to their model.

Parallelism analysis is applied to representative data from Hicks and Campbell in Figure 2.8. The lower right curve plots their paired-

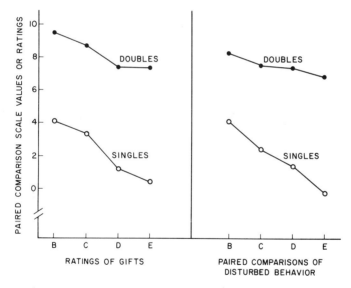

FIGURE 2.8. *Nonparallelism implies failure of Thurstonian adding model in reanalysis of data of Hicks and Campbell (1965). Right panel shows paired comparisons scale values measured in centimeters from Figure 1 of Hicks and Campbell. Lower curve plots values for four single items, B, C, D, and E, listed on horizontal axis; upper curve plots values for same four items when combined with Item A. Thurstonian adding model implies that distance between the two curves should be constant and equal to the value of Item A. Data were pooled over three groups of stimuli (context conditions), which all showed the same pattern. Left panel shows analogous plot for ratings of gifts from Hicks and Campbell's Figure 6, also pooled over the three context conditions and with items within context condition rank ordered by observed value.*

comparison scale values for single items B, C, D, and E, which are listed on the horizontal. The upper right curve plots their paired-comparison scale values for these same items combined with item A. The Thurstonian adding model implies that these curves should be parallel, separated by a constant difference equal to the value of item A. The clear nonparallelism thus implies that the adding model is incorrect or that the scale values are invalid. A second illustration, in the left panel, also shows nonparallelism of the same form for scale values obtained from ratings of birthday gifts. The other seven available sets of data showed similar patterns. Even without a formal test of goodness of fit it seems clear that the adding model does not hold within any group of stimuli either for ratings or for paired comparisons.

Still another reanalysis of the Hicks and Campbell data was presented by Jones (1967; Bock & Jones, 1968), who suggested that the variation of zero points across stimulus groups was not a true context effect, but meant that Case V scaling was not appropriate for these stimuli. Instead, Jones applied Case VI, which takes discriminal dispersion to be a linear function of scale value. In this scaling the zero points were essentially identical across stimulus groups, and Jones concluded that the Thurstonian adding model was correct.

The present view is that the adding model is not correct for any of these three classes of stimuli. For birthday gifts and other commodity bundles, Shanteau's (1970b, 1975a) analyses have found marked subadditivity. Shanteau's interpretation that his data reflect real discrepancies from the adding model rests on the assumption that his own response scale was linear. That assumption seems justified on the basis of response generality and by virtue of the constraint provided by the success of the multiplying model in these same judgments. Results with the behavior disturbance items (Anderson, 1962c, 1972b) and with criminal offenses (Leon et al., 1973) have favored an averaging rule. Although the evidence for averaging with these particular stimuli is not as definite as is desirable, both tasks are similar to others that have shown averaging.[b]

It should be noted that Jones's Case VI scaling would tend to eliminate the nonparallelism in Figure 2.8 as well as the subadditivity observed by Shanteau. In the present view, however, this is a numerical coincidence that stems from a law of diminishing returns implicit in Case VI. There seems to be no variant of the Thurstonian adding model that could account for the averaging results of Section 2.3.

In general, paired comparisons theory hardly seems meaningful for judgment tasks of the kind considered by Hicks and Campbell and by Bock and Jones. With the birthday gifts, in particular, scaling requires pooling data over individuals who have different rank orders of preference. It seems doubtful that such data have much bearing on additivity, which is a within-individual question (Section 5.3).

Other Models. Except for integration theory and the Thurstonian approach, only passing attempts have been made to give a quantitative account of the set-size effect. That is surprising in view of the many attempts to erect theoretical formulations on an adding hypothesis. Of course, a simple adding model is hardly plausible, for it predicts a linear set-size function. Some law of diminishing returns could presumably be adopted that would account for the nonlinearity observed in Table 2.2 and Figure 2.7. In fact, a reasonably good fit was obtained by Gollob, Rossman, and Abelson (1973), who employed a diminishing returns parameter together with a functional measurement method of analysis. However, their task actually obeys an averaging model, as shown in qualitative tests by T. Anderson and Birnbaum (1976).

Birnbaum (1972, 1973, 1974a; Birnbaum, Parducci, & Gifford, 1971) has considered a range model that specifies the response to two stimuli as

$$ws_1 + (1 - w)s_2 + \text{a range term,}$$

where the range term may be $w'|s_1 - s_2|$ or $w'|s_1 - s_2|/(s_1 + s_2)$. As one interpretation, the range between the stimuli could be considered as an additional stimulus itself, to be added onto the simple average of the two given stimuli. This range model does not account for the set-size effect. If $s_1 = s_2$, the stimulus range is zero, so the range model becomes a simple averaging model that predicts no set-size effect. Later work (e.g., Birnbaum & Stegner, 1979) has adopted the averaging model from integration theory, which does account for the set-size effect, and has suggested that a range term could be added on to that to account for certain configural effects.

A range formulation has been proposed by Warr (1974; Warr & Jackson, 1975), who assumed that the response to two-item sets is controlled by two quite different processes, depending on the range between the two stimulus values. When the range is small, the two items are treated as a single perceptual cue (Warr, 1974, p. 196). There is no integration, therefore, but the scale value of this compound cue is considered to be greater than the value of either separate item, a "pushover" effect, as though the two items emphasize and intensify each other. When the range is moderate or large, however, then the response to the two-item compound is assumed to be a simple average of the values of the two separate items, a "pullback" effect.

None of Warr's data present any problem for the present averaging model. The pushover effect is just the set-size effect, and the pullback effect is just the critical averaging effect illustrated in Table 2.2. The discontinuity assumed by Warr merely reflects this transition between the decremental and incremental effects in Table 2.2. The present averaging model provides a parsimonious, one-process account of these data. Warr's formulation has interest because it proposes to do without the concept of prior impression. However, it then requires the separate pushover process to handle the set-size effect. Moreover, the assumption that no integration is involved when the stimuli have equal value makes it awkward to account for serial integration.[c]

Wyer (1974a) has proposed that each trait adjective can be represented as a distribution of reference persons who possess that trait. Judged likableness, say, of a hypothetical person described by one trait is defined as the mean likableness value of the distribution of reference persons. Likableness of a hypothetical person described by two traits is defined as the likableness of the distribution of reference persons who possess both traits. This scheme could provide a testable account of set-size and averaging effects if the two-trait distribution was predictable from the two one-trait distributions, but that is not logically possible. Accordingly, Wyer suggested that the two-trait distribution might be approximated at each likableness value by a function of the product of the numbers of reference persons of that likableness value in each single-trait distribution. Arbitrary but reasonable assumptions allow directional predictions of both set-size and averaging effects, but the scheme is so flexible that exact tests seem difficult to obtain. One reservation about this scheme is that it is conceptually inadequate: It does not apply to inconsistent information; that is, when two single-trait distributions have no reference persons in common (Wyer, 1974a, p. 312). To handle this case, Wyer assumes either that the two trait values are averaged or that one trait is completely discounted. This does not accord with the data, which show partial discounting under inconsistency (Anderson & Jacobson, 1965). Moreover, the need for such ad hoc assumptions complicates an already complicated formulation.

Manis, Gleason, and Dawes (1966) adapted an averaging model by multiplying the average value of the given stimuli by an arbitrary function of the number of items in the set. This multiplier can account for the set-size effect while still allowing for the averaging effects of the kind discussed in Section 2.3. Despite these advantages, this model has not been pursued further, probably because the set-size dependence is only a curve-fitting procedure with no theoretical rationale.

An extended program of research by Byrne and his colleagues (e.g.,

Byrne, 1969; Clore & Byrne, 1974) has applied a reinforcement model or "linear law" to interpersonal attraction. As Kaplan and Anderson (1973a,b) point out, however, this reinforcement model predicts no set-size effect.

A two-variable regression analysis of the averaging model with equal weights has been suggested by Wyer (1969), who showed that the response to sets of two stimuli could be expressed as a linear function of the responses to the single stimuli. Let R_{Ai} and R_{Bj} denote the responses to the single stimuli, R_{ij} the response to their combination, and set the sum of the weights equal to unity ($w_0 + w_A + w_B = 1$). Then the averaging model can be written

$$R_{ij} = (w_0 + w_A)R_{Ai} + (w_0 + w_B)R_{Bj} - w_0 s_0.$$

In this form, R_{ij} is a linear function of R_{Ai} and R_{Bj}, and Wyer accordingly applied standard regression analysis. The averaging model yielded smaller deviations from prediction than three other models, but the estimates of s_0 were more positive than the response to the most positive single stimulus. This is theoretically unreasonable and would disallow the averaging-theoretical explanation of the set-size effect.

A possible resolution of the difficulty is that standard regression analysis is conceptually inapplicable in this situation. Standard regression is based on a prediction equation, whereas integration models must ordinarily be treated as functional regression equations (Anderson, in press, Section 4.3). With standard regression analysis, error in the predictors can bias the estimated regression coefficients and thus the estimate of s_0.

Hodges (1973) has considered a variant form of the averaging model that may be obtained by setting $s_0 = 0$ and $w_0 = 1$. The set-size equation for a single stimulus is then solved to yield $s_i = (1 + w_i)R_i/w_i$, which is substituted for s_i in the general averaging model. The theoretical expression for the response to a set of stimuli then becomes

$$R = \Sigma(1 + w_i)R_i/(1 + \Sigma w_i).$$

Hodges considers this to be an adding model because the sum of the relative weights, $\Sigma(1 + w_i)/(1 + \Sigma w_i)$, is not constant. That seems to reflect overconcern with the mathematical formalism. Inasmuch as Hodges's model is quantitatively equivalent to the averaging model, it also can account for the qualitative tests between adding and averaging discussed in Section 2.3. Those tests appear to disprove adding rules on conceptual grounds.

Hodges's equation is attractive because it avoids the concept of initial impression. However, it then becomes awkward to handle situations that require a representation for prior information (Section 4.3). Additional comments are given in Anderson (1973a).

2.5 Serial Integration

2.5.1 THE PROBLEM OF ORDER EFFECTS

In everyday life, information integration is a sequential process. Information is received a piece at a time and integrated into a continuously evolving impression. Each such impression, be it of a theoretical issue, another person, or a social organization, grows and changes over the course of time. At any point in time, therefore, the current impression looks both forward and back. In one perspective, the current impression is the cumulated resultant of all past information. In the other perspective, it is the initial impression into which future information will be integrated.

Basic to the study of serial integration is the problem of order effects. The very same set of stimuli can yield a different response depending on the order in which the stimuli are presented. Such order effects can be quite large, even with a set of only four or six adjectives. Moreover, order effects are found in many other tasks.[a]

A little reflection suggests many possible causes of order effects. Forgetting of the earlier stimuli would make the later stimuli relatively more important, a recency effect. On the other hand, the earlier stimuli might crystallize the response, causing a discounting of the later stimuli, hence producing a first impression or primacy effect. Again, primacy might be expected on the ground that initial stimuli are treated as more central and informative.

The key to the study of order effects is the ability to evaluate the influence of each successive piece of stimulus information on the overall response. Many integration tasks consider only the overall response after all the information has been integrated. However, this overall response is the resultant of a sequence of generally unobservable integrations of the successive stimuli. The following serial integration model can, under certain conditions, reconstruct this unobserved sequence of integrations knowing only the terminal response.

2.5.2 AVERAGING MODEL FOR SERIAL INTEGRATION

Serial Averaging Model. The averaging model helps clarify some conceptual problems connected with order effects. Moreover it becomes indispensable for getting beyond the customary crude index of primacy–recency to analyze the effect of each serial position.

The averaging model for the response to a sequence of N stimuli can be written

$$R = w_0s_0 + ws_1 + \cdots + w_Ns_N, \qquad (\Sigma w_X = 1)$$

where w_X and s_X are the weight and scale value of the stimulus at the Xth serial position. It is assumed that the scale value of any stimulus is constant, independent of its serial position and of previous stimuli. Although this independence assumption may not be true in general, it is well grounded for the personality adjective task, among others. For present purposes it is also assumed that all stimuli have equal natural weight. Accordingly, the weights in this serial integration model depend only on serial position per se.

On this basis, order effects can be interpreted in terms of weight parameters. Under the given assumptions, w_X represents the relative importance of serial position X. The serial position curve of the weight parameters thus constitutes the theoretically proper description of the order effects. These order effects will, of course, depend on attentional factors that can be experimentally manipulated in various ways.

Net Primacy–Recency Index. The traditional measure of order effects is an index of primacy–recency. Two stimuli, A and B, are presented in both AB and BA orders. The difference in response is defined as the order effect

$$\text{Order Effect} = R_{AB} - R_{BA}.$$

It is no restriction to assume that A has higher value than B. Hence a positive order effect indicates primacy; that is, greater effect of the stimulus presented first. Similarly, a negative order effect indicates recency.

This traditional definition of order effects is purely empirical, and its meaning may seem to be theory-free. That is not entirely true, however, as can be seen from the following model analysis. Under the given assumptions

$$R_{AB} = w_0s_0 + w_1s_A + w_2s_B,$$
$$R_{BA} = w_0s_0 + w_1s_B + w_2s_A.$$

Hence the theoretical expression for the order effect is

$$R_{AB} - R_{BA} = (w_1 - w_2)(s_A - s_B).$$

Primacy corresponds to greater weight at the first serial position; that is, $w_1 > w_2$. The theoretical order effect is then positive since $s_A > s_B$. Similarly, $w_1 < w_2$ represents recency and yields a negative order effect.

Under the given assumptions, therefore, the model analysis justifies the traditional interpretation of the order effect index.

However, the traditional index can be misleading if the given simplifying assumptions do not hold; in particular, if order of presentation affects weights or scale values. For example, suppose that order of presentation affects scale value so that A and B have values 7 and 3 in the AB order, 8 and 3 in the BA order. This represents a recency effect since s_A is greater when given second. Assume that weight depends only on serial position and that $w_0 = 0$. Then the model implies that

$$R_{AB} - R_{BA} = 4w_1 - 5w_2.$$

If $w_1 = w_2$, then the difference is negative, correctly indicating the recency effect associated with s_A. If $w_1 > w_2$, however, then there is also a primacy effect associated with the weights. What is actually observed, therefore, is a net difference between competing recency and primacy. This observed effect is mute about its true basis. Whenever there is stimulus interaction, therefore, the traditional index may be inadequate or misleading.

A similar conclusion follows even when the given assumptions all hold. With three stimuli, primacy and recency may occur together, as reflected in the pattern of weights, $w_1 > w_2 < w_3$. In this case also, a single index of net primacy–recency may be misleading.

The model approach has the advantage that it can evaluate the weight at each serial position to yield a complete serial curve. This evaluation rests on certain assumptions, it is true, but the model provides warning signs about violations of these assumptions. When these assumptions are not met the traditional index is also problematical, but it provides no warning sign. In general, the model approach seems almost necessary to get a conceptual understanding of order effects.

Two Processing Modes. Although the serial integration model is written out as an explicit average of the given stimuli, it could represent two extreme modes of processing. In the memory storage mode, the subject would store each stimulus as it was received, then make a single overall integration after the last stimulus was received. In the cumulative average mode, the subject would use a running average strategy, integrating each stimulus as it was received but retaining only the cumulated value of the impression in the relevant memory store.

Although this question has not been studied systematically, there is scattered evidence for the importance of the running average strategy. In the personality adjective task, for example, subjects who were asked for recall without warning immediately after making the impression recalled only two of eight adjectives (Anderson & Hubert, 1963). Moreover, these recalls were predominantly of the last three adjectives in the sequence, whereas the impression itself showed a primacy effect (see also Section 3.3). The running average mode seems to be important, therefore, and it is economical in its use of attention and storage capacity.

Hybrid modes are also possible and are perhaps the norm. For example, a request for judgment about another person often seems to evoke a general impression together with an aggregation of particular remembered incidents. The general impression represents previous cumulative integration, and it is a reasonable hypothesis that this general impression is augmented by present integration of the particular incidents from memory storage. This hypothesis is difficult to study, however, because the remembered incidents may already be largely represented in the general impression, having been integrated at previous times. The remembered incidents may have little or no effect on the response, therefore, a possibility that agrees with available evidence that allowing more time for thought has little effect on judgment in tasks of this kind (Section 2.2.3).

2.5.3 SERIAL CURVES OF RESPONSE

One function of serial curves is to exhibit how the impression develops over the sequence of information. The obvious way to get such curves is to ask for an overt response after each successive piece of information. In some tasks, however, this very responding affects the development of the impression. Before considering results from continuous responding, therefore, an alternative method will be illustrated that requires a response only at the end of the sequence.

Probe Technique for Serial Impression Curves. The probe technique is illustrated by the following seven sequences of three L and six H adjectives. The six H's are given in the same fixed order in each sequence, and the three L's are interpolated as a block at the seven possible positions.

```
LLLHHHHHH
HLLLHHHHH
HHLLLHHHH
HHHLLLHHH
HHHHLLLHH
HHHHHLLLH
HHHHHHLLL
```

This set of seven sequences is denoted (3L)6H. In the experiment, the subject was read each sequence of nine adjectives and then judged how likable a person with those traits would be. The data are shown in the upper right curve of Figure 2.9.

Every pair of sequences embodies an order effect comparison. For example, the first and last sequences, LLLHHHHHH and HHHHHHLLL, respectively, represent the usual low–high versus high–low comparison. The more favorable response to the latter sequence represents a primacy effect. As another example, the third and fifth sequences are identical over the first two and last two serial positions. They differ only in the order of the middle five adjectives, LLLHH versus HHLLL, respectively. Accordingly, the higher response to the fifth sequence also represents a primacy effect.

In fact, every such comparison yields primacy as shown by the upward trend of the curve: The later the (3L) probe, the less its effect on the impression. The complementary pattern in the lower (3H)6L curve, in which a (3H) probe was interpolated at the seven possible positions in a 6L sequence, also represents primacy. The greater regularity of the

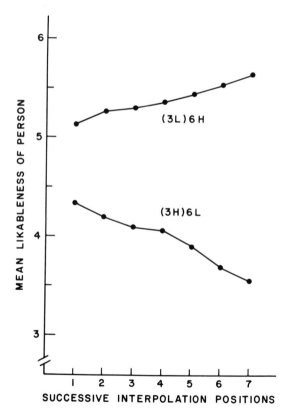

FIGURE 2.9. *Linear primacy effect. Mean person impression as a function of order of presentation. Upper curves represent responses to sequences of three L and six H adjectives, with the three L's interpolated as a block at each possible point within the sequence of six H's. Lower curves show response to complementary sequences. (After Anderson, 1965b.)*

curves for Experiment 2, which are shown in the figure, resulted from improved design that reduced the error variability from that of Experiment 1 (Section 4.1.5; see also Anderson, in press, Figure 1.5).

A notable feature of these curves is their linearity. This straight-line primacy had not been expected in Experiment 1, but it has a simple interpretation in terms of the serial averaging model. With initial impression neglected, the observed linear primacy is equivalent to a corresponding linear decrease in the weight parameter across serial position.

Linear primacy is not consistent with the verbal memory hypothesis (Section 4.2). Verbal memory for the adjectives would exhibit the usual bowed serial curve so that the three L adjectives would be remembered best when at the beginning or end of the sequence. If the impression response is based on verbal memory, therefore, it should be lowest for the first and last sequences, highest for the middle sequences. Indeed, the verbal memory hypothesis implies that the (3L)6H curve of Figure 2.9 should be the mirror image of the bowed serial curve of recall. The observed data thus provide strong evidence against the verbal memory hypothesis.

The linear primacy effect is consistent with the attention decrement hypothesis that primacy is caused by decreasing attention to the successive adjectives in the sequence. In this view, attention is a determinant of the weight parameter of the model. Furthermore, the linear primacy effect seems awkward to account for in terms of change of meaning. This issue is taken up in Section 3.3.

Serial Responding. Figure 3.5 shows the data resulting when subjects revised their responses after each successive adjective (Stewart, 1965). In the left panel, the top curve represents serial responding to a sequence of four H adjectives followed by four L adjectives. The curve shows the expected shape, rising as favorable information is integrated, then falling as unfavorable information is integrated. The bottom curve, in which the same adjectives are given in opposite order, shows the complementary pattern.

The crossover of these two curves at Serial Position 7 represents a recency effect. This recency contrasts with the primacy effect that is typically obtained when a response is given only at the end of the sequence (e.g., Figure 2.9).[a] Stewart's data show, regrettably, that serial responding itself sometimes affects the development of the response. Thus, the probe technique cannot be abandoned even though it provides less information than serial responding. An extension of the probe technique is considered next.

2.5.4 SERIAL CURVES OF WEIGHT

The averaging model may be used to measure the weight parameter at each serial position if the sequences are constructed according to factorial-type design. Experimentally, this can be done most simply by choosing two stimuli at each serial position so that the difference in their scale values is the same at each serial position. The difference D_X between the two marginal means at the Xth serial position yields an estimate of w_X on a ratio scale (Anderson, 1974d, p. 263). Two experimental applications will be given here.[a]

Person Perception. This model analysis was used in the two experiments summarized in Table 2.7. Subjects were read five-adjective descriptions and rated likableness of the person on a 1–10 scale after hearing all the adjectives. Each factor in the design was a polar, H–L pair of adjectives such as *cheerful–gloomy, trustful–suspicious,* etc. These were chosen with constant difference in scale value across pairs so that the model could be used to estimate the serial weights.

Table 2.7 gives the empirical values of D_X, which are proportional to the weight parameters. The main trend is a linear decrease in these weight estimates. Experiment 1 shows some irregularity but Experiment 2, which was run as a check, shows a steady decrease over serial position. These weight estimates thus agree with the linear primacy discussed in Figure 2.9.[b]

Intuitive Statistics. The number-averaging experiment presented here has historical interest as the first application of the serial integration model that provided a test of goodness of fit. The subject saw seven two-digit numbers one at a time and gave an intuitive running average of all previous numbers at each serial position in the sequence. Only the response at Serial Position 6 will be considered here.[c]

TABLE 2.7
Serial Position Curves of Weight Parameter for Person Perception

Experiment	Serial position				
	1	2	3	4	5
1	1.72	1.67	1.81	1.65	1.34
2	1.77	1.59	1.52	1.29	1.23
Mean	1.75	1.63	1.66	1.47	1.28

SOURCE: Data from Anderson (1973d). (Copyright © 1973 by the American Psychological Association. Reprinted by permission.)

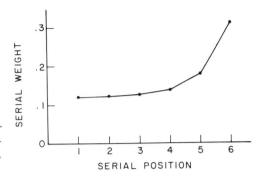

FIGURE 2.10. *Serial position curve for weight parameter in serial number-averaging task. (After Anderson, 1964d.)*

The model analysis yielded the serial weight curve shown in Figure 2.10. These weights show a uniform increase, or recency effect, across serial position. The main component of the recency is at the last serial position, which carries almost a third of the total weight. This disproportionate influence is not unreasonable since it represents the currently visible number in the sequence. Accordingly, this end effect in the curve suggests the operation of a short-term salience factor (see also Figure 2.11). This graph also illustrates how complete serial weight curves are more informative than an overall index of primacy–recency.[d]

Attitude Formation. Complete serial curves for weight in an attitude study are shown in Figure 2.11. Subjects received a sequence of paragraphs about a United States president and rated the president on statesmanship after reading each successive paragraph. The curve labeled R4 gives the relative importance of the paragraphs at each serial position on the final response after the last paragraph had been read. This curve has two noteworthy features. First, it is flat over the first three serial positions; the first three paragraphs each have equal weight in the final attitude. Second, there is a marked upswing at the last position, which has substantially higher weight than any of the first three.

A similar serial curve was constructed for the response after the third paragraph. This curve, labeled R3, shows the same pattern: flat over the initial serial positions with a strong upswing at the last position. Again, there is a recency effect, with the latest information having greatest influence.

Comparison of the two curves shows that the recency in the R3 curve has completely disappeared in the R4 curve, in which the weight at the third serial position is no greater than the weight at the first two. The recency effect is thus short term. The same conclusion follows from

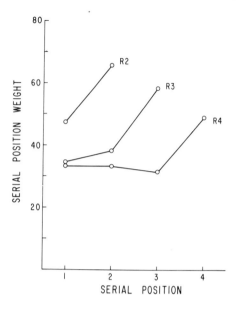

FIGURE 2.11. *Serial position curves for weight parameters in attitude formation. Curves labeled R2, R3, and R4 represent weights of given information in responses at second, third, and fourth serial positions, respectively. (After Anderson & Farkas, 1973.)*

comparison of the R2 and R3 curves. Very similar results have been reported for continuous responding conditions in person perception by Dreben, Fiske, and Hastie (1979).

The basal–surface hypothesis of attitude structure presented in an earlier experiment with jury trial materials (Anderson, 1959b) provides a theoretical interpretation of the present weight curves. The basal component is considered to be an enduring component of attitude that resists change once it has been formed. The surface component is considered to be an immediate reaction or overreaction to the latest information that is quickly dissipated. In the present experiment, the terminal upswing of the curves would represent the surface component, and the flat segment the basal component, which appears to develop at a uniform rate across serial position.

The basal–surface conception of attitude structure is important but it also complicates theoretical analysis. Both components will be present in any observed attitudinal judgment, and some method is needed to separate them and to determine their relative magnitudes. The separation of the two components in the jury trial experiment depended on a lucky correspondence between the design and the locus of attitude formation that could not be expected to hold in general. The theoretical development of the serial weight analysis has provided a tool that makes feasible further study of this aspect of attitude structure.[e]

2.5.5 COMMENTS ON SERIAL INTEGRATION

The most important outcome of the work summarized in this section is the support for the serial integration model. Despite the complexities introduced by order effects, serial presentation seems to obey the same integration rules as simultaneous presentation. In addition, the model-based capability for obtaining complete serial curves of weight is a useful tool for further analysis.

Moreover, the serial integration model has conferred theoretical unity on the welter of conflicting results on order effects in the personality adjective task. Seemingly similar experiments find primacy, recency, neither, and even both, with little visible pattern. The primacy, or first impression, effect can be substantial, as shown in Figure 2.9, but it is also quite labile. Seemingly minor changes, such as having the subject respond after each adjective is presented, can change primacy into recency.

These results, despite their lack of empirical regularity, have been found to make theoretical sense. The standard primacy effect appears to result from a progressive decline in attention across serial position. Experimental manipulations that act to eliminate attention decrement also eliminate primacy. The evidence is discussed in Section 3.3, which takes up alternative theoretical interpretations, especially the change-of-meaning hypothesis. These results also bear on the validity of the serial integration model.

The distinction between process generality and outcome generality (Section 1.8.2) is well illustrated in the history of the order effect problem. The traditional approach was oriented toward outcome generality and was concerned with the surface pattern of primacy and recency. Much of this research was imbued with the idea that there should be a "law of primacy," or at least that some empirical rule should be sought that would predict whether primacy or recency would be found. However, this search for outcome generality was not successful. In contrast, the search for process generality in valuation and integration has had some success in unifying the data. This search for process generality has advanced the conceptual formulation of the order effect problem and has also produced new analytical tools.

The serial integration model has also done well in other areas of psychology. These include psychophysical integration (e.g., Anderson, 1967a; Anderson & Jacobson, 1968; Weiss & Anderson, 1969), probability learning (Anderson, 1969a; Friedman, Carterette, & Anderson, 1968), decision theory (Shanteau, 1970a, 1972, 1975b), level of aspiration (Lopes, 1976a), and social attitudes (Anderson, 1959b; Anderson & Farkas, 1973).

There is much that is not yet understood about order effects. The area of attitudes presents an unclear picture that may result from the more extended verbal stimulus materials that are common in attitude research. Psychophysical integration seems to show uniform recency that may result from the use of sensory, nonsymbolic stimuli. And even in the personality adjective task, the concept of attention decrement is not satisfactorily specific. It seems fair to say, however, that the serial integration model has provided an effective approach to the study of these problems.

Notes

2.1.1a. Some of the material in Chapters 2–4 was used in technical report form (Anderson, 1974e,f,g,h) for a Workshop in Mathematical Approaches to Person Perception, directed by Seymour Rosenberg and myself, and held at the Center for Human Information Processing, University of California, San Diego in August 1974.

2.1.1b. A number of early investigators (e.g., Bruner, Shapiro, & Tagiuri, 1958; Feldman, 1962; Hammond, 1955; Hays, 1958; Hoffman, 1960; Osgood, Suci, & Tannenbaum, 1957; Podell, 1961; Rimoldi, 1956; Willis, 1960; Wishner, 1960) have also considered algebraic models for combinations of traits and similar stimuli. However, only one or two begin to come to grips with the basic question of model validity. In most cases, no more is shown than that the model provides high correlations between predicted and observed. That is seldom informative because high correlations can readily be obtained from seriously incorrect models. Accordingly, these early reports will not be considered here.

Reliance on such weak inference methods based on correlation or scatterplot analysis has continued to trouble theoretical approaches to social judgment. One illustration of weak inference has been given in the discussion of Figure 1.4. A similar illustration is provided by the date-rating data of Figure 2.5. The divergence of the solid curves is a strong violation of adding or summation models. Yet an adding model fit to these data yields a correlation between predicted and observed of .985 (Anderson, 1971a, p. 193). An investigator who relied on correlations would conclude that the adding model did very well, which is not so.

The data of Table 2.2 provide another illustration. With normative values of -3, -1, 1, and 3 assigned to the L, M$-$, M$+$, and L traits, respectively, the simple adding model may be tested without estimation of any parameters. This yields a correlation of .983, which might be taken as support for the model. The fact is, of course, that this analysis misses everything of interest in these data, including the diminishing returns in the set-size effect and the evidence for averaging.

The problem of weak inference is basically a problem of experimental design, not of statistical analysis per se. In the examples that have been cited, the inadequacies of correlation-scatterplot analysis could be made clear because the design allowed application of alternative methods. However, most studies that have relied on correlation-scatterplot analysis have design limitations that disallow the stronger methods of inference. The source of the difficulties is thus in the design stage.

2.2.1a. Some details of design and method in this experiment (Anderson, 1962a) are worth noting (see also Appendix A on procedure). To help ensure a linear response scale, end-anchor stimuli of extra-high or extra-low value were included. This is a standard procedure, intended to tie down the ends of the response scale and reduce potential end effects in the main data. A 20-point scale was used to allow ample room in the interior of the scale and to minimize effects of the equal-frequency tendency that can be prominent in rating scales with only a few categories.

Several precautions were taken to ensure equal weighting for each adjective in the description. The instructions emphasized that all adjectives were accurate and of equal importance. Furthermore, each adjective was attributed to a different acquaintance of the person in order to minimize minor inconsistency and redundancy. These instructions are still standard. Although later experience has shown that subjects are not very sensitive to inconsistency or redundancy, the equal weighting condition is even more important than was recognized when this experiment was run (Section 4.4).

For each stimulus person, the experimenter first read aloud the adjectives from bottom to top, then handed the card to the subject, who read them aloud from top to bottom before making his response. This was intended to equalize attention to all the adjectives and to minimize order effects. Later work has indicated that this procedure is unnecessarily conservative and tedious, at least for most purposes.

The first two sessions were treated as practice, and only the data from the last three sessions were analyzed. Later work has shown that a small amount of practice at the beginning of a single session is often adequate, at least for group analysis. For individual analysis, more extended practice may be necessary to iron out minor nonlinearity in the response scale.

Repeated sessions with the same set of stimuli carry the risk of response stereotyping due to memory effects. The data were studied carefully for such stereotyping effects, but no sign of any was found.

The six different sets of adjectives used were chosen randomly from a larger list. This procedure extends the generality of the results and eliminates the possibility that the parallelism reflects some unwitting selection process that avoided the interactive combinations (see also Section 3.4).

Some aspects of the statistical analysis also deserve mention. In this experiment, all three factors were equivalent and so all interactions were pooled for summary purposes. The complete analysis tested all four interactions for each subject, but the overall picture seen in Table 2.1 was virtually identical.

The values of the error mean squares in Table 2.1 are worth thinking about. They vary by a ratio of 10:1 across subjects, a range found in other work as well (Shanteau & Anderson, 1969). Why some subjects show large error variance, or whether it can be reduced in some way, is unknown. A subject with large error variance does not give a very strong test of the model since the discrepancies would need to be very large to become significant.

Finally, the correlations in Table 2.1 between the observed values and those predicted by the linear model also deserve notice. The mean correlation is .967, which indicates that the model has good predictive power. But this predictive power is trivial. The correlations are equivalent to the F ratios for additivity; that is, to the main effects of the design. Large correlations are thus a trivial consequence of the choice of stimuli. These correlations were included in the original report in accord with prevailing practice, but it was pointed out that correlations do not test the deviations from the model, which is what is important in the search for understanding of the integration processes (see Figure 1.4).

2.2.1b. As a historical footnote, I should say that I did not have a firm a priori hypothesis that the model would succeed. Indeed, I more than half expected that it would fail but would still serve as a baseline so that deviations from parallelism could be interpreted in terms of semantic interaction among the adjectives. After living with these results for a time, I began to feel that they had some generality. At that time (1963), an undergraduate named Penny Brooke came by looking for a research project and I told her about this work. A few days later she came back with a very nice hypothesis about certain specific adjective interactions. For example, *thoughtful* could be expected to have different implications when combined with *honest* and *dishonest*. I felt the force of her examples and encouraged her to test them. Her subjects, however, did not show the expected effects. Their data did show some interactions, it is true, but they had no particular relation to the critical combinations. Later evidence has supported this outcome and indicated surprisingly little stimulus interaction in this person perception task, either for inconsistency (Section 3.4) or for redundancy (Table 2.3).

2.2.2a. Guilford (1954) refers to bimodal distributions in ratings of colors, odors, poems, and so forth, but treats bimodality as faults of the rater and the rating scale: "There is no doubt that there is considerable distortion of the scale going on [p. 291]." Guilford's argument is based on the assumption that the true distribution should be unimodal, which seems plausible but could not be tested with the methods available to him. Functional measurement provides a way to test this assumption, as noted in the text, and it establishes the bimodality as psychologically real.

2.2.3a. Since this conclusion rests on acceptance of a null hypothesis, it should be noted that there were 48 subjects in each condition, which would seem to provide adequate power. Moreover, a similar initial experiment also showed nearly identical results in the Paragraph and No-paragraph conditions. It may be added that the person impression data in the initial experiment showed significant nonparallelism, an outcome that was attributed to nonlinear bias in the impression response scale. The main concern in the initial experiment was with judgments of individual component adjectives (Section 4.1). To reduce similarity between the two response scales, the impression response was assessed on a rating scale with nine verbal categories and without end anchors. In the second experiment, shown in Figure 2.3, the impression response was assessed on a graphic scale that seems to have cleared up the nonparallelism. This experiment provided another reminder of the importance of procedural details in the rating method (Anderson, in press, Chapter 1).

2.2.3b. Surprisingly, no systematic analysis of subjects' self-reports of integration processes in the personality adjective task seems to have been presented. Some subjects say that some descriptions remind them of specific people, an idiosyncratic factor that would tend to produce deviations from any simple integration rule. An interesting line of attack would be to investigate the paragraphs that subjects write when they describe the person in their own words. Aside from their relevance to person perception, these paragraphs have great interest as verbal behavior. Asch (1946) apparently made extensive use of such paragraphs but gives only illustrative, anecdotal examples.

2.2.3c. The equivalence of the Paragraph and No-paragraph conditions in Anderson (1971c) also bears on the interpretation of procedures in which subjects are asked to list their thoughts when making judgments. If these thoughts had a causal role in the judgment, as is often assumed, then the person impression would be based on more information—not just the three given adjectives, but also the additional thoughts and inferences generated in writing the paragraph. By virtue of the set-size effect, therefore, the impression should be more extreme in the Paragraph condition. In fact, the impressions were a little less extreme. This result has been corroborated by Simpson and Ostrom (1975) and by Burnstein and Vinokur (1975, Condition III); see also Anderson (1981b).

2.3.1a. The only previous application of this idea that I have found is in a study of affective value of combinations of odors by Spence and Guilford (1933). Their results appeared to favor averaging.

2.3.2a. Here, as elsewhere, the psychological zero lies in the middle of the response scale, which is near the nominal value of 50 in this experiment. Psychologically, therefore, the response of 23.7 to the LL description is negative. Adding the M^-M^- adjectives yields a mean response of 25.7, which is less negative.

This test between averaging and adding requires that the M^+ traits be positive and the M^- traits negative. That this condition is satisfied is shown by the set-size effect: Four traits yield more extreme response, positive or negative, than two. Similar evidence that the M^+ traits are truly positive and the M^- traits truly negative is given in Anderson and Alexander (1971).

2.3.4a. A doubt about this social pressure interpretation of the desirability of unattractive males as dates has been raised by Hagiwara (1975). These data showed nonparallelism similar to that of Figure 2.5 when similar stimuli were used; namely, photographs and trait adjectives. When the photographs were replaced by verbal descriptions of physical appearances, however, the curves were essentially parallel. It is possible, therefore, that the visual form of the information is important in producing differential weighting.

2.3.4b. I am indebted to Joseph Kruskal for correspondence that helped clarify some issues involved in the interpretation of this experiment.

2.3.4c. The main intended purpose in Oden and Anderson (1971) was to get a quantitative test of the averaging model with differential weighting. Unfortunately, the requirement of equal weighting on one factor of the design was not fulfilled in these data, so an exact test was not possible. Subsequent developments in statistical procedure have led to exact tests for the general averaging model.

2.3.5a. The congruity model is an averaging model in which weight is a specified function of scale value. With two stimuli, for example, $w_1 = |s_1|/(|s_1| + |s_2|)$, and $w_2 = |s_2|/(|s_1| + |s_2|)$. The congruity model thus requires a neutral stimulus to have zero weight, which is clearly wrong. Also, it cannot account for the set-size effect. For these and other reasons, it has done exceptionally poorly although it still retains straw-man popularity.

2.3.5b. The nonlinear bias in Birnbaum's rating response presumably resulted from procedural causes including lack of end anchors, use of only nine response categories, and running subjects in classroom batches (see Anderson, in press, Chapter 1).

2.3.6a. As far as it goes, the theoretical interpretation of the paradoxical averaging effect considered by Nisbett and Ross (1980) and by Nisbett, Zukier, and Lemley (1981) seems much the same as an averaging process considered in integration theory. The authors assume an element integration like that in Anderson (1974d); indeed, their term *dilution effect* agrees well with the idea of mixing elements, with nondiagnostic information corresponding to neutral elements that dilute the concentration (see also Section 4.5.4). They also assume a theory of similarity judgments like that considered in Anderson (1974a); Section 5.6.5 compares this theory of similarity with later developments by Tversky (1977). This and other averaging processes are discussed further in Section 4.5.4.

2.4.1a. It hardly needs saying that "simultaneous presentation" generally involves serial processing. With a few trait adjectives, however, it is assumed that serial position effects are largely eliminated by simultaneous presentation, especially with a forward–reverse presentation procedure (Note 2.2.1a). Residual effects may sometimes be appropriately confounded with serial position.

2.4.3a. Experiment 1 of Figure 2.7 on the set-size effect used two groups of H and two groups of L adjectives, chosen to avoid obvious redundancy within each group: (*thoughtful, wise, considerate, good-natured, reliable, mature*); (*warm, earnest, kind, friendly, happy, interesting*); (*spiteful, annoying, conceited, narrow-minded, disrespectful, greedy*); (*rude, thoughtless, vulgar, heartless, selfish, loudmouthed*). Each group of adjectives was used to construct 15 sets such that each adjective appeared equally often within each set size (one, two, three, four, and six adjectives per set). Besides these 60 experimental sets, 28 filler sets of varied size containing only M^+ and M^- adjectives were included.

Subjects were told that all adjectives were equally important and should be given equal attention. The adjectives of each set were typed on an index card that was handed to the subject at the start of each trial. The subject read the adjectives aloud slowly and looked them over for 5 more seconds before responding. This procedure was used to minimize possible inattention to adjectives in larger sets.

2.4.3b. Later experience has indicated that the -50 to $+50$ rating scale is not the best. It was used in this set-size experiment to facilitate the assumption for parameter estimation that $C_0 = 0$, that is, that the nominal zero of the rating scale was the true psychological zero.

2.4.3c. The w-estimates reported by Kaplan (1971e) showed a pattern similar to that of Experiment 1 in Table 2.6: w-estimates roughly constant for sets with two, three, and four adjectives but larger for sets with six adjectives.. Also included was a redundant condition, in which adjectives within each set were chosen to have high implicational relatedness. In this redundant condition the w-estimates were generally smaller for the larger set sizes, in accord with theoretical expectation. These estimates were made from the group means, however, so that significance tests were not available.

Two tests of the set-size equation have been reported by Takahashi (1970, 1971c), who used Japanese words. The first report agreed fairly well with the above results; the w-estimates were roughly constant over set size, though perhaps with a tendency to be too large for the largest sets of positive adjectives. The second report showed a significant dependence of w on set size for negative adjectives. However, that seemed to represent an inversion in the set-size function itself, possibly as a result of confounding adjectives with set size.

2.4.4a. Clear evidence is lacking on these two interpretations of the relation between confidence and the set-size effect. Anderson (1968a) noted that the causal view of confidence would have difficulty explaining why adding M^+ adjectives to H adjectives can decrease the impression because the added information would presumably increase confidence. However, this latter assumption has never been tested.

An interesting experiment by Lee and Ostrom (1976) obtained likableness judgments of persons and then asked for likableness judgments of a single trait in the description and also for certainty ratings of that single trait judgment. If set size is the cause of confidence, then these certainty ratings should be constant because set size is always one. However, certainty varied directly with the extremity of the likableness response to the single trait as manipulated by the other traits in the description (the positive context effect of Section 4.1). Lee and Ostrom interpreted this to mean that extremity of the likableness response has a direct effect on certainty when set size is constant. An alternative interpretation of their result is that the certainty rating of the component reflects the certainty of the overall person impression, in line with the halo model of Section 4.1.4. However, these latter certainty judgments were not obtained.

2.4.4b. The cited study of behavior disturbance judgments did not include an averaging–adding test. Such a test was included in the cited study of criminal offenses and it supported averaging. It could be argued, however, that criminal offenses are qualitatively different from mere traffic violations.

2.4.4c. Warr's range adjustment model makes a quantitative prediction that can be tested. Let S and L denote the responses to the less and more polarized of two single items, and let C denote the response to the two-item set. As long as the range $L - S$ is greater than some minimum, typically assumed to be 1 point on a 10-point scale, a simple averaging model is assumed to hold. Thus $C = kS + (1 - k)L$, which may be solved for the weight parameter k to obtain $k = (L - C)/(L - S)$. Warr requires that k be constant for fixed S, independent of the range $L - S$, which would be contrary to averaging theory. Warr and Jackson (1975, Table 2) present data to support this constancy prediction, but these ratio estimates of k are subject to unusually severe statistical difficulties: First, since the range is defined in terms of the observed responses, S and L, it will be subject to regression artifacts that tend to make k too small for small $L - S$, too large for large $L - S$. Second, unreliability in $L - S$ will cause a ratio bias in the estimate of k; the magnitude of the bias depends on $L - S$. Third, estimates of k for different values of $L - S$ were based on whatever data were available pooled across 10 stimulus pairs, 14 different adjectival response dimensions, and 39 subjects; different estimates of k are based on different numbers of instances, ranging from 2 to 139, and so are confounded with different aggregates of heterogeneous data. Fourth, ratings of the single items and the two-item combinations were obtained in different sessions, which renders their comparability doubtful. Last, all analyses depend on the assumption that the response scale is linear, but no evidence is given to support this assumption. Because of these statistical problems (Anderson, in press, Sections 1.1.1, 1.1.8, and 7.6), these data on the constancy prediction of the range model are difficult to interpret.

2.5.1a. More extensive discussions of order effects in social judgment are given by Hovland (1957), Insko (1967), Jones and Goethals (1971), and Anderson (1974b, pp. 68–81). Many treatments in the literature rest on implicit assumptions that primacy and recency require interactive explanations of the kind considered in Section 3.3. When direct tests have been made for these interactive processes, however, little supportive evidence has been found. Many theoretical interpretations of primacy and recency thus rest on untested assumptions that have become increasingly doubtful in the light of the data.

2.5.3a. Continuous responding curves obtained by Anderson (1968a, Figure 1) had a somewhat different shape than those obtained by Stewart (1965). The change in response in midsequence was more abrupt, and two curves showed a trend reversal at the last adjective. These differences may result from the use of individually selected adjectives, which would make salient the shift in adjective value, or the use of a pronouncing condition, or the rapid presentation rate. In this experiment, however, the final response was not much different for continuous, intermittent, and end responding (see Figure 2.4).

2.5.4a. This serial integration model originated in the distance-proportional model used by Anderson and Hovland (1957) and in the extension of the linear operator models of Bush and Mosteller (1955) used by Anderson (1959a) to handle sequential dependencies. Relevant later articles are Anderson 1961b, 1964a,b,c,d, and 1969a; see also 1974d, Section 6. This work also generalized standard learning models by allowing learning rate to vary over trials.

2.5.4b. A complication of some interest was that the two-way interactions were generally significant (Anderson, 1973d). The shape of these interactions was uniformly consistent with a negativity effect, that is, with greater weighting of negative than positive adjectives. Also, there were no significant differences among the magnitudes of the various two-way interactions. Since the design factors are the serial positions, each two-way interaction corresponds to a pair of serial positions. The fact that these two-way interactions were as large when the serial positions were separated as when they were adjacent argues against interactive interpretations such as inconsistency discounting and supports the negativity weighting interpretation.

2.5.4c. An interesting note on method in serial integration arose in the pilot work on the number-averaging task in which the experimenter read the numbers one at a time. The pilot subjects were graduate students in psychology but nevertheless found the task very difficult, seemingly because of fluidity of the cumulated impression. This difficulty was eliminated by writing each number on a card and presenting the task as that of determining the average of the deck of cards. The task became easy when it was made physically concrete in this way.

2.5.4d. Many real-life situations appear to exhibit similar short-term recency due to the salience of the most recent information. Predictions about the stock market by pundits in the television program "Wall $treet Week" give great attention to minor business and political fluctuations and little attention to economic fundamentals. Affective content of relations between nations (and other groups) exhibits similar reaction to temporary factors. This is a considerable handicap to the United States, which is gravely deficient in long-term policy and so is controlled by short-term fluctuations. As a consequence, the United States is retaining little permanent benefit from its lucky preeminence in natural resources.

2.5.4e. The following questions illustrate aspects of the basal–surface hypothesis that deserve investigation. The first concerns the developmental structure of the basal component. Formation of the basal component was uniform across serial position in the attitude experiment (Anderson & Farkas, 1973), but occurred mainly in the middle positions in the jury trial experiment (Anderson, 1959b). An associated question concerns the shape of the serial curve for longer sequences. The cause of the short-term recency is also unknown. The two obvious explanations are passage of time and active interference from other material, but existing data do not distinguish between them. Finally, there is the question of the relations between the surface and basal components and the analogous concepts of short- and long-term memory.

Meaning Constancy in Person Perception

Many problems of cognitive interaction arise in judgment-decision theory and interactional processes have been considered by many writers. Among the more prominent are theories that base themselves on one or another postulate about the mind's need for cognitive consistency. The popularity of this approach may be seen in the 80-odd chapters by 60-odd authors compiled in *Theories of Cognitive Consistency: A Sourcebook,* edited by Abelson, Aronson, McGuire, Newcomb, Rosenberg, and Tannenbaum (1968).

Such consistency postulates, as well as other kinds of cognitive interaction, can be most attractive on phenomenological grounds. But before such processes can be accepted as explanatory concepts, they must be shown to be operative. Because of their phenomenological attractiveness, however, they have generally been taken for granted, not in need of proof. As one consequence, conceptual analysis and experimental study of interactional processes have both suffered neglect.

This chapter presents a case history of one interaction problem. This concerns the competing assumptions of meaning constancy and meaning change in the personality adjective task. One main outcome is the failure of the phenomenological approach and the success of an approach based on informational principles.

3.1 Problem and Initial Evidence

3.1.1 THE PROBLEM OF STIMULUS INTERACTION

The idea that the meaning of a word depends on its context may seem too obvious to require proof. Phenomenologically, the quality of patience seems different in a *resourceful, patient* person than in a *submissive, patient* person. An *irresponsible mother* seems worse than an *irresponsible aunt*. When the separate words are combined as a unit, they seem to change each other's meanings. Such interaction processes are obviously fundamental to any attempt to build a theory of information integration.

Two phenomena are of primary concern in this chapter. One is the first impression, or *primacy effect*. When the same few adjectives yield different impressions in different sequential orders, the same adjective must have variable effects. A natural explanation is that the same adjective has different meanings in different orders, as though earlier adjectives influence and control the meanings of later adjectives.

The other phenomenon is the *positive context effect*. When asked directly about the meaning of a single adjective in a person description, the subject's response shows an apparent shift toward the other adjectives in the description. Such responses provide direct measures of meaning change whose interpretation seems hardly open to doubt.

These two phenomena are discussed in Sections 3.2 and 3.3, but in counterhistorical order for expositional convenience. First, however, a preliminary view will be given of the initial evidence on the problem.

3.1.2 ASCH'S EVIDENCE

The problem of change of meaning in the personality adjective task was pursued by Asch (1946) in an extensive series of experiments, nearly all variations on one experimental procedure. Subjects received a set of adjectives that described a hypothetical person about whom they were to form an impression, usually by writing a paragraph about the person. Most experiments also included some more quantifiable response measure, especially an adjective checklist on which subjects checked pairs of polar opposites (e.g., *generous–ungenerous, reliable–unreliable*) according to which better described the person. The standard procedure compared responses to two person descriptions that had some common, some noncommon adjectives.

In Asch's Experiment I, for example, two groups of subjects, A and B, received the following descriptions, which were identical except for the middle term, *warm* or *cold:*

A: Intelligent–skillful–industrious–*warm*–
    determined–practical–cautious
B: Intelligent–skillful–industrious–*cold*–
    determined–practical–cautious

The two groups showed substantial differences on certain checklist pairs, such as *generous–ungenerous*, but not on others, such as *reliable–*

unreliable. From several such experiments, Asch (1946) concluded that the adjectives change meaning and "do not contribute each a fixed independent meaning [p. 268]" to the impression. Instead, the processes of person perception involve intricate interaction among the given adjectives that cannot be reduced to any simple rule. If this is true, it needs to be established and studied.

It should be evident, however, that these checklist results have no bearing on the issue. The two descriptions contain noncommon words so they must produce different responses. All differences in the data of Groups A and B can be understood in terms of the noncommon words, *warm* and *cold*. A warm person is expected to be more generous than a cold person but not to be particularly more reliable than a cold person. This has nothing to do with any relations between *warm* or *cold* and the other stimulus words in the descriptions; it merely reflects the relations between *warm* and *cold* and the response words in the checklist. Hence the same checklist results would be expected under the opposite hypothesis; namely, that each adjective does indeed contribute a fixed, independent meaning to the overall impression.

A different procedure was used in some experiments, with subjects being asked to respond to a single trait in the description. In Experiment V (Asch, 1946), for example, each subject received one of the following two descriptions:

Person A: Kind–wise–honest–calm–strong
Person B :   Cruel–shrewd–unscrupulous–calm–strong

Subjects were instructed to write synonyms of the given terms, and these were quite different for the two descriptions. For example, Asch's Table 6 includes the following nine words that occurred a total of 20 times as synonyms to *calm* for person B:

Shrewd–scheming–nervy–conscienceless–
cold–frigid–icy–cool–calculating

None of these words occurred as synonyms to *calm* for person A, for whom the given synonyms were much more desirable. Accordingly, Asch concluded that *calm* had a different meaning in the two cases.

This experiment makes better sense than the first because the subjects are judging the meaning of the same word in different contexts. This and other such results were taken to support Asch's (1946) general thesis that "a given trait in two different persons may not be the same trait [p. 278]."

But Asch's reasoning becomes suspect on closer inspection of the listed synonyms. At least half of them appear to be synonyms not of *calm,* but of other traits of person B: *Conscienceless* is evidently a synonym of *unscrupulous* rather than *calm; shrewd* is an extreme case in which the response word actually is one of the other words in the list (Anderson & Norman, 1964).

Evidently, Asch's instructions to his subjects cannot be taken at face value. The difficulty is obvious in this example, but it is just as serious in Asch's other experiments. The conclusion that "a given trait in two different persons may not be the same trait" does not follow from such data.

Asch's evidence is unsatisfactory because nearly all of it is readily predicted by the opposite of Asch's hypothesis. A more satisfactory kind of evidence, namely, the primacy effect, is taken up in Section 3.3, and a more detailed treatment of Asch's position is given in Section 3.6.3.

3.1.3 EVIDENCE FROM PARALLELISM

According to the meaning-constancy hypothesis, each adjective has a fixed meaning that is independent of the other adjectives in the description. Strong evidence for this meaning-constancy interpretation emerged from the study of integration rules in the personality adjective task.

If each adjective has a fixed meaning in all person descriptions, then it will also have a fixed scale value; if its meaning varies with the other adjectives in each description, then its scale value will in general also vary. The parallelism test will tend to fail, therefore, because the parallelism theorem rests on the independence assumption that each stimulus adjective has a constant value across different descriptions (Section 1.2.5). If Asch's hypothesis was correct, the tests of the integration model should have provided evidence for it—by obtaining deviations from parallelism that reflected semantic relations among the adjectives. Conversely, the success of the model cast doubt on Asch's change-of-meaning position.

The initial studies of integration processes in the personality adjective task had thus led to the following position. Most results presented by Asch had little or no bearing on the issue for reasons already indicated. His interpretation of the primacy effect as change of meaning was reasonable, but it had failed to find support in several studies (Section 3.3). Moreover, the observed parallelism in the person impression task provided strong support for the alternative hypothesis of meaning constancy. The issue of meaning constancy seemed fairly settled, therefore, when the discovery of the positive context effect resurrected it in a form that is not entirely resolved even today.

3.2 Studies of the Positive Context Effect

3.2.1 THE POSITIVE CONTEXT EFFECT

One of the most substantial effects in person perception is the positive context effect. In this effect, each single trait of the person seems to shift in value toward the person's other traits (Anderson, 1966a; Anderson & Lampel, 1965). A typical case is shown in Figure 3.1. The subject received person descriptions that contained three trait adjectives, gave an overall impression of the person, and then rated one trait in the description according to the likableness of "that particular trait of that particular person." For convenience, the single rated trait is called the *component test trait*, and the other two traits are called the *context* (see also Section 4.1).

The positive context effect is visible in Figure 3.1 as the upward slope of the curves. The open-circle curve at top left shows how judgments of the same component, M$^+$ adjective increase as the context traits listed on the horizontal increase in value. This upward slope means that the judgment of the test trait is displaced toward the value of the other traits in the description. The other curves show similar positive context effects.

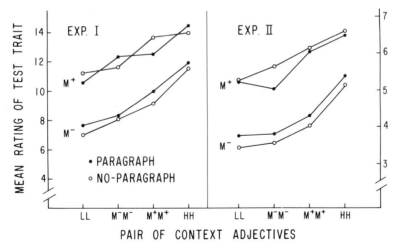

FIGURE 3.1. *Positive context effect. Rated likableness of one trait of person (M$^+$ or M$^-$) increases with value of the other two traits in the person description (horizontal axis). Filled circles represent data from conditions in which subjects wrote a paragraph describing the person in their own words before rating the one test trait. There were 48 subjects in each of the Paragraph and No-paragraph conditions in each experiment. (From Anderson, 1971c. [Copyright © 1971 by Academic Press. Reprinted by permission.])*

Change of Meaning or Halo Effect? At face value, the positive context effect *is* change of meaning, for subjects were told to judge the value of the component trait. If that judgment depends on the other traits in the description, it is natural, almost inevitable, to conclude that the other traits influence the meaning and value of the component trait. However, an alternative interpretation is also possible that involves no meaning change.

For conceptual clarity, it is important to recognize that two successive response processes are involved. The first refers to the integrated impression of the person; the second refers to the judgment of the single component trait of that person. Implicit in the change-of-meaning interpretation is the assumption that the second response truly measures the effective value of the component trait in the first response.

Under the alternative meaning-constancy hypothesis, the adjectives do not influence one another's meanings as they are integrated; each adjective has the same fixed meaning and value regardless of context. However, the second response process is affected by the outcome of the first: The judgment of the component is influenced by the integrated person impression. In other words, the overall person impression exerts a kind of halo effect.

The distinction between the two interpretations of the positive context effect can be explicated in terms of the integration diagram, Figure 1.1. Figure 3.2 gives the diagrammatic representation of the two interpretations, somewhat simplified and modified to make explicit the presence of the two different response processes.[a]

The meaning-constancy view appears in the left panel of Figure 3.2. The upper part represents the integration of two adjectives, s_1 and s_2, to obtain the person impression, I. The lower part, which is the focus of present interest, represents the judgment of the second adjective as a component trait. The diagram shows that this component judgment results from a second integration in which the context-free value, s_2, and the person impression, I, are combined to produce the component judgment, denoted by s_2^*. This diagram provides a straightforward explanation for the positive context effect: The value of the first adjective, s_1, influences the impression, I, which in turn influences s_2^*. By virtue of this causal chain, s_2^* moves toward the value of the context, in this case s_1.

The corresponding diagram for the meaning-change view, given in the right panel of Figure 3.2, is more complex in one respect, less complex in another. This diagram requires a more complex valuation operation in which the stimulus adjectives undergo mutual interaction. This interaction, denoted by the dotted lines, causes the adjective values to

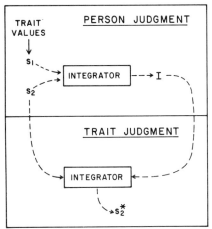
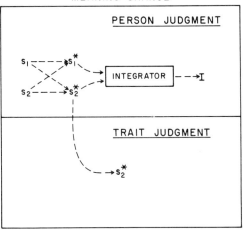

FIGURE 3.2. *Integration diagrams for halo and meaning-change interpretations of positive context effect. Upper panels represent judgment of person described by two traits, lower panels represent in-context judgment of one trait of that person.*

change to s_1^* and s_2^*; these changed values are integrated to obtain the person impression, I. The lower part of the diagram, which represents the judgment of the second adjective as a component trait, is simplicity itself. The subject is assumed to have introspective access to the changed stimulus values, and they are reported when the component judgment is requested.

These diagrams exhibit the distinctive differences between the two interpretations. The meaning-constancy interpretation assumes no interaction among the adjectives in forming the person impression; the positive context effect appears because that person impression affects the subsequent judgment of the component trait. The meaning-change interpretation assumes interaction among the adjectives in forming the impression; that interaction yields the positive context effect. There is thus a sharp conceptual difference between these two interpretations.

Initial Evidence. The initial experiments on the positive context effect (Anderson, 1966a; Anderson & Lampel, 1965) contained two results uncongenial to the change-of-meaning interpretation. These were the linearity of the effect and its large magnitude.

The linearity of the positive context effect, represented by the parallelism of the M^+ and M^- curves in Figure 3.1, suggests that it depends on the affective value of the adjectives rather than their semantic interrelations. If the words did interact in terms of their specific meanings, then the judgment of a given M⁺, for example, would depend on which particular H it was paired with. Hence deviations from parallelism would be obtained, contrary to the obtained results.

The very magnitude of the positive context effect is also troublesome to a change-of-meaning interpretation. In the left panel of Figure 3.1, two context adjectives shift the rating of the test adjective approximately 3 points on the 20-point scale. If this shift is true change of meaning, it points to a surprising lability of word meanings.

Both considerations are only suggestive. More cogent tests of the positive context effect are given in the following sections.

3.2.2 COMPONENT JUDGMENTS WITH PARAGRAPH WRITING

If change of meaning arises from interaction among the specific meanings of the words, then requiring greater interaction should produce greater meaning change. The present experiment was based on this rationale.

Two main experimental conditions were used. The No-paragraph subjects were run under the usual condition: They formed an impression of the person, rated the person, and then rated one adjective in the description of that person. The Paragraph subjects were treated similarly, except that they wrote a paragraph describing the person in their own words before the ratings (see also Section 2.2.3). Two similar experiments were run and both yielded positive context effects, as already shown in Figure 3.1.

This positive context effect should be larger for the Paragraph than the No-paragraph condition—according to the change-of-meaning rationale. In writing their descriptive paragraphs, subjects are required to work over the adjectives in more detail, thereby becoming more aware of inconsistencies and alternative meanings. More meaning change should occur in the Paragraph condition, therefore, and so the positive context effect should be larger.

This prediction is not supported by the data in Figure 3.1. It requires the component judgments to be more extreme in the Paragraph condition, in the direction of the context. At the lower left, therefore, the filled circles should lie below the open circles, whereas in fact they lie above them. Moreover, this filled-circle curve should show greater vertical rise and cross over the corresponding open-circle curve, also contrary to the data. The three other pairs of curves in Figure 3.1 tell much the same story. Overall, the positive context effect is slightly less in the Paragraph condition. These results suggest that the positive context effect does not represent change of meaning.[a]

This test is notable because it applies even if the positive context effect is only partly true change of meaning. The two conditions will produce different effects to the extent that there is any differential change of meaning between them. Accordingly, the results suggest that not even part of the positive context effect represents change of meaning. It could still be argued, of course, that the maximal change of meaning already occurred in the No-paragraph condition so no more could be produced by the greater cognitive organization required in the Paragraph condition. Subject to this reservation, however, these results weigh against the change-of-meaning hypothesis.

These results have been confirmed by Simpson and Ostrom (1975), who also found little or no effect of a paragraph condition, either on the person impressions or on the component ratings. Their experiments varied the Paragraph–No-paragraph condition on a within-subject basis, thereby providing a more powerful test. By the reasoning described previously, their results also suggest that the positive context effect cannot be attributed to stimulus interaction, even in part. This result, like others noted later, is important because it avoids the simplistic, either–or view that only one kind of process can be operative.

3.2.3 CONNOTATIVE VARIABILITY AND THE POSITIVE CONTEXT EFFECT

Theoretical Rationale. If the positive context effect represents true change of meaning, it should be greater for words that have greater natural range or variability of meaning. The value of *nice,* for example, seems less crisp and well defined than the value of *able,* and so would be more susceptible to contextual influence. This idea was independently proposed and tested by Wyer and Watson (1969) and by Kaplan (1971a). It has the advantages of simplicity and directness. And like the test described in Section 3.2.2, it has the important advantage of being sensitive to meaning change even if the positive context effect itself includes other components besides meaning change.

To implement this test requires selecting pairs of words that are equated in value but high or low in variability of meaning. The several experiments discussed in this section used different procedures for selecting such pairs of words. They were otherwise similar in that they employed the standard task of component judgments. If meaning change does occur, then the positive context effect should be greater for the high-variable traits.

Kaplan (1971a). Kaplan began by selecting M^- and M^+ adjectives that had high or low between-subject variance in the standard list of 555 words (Anderson, 1968d). He then showed that these same traits were high and low, respectively, on within-subject variability as measured by the difference between ratings on two separate occasions by the same individuals.

The critical test is between the component judgments of the high- and low-variable traits. This difference did not approach significance, contrary to the change-of-meaning interpretation.

Kaplan's test was reasonably powerful, for it was based on within-subject comparisons from 24 person descriptions from each of 76 subjects. Moreover, the positive context effect itself was quite large, so any true change of meaning had ample opportunity to exhibit itself. Kaplan's null result thus creates serious difficulties for the change-of-meaning interpretation.

One criticism of Kaplan's experiment is that it, like most others cited here, failed to set aside those few adjectives such as *discriminating* and *sensitive* that have clearly bimodal meanings (Anderson, 1968d, p. 277). Including such adjectives would inappropriately favor the change-of-meaning interpretation. Of course, the obtained null result nullifies the practical effect of this criticism.

Kaplan (1975a). In his 1975a experiment, Kaplan used the same stimulus materials as in his 1971a report, but the components were not rated on likableness. Instead, each component was rated on two scales of denotative meaning, one semantically related to the component adjective, the other semantically unrelated. Variability of the component had no apparent effect on either scale. This result supports and extends Kaplan's earlier work.

Schümer (1973). Schümer questioned Kaplan's (1971a) work on the ground that variability may not be an adequate index of lability of meaning. Accordingly, Schümer selected words on the basis of direct ratings on an ambiguity scale.

Schümer also obtained a null result. The component ratings were nearly the same for traits that were high or low in ambiguity. Schümer's experiment was also reasonably powerful since it was based on within-subject comparisons from 64 person descriptions from each of 32 subjects. Schümer's confirmation of Kaplan's result is interesting because of the difference in the index of range of meaning, and because Schümer used German words with German subjects.

Wyer and Watson (1969). Wyer and Watson (1969, Experiment III) used a direct index of range of meaning obtained by asking subjects to rate not only expected likableness of a person described by a given trait, but also most and least probable likableness. Both the upper half-range and the lower half-range were used as indexes of range or variability of meaning.

When the lower half-range was used, there was a marginally significant effect in the direction predicted by change of meaning; when the upper half-range was used, the effect was in the opposite direction and a little larger. Wyer and Watson dismissed the latter result and concluded that their data favored the change-of-meaning interpretation. As both Kaplan and Schümer have observed, this interpretation does not seem consistent with the data.

From other data, Wyer and Watson (1969, p. 31) concluded that change-of-meaning also caused the positive context effect observed under two other instruction conditions. In the word condition, subjects were told that the adjectives did not apply to a single person, but that they were nevertheless to judge how much they would like the traits in combination and also to judge one single trait. In the group condition, subjects were told that each adjective described a different person and that they were to judge how much they would like the group as a whole and also one person in the group.

The word condition is problematical because the instructions not to treat the adjectives as describing a person seemed to have just the opposite effect (see Section 4.1.6). The result from the group condition may perhaps be taken as better evidence for a halo interpretation than for change of meaning. Since the group existed only by the arbitrary juxtaposition of three hypothetical persons, each described by one adjective, these adjectives are presumably under no consistency constraint such as when they describe a single person. Without such a consistency constraint it is not clear that there is any causal basis for change of meaning. The result is consistent with the halo interpretation, as noted by Anderson, Lindner, and Lopes (1973), who obtained similar effects in group attractiveness. Moreover, the halo model provides a good account of a complex pattern of data obtained by Takahashi (1971a) for a similar group condition (see Section 4.1.6). On the other hand, it would be difficult, using the halo hypothesis, to explain why the positive context effect was as large in the group condition as in the standard person condition.

Wyer (1974b). In his 1974b report, Wyer used a different index of meaning variability, namely, a dispersion measure of a subjective distribution of likableness values of a given adjective. The component ratings in the first experiment showed a significant interaction between context value and the variability index of the component adjective, with ratings of the high-variable adjectives more influenced by context. At face value, this seems to support the change-of-meaning hypothesis, but this interpretation is clouded by an interaction of similar shape in the person impression. By the halo model, the component rating depends on the overall impression; hence the interaction in the person impression implies a corresponding interaction in the component ratings. Hence the halo interpretation also accounts for the critical result.

Complications arose in Wyer's second experiment, in which subjects not only rated the component test adjective, but also performed a meaning-selection task, checking those of 15 given words and phrases that would be possible interpretations of the test adjective. The mean normative rating of the selected words provides a second response measure with the same significance as the test adjective rating. The two response measures should therefore show similar trends, and that was verified when the set of three adjectives described a person. But when each adjective described a different person in a three-person group, the rating of the test adjective showed strong context effects, whereas the mean value of the selected words showed little or none. This marked discrepancy between two presumably equivalent response measures is a source of concern.

In Wyer's (1974b) interpretation, the mean value of the selected words was taken to be the valid index of meaning. To account for the discrepancy between this index and the component ratings in the group condition, a halo-type interpretation was assumed in which likableness of the specified person was directly affected by the adjectives describing the other two persons. However, this halo interpretation creates an unexplained inconsistency with Wyer's earlier claim that the positive context effect in the group condition represented true meaning change. Furthermore, the discrepancy between the two response measures creates two difficulties within the two experiments under consideration.

The first difficulty concerns the first experiment, in which the critical interaction in the component judgments was similar in both person and group conditions. To reinterpret this interaction as a halo process in the group condition vitiates its interpretation as change of meaning in the person condition. The second difficulty arises because any change in the likableness judgment of the specified person in the group, even though it is a halo effect, should affect the selection of the corresponding words. The difference between the two indexes thus seems as much a problem for meaning constancy as for meaning change.

In sum, it seems difficult to reach any clear conclusion from these experiments. However, they do contain results that, if verified, could cause difficulties for the meaning–constancy hypothesis. Accordingly, the line of inquiry opened up by Wyer deserves further investigation.[a]

3.2.4 OTHER RESULTS ON THE POSITIVE CONTEXT EFFECT

Other Response Dimensions. Theoretically, the positive context effect should be obtainable on any response dimension. This follows from the generalized halo interpretation (Section 4.1), in which the general impression of the person constitutes one source of information that is relevant to diverse judgments about the person.

The experiments on this issue have all treated the halo effect in its traditional sense; that is, as a purely evaluative process without specific semantic content. This differs from the integration view, which allows for semantic structure both in the general impression and in particular response dimensions. As a consequence, none of these experiments has much bearing on their main purpose of testing the meaning-change and meaning-constancy hypotheses, for both hypotheses make the same predictions. Since this matter has caused confusion, a short discussion is included here.

The essential idea of Hamilton and Zanna (1974) can be illustrated by the two person descriptions, (*proud, friendly, well-spoken*) and (*proud, boring, rude*), in which the component test adjective, *proud,* appears in positive and negative contexts, both chosen to be semantically unrelated to the test component. Subjects first judged likableness of each person as well as likableness of *proud* in that person. Then they judged *proud* on an 8-point scale of connotative meaning defined by the endpoints *conceited* and *confident.* The main result was a positive context effect on the "conceited–confident" dimension: *Proud* was judged to be more toward the *confident* end of the scale when it appeared in the positive context (*friendly, well-spoken*) than in the negative context (*boring, rude*).

Hamilton and Zanna (1974) interpreted this result as change of meaning, on the grounds that the conceited–confident scale ostensibly measures connotative meaning. As they recognized, however, even the traditional evaluative halo interpretation could account for the result. The conceited–confident scale contains a substantial evaluative component of social desirability. Accordingly, the evaluative component of the general person impression could produce the observed effects on the conceited–confident dimension.

In Zanna and Hamilton (1977, Experiment 2), two-word descriptions were used in which the context word was either semantically related or semantically unrelated to the test word (e.g., *proud–dignified* or *proud–religious*). The two context words, *dignified* and *religious,* were matched on likableness value in order to equate evaluative halo. The central result concerns the in-context component rating of the test word *proud* on its own connotative meaning scale of conceited–confident.

In the integration-theoretical view, this component rating results from an integration of two relevant pieces of information, namely, the specified stimulus word, *proud,* and the general person impression. Because the response is on the conceited–confident dimension of judgment, both pieces of information must be evaluated with respect to that dimension. *Dignified* obviously has a higher value than *religious* on the conceited–confident dimension. Therefore, the proud–dignified person will have a higher value than the proud–religious person on the conceited–confident dimension. By virtue of the halo process diagrammed in Figure 3.2, the same applies to the component rating of *proud,* which agrees with the obtained results.

Zanna and Hamilton thought that the halo formulation could not account for the data and that a change-of-meaning interpretation was required. This view appears to rest on the traditional concept of the halo as a purely evaluative process. However, integration theory allows for nonevaluative halos by the logic of the preceding paragraph.[a]

The issue may become clearer by considering the related task in which the subject initially judges the person on the conceited–confident dimension, then judges the component trait on that same dimension. This task is isomorphic to the task considered in Figure 3.1, except that the response dimension is changed from dislikable–likable to conceited–confident. Although this experiment has not actually been run, it must also yield a positive context effect, quite like that observed by Zanna and Hamilton. Their data go no further than an empirical reflection of the positive context effect on other response dimensions. That has no bearing on the theoretical issue, which is to discriminate between the two hypothesized causes of this positive context effect.

A Correlational Test. An interesting implication of the change-of-meaning hypothesis has been pointed out and studied by Bryson. Owing to individual differences, ratings of a component trait will be correlated across contexts. If the trait does change meaning across contexts, it will be changed by different amounts for different subjects, thereby reducing the correlation.

To implement this idea, Bryson and Franco (1976) employed a 2 × 2 design that provided the necessary comparisons. Each set of three adjectives described a single person in the person condition but were unrelated traits in the word condition. Each condition was tested with nonambiguous component traits (e.g., *confident, obstinate*) and ambiguous or bivalent component traits (e.g., *discriminating, sensitive*). The context traits were chosen to bring out one meaning or the other of the ambiguous traits.

The bivalent traits can certainly take on different meanings in different contexts. However, this change of meaning should be restricted to the person condition, for the word condition imposes no contextual constraint. Therefore, the correlation across contexts should be lower in the person than in the word condition. This was the case; the two respective correlations were .23 and .60.

The critical result comes from the nonambiguous traits, for just these are at issue in the question of meaning change. As already noted, the change-of-meaning hypothesis implies lower correlation in the person condition than in the word condition. In fact, the correlations were virtually identical, .47 and .48, respectively. That this critical test had adequate power is demonstrated by the difference for ambiguous traits already given. Like certain of the previous tests, this correlation test applies even if only part of the positive context effect is change of meaning. The observed outcome thus argues against change of meaning and for meaning constancy.

This correlational test is interesting because it looks at the data in a novel way. It can capitalize on individual differences and analyze cases in which the response means do not vary across conditions. However, the judgment of the component is more direct and meaningful as a descriptive statistic and has the advantage over the correlation of allowing analysis for single individuals.

Averaging Model for the Positive Context Effect. The idea of the halo interpretation is that the positive context effect results from an integration of the overall impression and the component being judged. Although the exact form of integration is not important to the previous discussion, it is an attractive assumption that this integration obeys an averaging model (Section 4.1.4). The following two subsections take up implications of this model.

Component Judgments and the Set-Size Effect. The halo averaging model implies that the number of adjectives in the description will affect how any one component adjective is judged. If the number of adjectives is increased and their value kept constant, then the person impression will in general become more extreme by virtue of the set-size effect. That will produce a corresponding trend in the component judgment because the person impression exerts a direct effect on it. Moreover, the weight of the impression might increase with set size, thereby making the component judgment more extreme.

This implication of the averaging formula is not critical for the change-of-meaning hypothesis. Instead, the effect would be relevant to a change-of-meaning approach because it would impose constraints on the underlying process. An interpretation in terms of affective assimilation would not work when all adjectives have the same value. However, a more molecular mechanism could account for such an effect by discounting less polarized element meanings. Indeed, by assuming that the more polarized meanings are discounted, the change-of-meaning hypothesis could even account for a result opposite to that predicted by the averaging formula (Anderson, 1971c, pp. 81–82). For the halo interpretation, of course, failure to obtain the predicted effect would be serious.

The first test of this prediction (Anderson, 1971c, Experiment III) is portrayed in Figure 3.3. Subjects received sets of two or six trait adjectives, all of equal value, rated the person, and then rated a specified

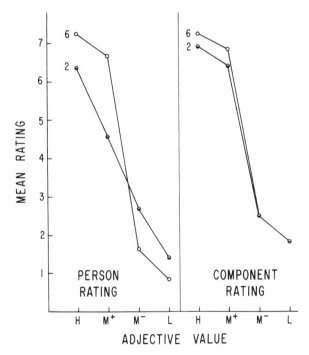

FIGURE 3.3. *Effect of set size on person impressions and on positive context effect. Left panel shows likableness rating of persons described by two or six traits of equal value as listed on horizontal. Right panel shows in-context likableness rating of one trait in each description. (From Anderson, 1971c.* [*Copyright © 1971 by Academic Press. Reprinted by permission.*])

component trait of the person. Person likableness, in the left panel, shows the expected set-size effect. Component trait likableness, in the right panel, shows the predicted effect for positive traits but not for negative traits.

In the design of this experiment, it was expected that the weight of the impression in the component judgment would be constant over set size. The observed dependence of the positive context effect on set size was therefore expected to be small, a fraction of the size of the set-size effect in the person impressions themselves. Accordingly, care was taken to reduce variability by preselecting adjectives equal in value for each individual subject.

The success of this methodological precaution is reflected in the highly significant $F(1,39) = 9.29$ for the main effect of set size in the component judgments. Since the mean effect was only .15 scale points, its significance testifies to the precision that can be obtained from within-subject, within-word design. However, the small size of the effect suggested that this was not an attractive line of inquiry.

Other reports on this set-size prediction differ in an important way, namely, in the use of context and test traits of unequal value. This value difference can be amplified by the set size so that substantially larger effects can be obtained. Kaplan (1971a) obtained large effects that conformed well to several quantitative predictions from the model (Section 4.1.6), and large effects were also obtained by Kaplan (1975a) on response dimensions other than likableness.

Lee and Ostrom (1976) obtained data both for an isovalent condition, which was similar to that of Figure 3.3, and for a contravalent condition, which was similar to that employed by Kaplan (1971a). The isovalent data yielded regular effects in the predicted direction, but these were small and apparently not significant. The contravalent data yielded significant effects in the predicted direction, but these were somewhat irregular and not as large as those obtained by Kaplan. Finally, Schümer (1973) obtained small, nonsignificant effects. The reason for the apparent disagreement among these two reports and those of Kaplan is unclear. The small effect in Schümer's data is a cause of concern because his procedure seems to have been careful and because the halo theory would presumably predict a substantial effect in this case.[b]

A Disguised Set-Size Effect in Component Judgments. An interesting experiment by Ostrom (1971) seems at first glance to contradict the halo interpretation and to demonstrate true change of meaning. In the person condition, a person was described by 15 traits, 12 positive followed by 3 neutral. Subjects rated the person and then the likableness of each trait of that person. In the word condition, the traits were presented as descriptions of 15 different people, and subjects rated likableness of each person.

The main result was that the positive traits were rated more positively in the person condition than in the word condition. Ostrom pointed out that this result could be predicted by a change-of-meaning hypothesis with a meaning elements rationale in which each adjective has various shades or elements of meaning. Incompatible meaning elements may be eliminated when two or more adjectives are integrated so that each adjective takes on a more extreme meaning in the person condition. No such effect occurs in the word condition because each description contains only one word. This change-of-meaning rationale implies that the rating of single positive traits will be more extreme in the person condition than in the word condition.

This result might seem to disagree with the halo interpretation. By the halo averaging model, the component judgment is an average of its context-free value, s, and the person impression, I. Since the same word is being rated in both conditions, s is the same, and the prediction depends on the value of I. Under a simple averaging model, the person impression would be more positive when based on one positive adjective than when based on 12 positive and 3 neutral adjectives. By this reasoning the observed result would disagree with the halo averaging model.

However, this reasoning neglects the role of the initial impression in the averaging model for person impressions. When the initial impression is included, the halo model can account for the result.

For a numerical illustration, the following assumed parameter values would be empirically reasonable. The neutral traits and the initial impression, I_0, have the value 0, the neutral point on the rating scale. Each positive trait has the value 4, the endpoint on the rating scale. Both w and w_0 have the value 1. With these values, the person impression for a single positive trait has the value $(1 \times 0 + 1 \times 4)/(1 + 1) = 2$. Similarly, the person impression based on the twelve positive and three neutral traits has the value $(1 \times 0 + 12 \times 4 + 3 \times 0)/(1 + 12 + 3) = 3$. The person impression is thus considerably more extreme in the person condition than in the word condition, 3 versus 2, respectively. Accordingly, the halo interpretation implies that the judgment of the single positive trait must also be more extreme in the person condition, in agreement with the data.

In addition to illustrating the role of the initial impression, this experiment also contrasts the predictive precision of averaging theory and the vagueness of the change-of-meaning hypothesis. Although it is plausible that elimination of incompatible meaning elements would produce more extreme meaning, the opposite effect could be claimed just as easily, as already noted in the subsection, "Component Judgments and the Set-Size Effect." At the same time, the halo averaging model has survived a test that could have caused serious theoretical difficulty.[c]

3.3 The Primacy Effect

When the adjectives in a person description are presented in sequence, it makes a difference which adjectives come first. Under a standard set of conditions, persons will seem more likable if their good traits are presented first, less likable if their bad traits are presented first. This represents a first impression, or primacy effect.

Primacy has theoretical importance because the very same adjectives produce different effects when presented in different orders. It follows that the effect of any one adjective is not constant but variable, depending on its position relative to the others. An obvious explanation of this variable effect is that the meaning is variable.

Moreover, the primacy effect is large. The same six adjectives can produce responses that differ by as much as 1 point on a 10-point scale. The primacy effect thus promises to be a useful tool for analysis of integration processes. Beyond its bearing on the problem of meaning constancy, therefore, the primacy effect is important to cognitive theory.

3.3.1 THREE THEORETICAL INTERPRETATIONS OF PRIMACY

Three main theoretical interpretations of primacy have been at issue: *inconsistency discounting, attention decrement,* and Asch's concept of *directed impression.* The nature of these three interpretations and the experimental evidence that bears on them are summarized in this section (see also Section 2.5).

Inconsistency discounting refers to a reduction in weight parameters of the later adjectives that results from their inconsistency with the earlier adjectives. It is easy to see how such inconsistency discounting would reduce the influence of the later adjectives. In a HHHLLL sequence, the first L is received and integrated into an impression based on the three preceding H's. Any felt inconsistency would work against the single L because it is informationally less weighty than the aggregate of three H's (Anderson & Jacobson, 1965). Much the same argument would apply to each succeeding L. If the effective weights of the L adjectives are reduced in this manner, primacy will result.

Attention decrement refers to a decrease in the weight parameter of the later adjectives corresponding to a progressive decrease in attention over the course of the sequence. This differs from discounting because it does not depend on inconsistency between the earlier and later adjectives. Instead, attention is considered to decrease across the sequence, regardless of any relation among the adjectives. In contrast to inconsistency discounting, therefore, attention decrement is a noninteractive interpretation.

Directed impression can be considered as a set established by the initial adjectives and affecting the meanings of the later adjectives. Like discounting, it represents an interactive cognitive process. Unlike discounting, it refers to changes in the meanings and scale values of the later adjectives, not merely to changes in their weight parameter.

It should perhaps be added that Asch was vague in discussing the concept of directed impression. However, the essential idea seems to be as indicated—that the earlier adjectives set up a preliminary impression that controls and changes the meanings of the later adjectives. In a more specific version (Anderson, 1965b), each adjective would be considered to have a population, or distribution, of meanings. Primacy would then result from a tendency to select those shades of meaning of the later adjectives that fit best with the earlier adjectives.

3.3.2 INITIAL EVIDENCE FOR PRIMACY

The primacy effect in the personality adjective task was first reported by Asch (1946) who claimed that person A, described by six adjectives read in the order:

A: Intelligent–industrious–impulsive–critical–stubborn–envious,

was more likable than person B, described by the same six adjectives read in the opposite order:

B: Envious–stubborn–critical–impulsive–industrious–intelligent

The reliability of this effect was uncertain, however, for Asch's procedures and data analyses were casual. Indeed, when Asch repeated the experiment with a second sequence of adjectives, the effect vanished.[a]

Moreover, a substantial body of attitude research not only failed to support a "law of primacy" enunciated in 1924, but found a mixed array of primacy, recency, and nonsignificant effects that defied coherent summary (e.g., Hovland, 1957). The idea that six single words would produce primacy was thus dubious on empirical grounds. The experiment discussed next was designed to assess the reliability of the primacy effect with extended trials in the personality adjective task.

In this experiment (Anderson & Barrios, 1961), the experimenter read six adjectives of a person description and the subject rated likableness on an 8-step scale. In Experiment 1, each of 64 subjects received 61 or 62 such descriptions. The adjectives were chosen randomly from a larger pool to avoid bias, and 48 different sequences of adjectives were used to assess generality. In half these descriptions, the value of the adjective changed gradually across the sequence; in the other half, there was an abrupt change in normative value, from high to low, or vice versa, midway in the sequence.

Primacy was measured as a difference score: the rating for a sequence of adjectives in the favorable–unfavorable order minus the rating for the same adjectives in the opposite (unfavorable–favorable) order. This direction of difference is such that a positive score reflects primacy. A negative score would reflect a recency effort (Section 2.5.1).

The results showed substantial primacy, .69 averaged over all conditions, which was statistically reliable. Of the 48 paired descriptions, all but 5 yielded primacy. Thus, the primacy effect was found to be reliable and large.

Figure 3.4 presents the primacy effect as a function of trials. There is a gradual decline from about 1.1 points in the first block of trials to about .5 points in the sixth block of trials. Similar declines have been seen in later studies, but nothing is known about their cause. Despite the decline, primacy is not transient, but stabilizes at a substantial level.[b]

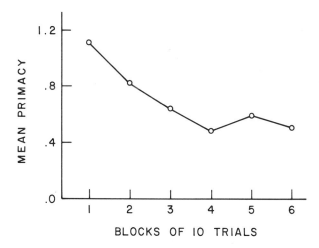

FIGURE 3.4. *Primacy in person impressions as a function of trials. (After Anderson & Barrios, 1961.)*

One further result was that mean primacy was approximately equal for both kinds of sequences, those in which adjective value changed gradually and those in which there was an abrupt change. If primacy results from cognitive interaction, larger effect might be expected from the abrupt change on the ground that it would make any inconsistencies more salient. The lack of difference between the two kinds of sequences thus provides no support for an interactive interpretation.[c]

A second experiment was also run to provide a more direct test of Asch's concept of directed impression. Each description contained only two adjectives, one high and one low, with interadjective intervals of 0, 2, or 4 seconds. Each of 24 subjects judged 20 descriptions at each time interval in latin square order. If the first adjective does set up a directed impression, as Asch claimed, it should become more solid and effective with a longer time interval and hence cause greater primacy. Small effects would be detectable in this within-subject design.

In fact, there was no primacy at all. The 95% confidence interval about the mean was .04 ± .12. This null result is a problem for the directed impression interpretation because it should operate with two adjectives as well as with six (see also Hendrick et al. [1973] in Section 3.3.5).

These initial results showed the primacy effect in the personality adjective task to be substantial and reliable. However, they gave no support to interpretation in terms of directed impression. None of these results raised any serious difficulty for directed impression, of course, but they did suggest the alternative attention-decrement interpretation that was explored in the studies reported next.[d]

3.3.3 ATTENTION DECREMENT HYPOTHESIS

The initial studies of primacy in the personality adjective task were mainly concerned with developing an integration model and with comparing the attention decrement and change-of-meaning hypotheses. The discounting interpretation grew out of later work with simultaneous presentation and was explicitly introduced into the primacy–recency issue only toward the end of this phase of the research (see Section 3.3.4). Because discounting is similar to change of meaning in assuming cognitive interaction, most of the experimental tests apply to both, vis-à-vis attention decrement. Accordingly, the two interactive hypotheses will be treated together here.

The rationale of the two tests considered in this section is straightforward. An experimental manipulation is chosen that is expected to equalize attention to all the adjectives. That should eliminate the primacy effect—if it is caused by attention decrement. In contrast, both interactive hypotheses imply that primacy is robust and should not be much affected by such attentional manipulations.

Primacy–Recency with Verbal Recall. In this experiment (Anderson & Hubert, 1963), subjects were read a sequence of personality trait adjectives and rated person likableness in the usual way. In certain conditions, subjects were also asked for casual recall of the adjectives in the description after rating likableness. The instructions placed a casual aspect on the recall task that was evidently successful since recall averaged only 3.9 words for descriptions with either six or eight adjectives.

The rationale was that the added recall task would eliminate the attention decrement—if it really existed—thereby also eliminating the primacy effect. Failure to eliminate, or at least markedly reduce, primacy would therefore cast doubt on the attention-decrement hypothesis. But if primacy was caused by change of meaning or inconsistency discounting, it should remain despite the added recall.

The results were clear. Inclusion of the recall task eliminated the primacy or even changed it to recency. The trend can be seen most clearly in Experiment 2 in Table 3.1. In the first half of the experiment, the No-recall condition shows a primacy effect of .50, whereas the Recall condition shows a recency effect of −.22. In the second half, both groups of subjects received recall instructions. The primacy effect in the first condition decreases to near zero, and the recency effect in the second condition becomes even larger. Both experiments show similar results. Not all the detailed tests were significant, and the need to allow for the usual decrease in primacy over trials should be kept in mind. Nevertheless, the main pattern of primacy and recency follows the predictions of the attention decrement hypothesis. This pattern has been corroborated by Riskey (1979; see Section 4.2.3).

This outcome does not support either interactive interpretation. If the initial adjectives produce a judgmental set that causes the later adjectives to change meaning in the standard No-recall condition, that same set should operate in the Recall condition. Inconsistency discounting might even imply increased primacy on the ground that more attention to the later adjectives would increase the salience of the hypothesized inconsistencies. Neither interactive interpretation is very definite, of course, and it could be argued that the concomitant recall weakens the process that produces primacy. However, that would leave the recency unexplained. On the other hand, the attention-decrement hypothesis would have found it awkward to explain no change or an increase in primacy. The attention-decrement hypothesis has thus done well in an important test.

TABLE 3.1
Mean Order Effect in Person Impression

	First half adjectives		Second half adjectives	
	Six	Eight	Six	Eight
Experiment 1 condition				
No recall–No recall	.57	.55	.40	.10
No recall–Recall	.55	.55	.07	−.13
Experiment 2 condition				
No recall–Recall	—	.50	—	.02
Recall–Recall	—	−.22	—	−.33

Source: Data from Anderson and Hubert (1963).
Note: The main design is explicit in this table, which specifies set length and recall conditions. For example, subjects in the No recall–Recall condition gave only the person impression in the first half of the experiment but also recalled the adjectives in the second half. The 80 sets of Experiment 1 were divided into 10 blocks, each with two HL, two LH, and six Random sets. The six-adjective and eight-adjective HL sets had the forms HHHLLL and HHHHLLLL. The Random sets consisted of equally many H and L adjectives in random order. The 30 sets of Experiment 2 were divided into six blocks, each with two MHL, two MLH, and one Random type set of eight adjectives (M represents two Medium adjectives). Sets were constructed by random selection of adjectives, and blocks of sets were balanced over serial position by latin square sequences. There were 64 and 72 subjects in the two respective experiments. Each subject served in only one condition and hence is represented in only one score in each half of the experiment. Responses were rated on a 1–8 scale of likableness. Entries are difference scores, positive for primacy, negative for recency.

One incidental aspect of the recall data provides a further argument against either change of meaning or inconsistency discounting. The adjective sets used in these experiments were of the HL type, with abrupt changes in normative value midway through the sequence. If the subject is sensitive to this change, as might be expected under either interactive hypothesis, greater recall would be expected at that point. No sign of such enhanced recall was seen in the data (Anderson & Hubert, 1963, Tables 2–4).[a] A more direct comparison was also available by comparing recall for the HL-type sets with the Random sets in which the adjectives were given in random order. At the position where the HL-type sets changed polarity, recall was slightly better for the Random sets. This result also argues against both the inconsistency discounting and the change-of-meaning interpretations of primacy.

Cumulative Responding. As a further test of the attention-decrement hypothesis, Stewart (1965) included a cumulative responding condition, in which subjects revised their responses to take each successive adjective into account. Subjects are thus required to give full attention to each adjective, which should eliminate any attention decrement. If the attention-decrement interpretation is correct, cumulative responding should eliminate primacy.

Opposite predictions seem to follow from change of meaning as well as from inconsistency discounting. Any felt incongruity at the shift between high and low adjectives should be accentuated and made more salient under cumulative responding. The two interactive interpretations thus seem to imply that primacy should increase under cumulative responding.

The experimenter read descriptions with four, six, or eight adjectives and subjects rated likableness under two response conditions. One was the usual condition in which subjects responded only after all adjectives had been read. The other was the cumulative condition in which subjects revised their opinions after each new adjective.[b]

The usual primacy effect was obtained under the standard condition in which the subject responded only at the end of the sequence of adjectives. Its mean value was .38, and it was nearly equal for sequences of length four, six, and eight adjectives.

The more interesting data are shown in Figure 3.5, which plots the cumulative impression as a function of serial position. The left panel illustrates the general pattern for eight-adjective sequences. The upper curve gives the response to the HHHHLLLL sequences. The impression rises steadily as each successive H is received, then falls steadily as each

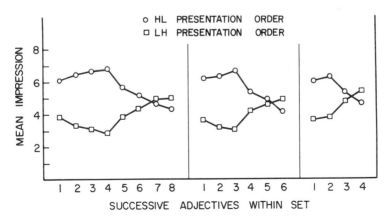

FIGURE 3.5. *Serial curves for cumulative response in personality adjective task. Upper curves for sequences with favorable adjectives first, followed by unfavorable adjectives; lower curves for opposite order of presentation. Crossover in each panel represents recency effect. (After Stewart, 1965.)*

successive L is received. The lower curve shows the response to the reversed (LLLLHHHH) sequence.

The critical feature of these data is the crossover of the curves. That corresponds to a recency effect; persons are less likable when their good traits are learned first. Thus, cumulative responding eliminates primacy and produces recency. This outcome agrees with the attention-decrement interpretation but disagrees with the change of meaning and discounting hypotheses.[c]

Linear Primacy Curves. Primacy–recency studies have traditionally used a gross index, namely, the difference between the responses produced by the forward and reverse orders of the same information (Section 2.5.1). However, the information at every serial position contributes to the response, and a complete serial curve would be desirable to reveal just where in the sequence the primacy and recency effects come from.

The obvious way to study the sequential development of the impression is to ask the subject to respond cumulatively after each piece of information. That is what Stewart did in the experiment shown in Figure 3.5. Unfortunately, as Figure 3.5 shows, this cumulative responding can markedly affect the outcome. Cumulative responding does not always have such effects (Anderson, 1959b), but Stewart's results showed that this direct approach had serious limitations and that some other approach was needed. Fortunately, collateral work on serial integration had led to methods for determining the influence of each successive piece of information knowing only the response at the end of the sequence.

These methods for obtaining complete serial curves were presented in Section 2.5, and one result deserves mention here. Linear serial curves were found in two experiments, which means that the influence of each adjective decreases as a straight-line function of serial position (see Figure 2.9 and Table 2.7). That makes ready sense in terms of attention decrement, as though attention itself showed a corresponding linear decrease.

One oddity in the experiments of Figure 2.9 is that the six-adjective curves (not shown in the figure) had markedly greater slope than the nine-adjective curves. This slope difference seemed consistent with the finding of equal primacy and equal recency for different sequence lengths in the experiment shown in Figure 3.5. The implications of this result have not been considered, but it suggests the possibility of a fixed amount of attention decrement that is distributed uniformly across the sequence.

The experiments of Table 2.7 have interest because the analyses evaluated the statistical interactions between every two serial positions. Change of meaning and inconsistency discounting both imply larger interactions for adjacent than nonadjacent serial positions because adjective–adjective interrelations would be more salient for adjacent adjectives. Contrary to both interactive hypotheses, the interactions had the same size for every pair of positions. That pattern is what would be expected from the noninteractive integration model, with negative adjectives having greater weight than positive adjectives (Section 4.4.2).

3.3.4 A FALSE STEP

The discounting interpretation of primacy, as already mentioned, only began to receive explicit consideration after inconsistency discounting had been studied with simultaneous presentation (Anderson & Jacobson, 1965; Section 3.4.1). The first explicit experimental test of discounting with serial presentation was included within a larger experiment designed to provide quantitative tests of the serial integration model (Section 2.3.3, Figure 2.4). Only the data comparing discounting to attention decrement will be noted here.

Virtually all research on primacy–recency has used information of opposite polarity presented in positive–negative and negative–positive orders. In each sequence, therefore, a shift occurs in affective value that would potentiate inconsistency discounting. The results of Anderson and Jacobson (1965) emphasized the importance of differences in affective value for inconsistency discounting. However, sequences that contained only positive information would presumably avoid any such inconsistency reaction. If primacy is caused by inconsistency discounting, therefore, it should not be found when only positive information is used. In contrast, attention decrement still implies primacy.

This test is simple in principle, requiring only the use of very favorable H and mildly favorable M^+ adjectives. But using H and M^+ unfortunately reduces the expected effect. Theoretically, the effect depends on the difference in scale value of the two halves of the sequence, and that difference is much smaller for H and M^+ than for the usual H and L adjectives. For this test to stand a reasonable chance of success, therefore, some way had to be found to increase the power of the experiment. That was done by preselecting adjectives that were equated in value for each individual subject, constructing the sequences separately for each subject, and using within-adjective design. This procedure also avoids the objection that selecting M^+ adjectives from the normative scaling would surely include some that are negative for some subjects.

In the experimental situation, the adjectives were presented visually and the subject read each adjective aloud as it was exposed. Sequences of six adjectives were used, of the form HM⁺ or M⁺H, and LM⁻ or M⁻L. The only condition of present interest is that in which subjects responded at the end of the sequence.

These data yielded a recency effect, .33 for the sequences with positive information and .66 for the sequences with negative information. This recency contradicted the prediction from the attention-decrement interpretation. It was suggested, therefore, that a discounting process was involved in the primacy effects that had previously been obtained. This suggestion was a false step, as will be seen next.[a]

3.3.5 THE WORK OF CLYDE HENDRICK

The work of Clyde Hendrick had an important role in the development of information integration theory. It includes the first replication of the averaging-versus-adding test outside the writer's laboratory, which was important in demonstrating generality of the averaging hypothesis (Hendrick, 1967, 1968a; see Section 2.3.3). It also includes the first careful studies of redundancy in implicit personality theory (Hendrick, 1968a,b, 1969). Of present concern is a cogent series of articles that did much to finalize the attention decrement interpretation of primacy–recency.

Direct Test of Inconsistency Discounting. The first report in this series (Hendrick & Costantini, 1970a) was designed as a direct test between attention decrement and inconsistency discounting. This was accomplished very neatly by varying the consistency between the positive and negative traits. Sequences were chosen in which the positive and negative traits had either a high or low probability of going together in a single person. Scale value was equated across these two degrees of consistency.

The theoretical hypothesis is straightforward. If primacy is caused by inconsistency discounting, then low consistency sequences should produce more primacy. In contrast, attention decrement predicts equal primacy for high and low consistency. The results unequivocally favored attention decrement.

This outcome deserves detailed consideration and the complete data are summarized in Table 3.2. The first rows in the upper and lower halves of the table represent high and low consistency sequences, respectively. In both cases, the three positive adjectives are *energetic–vigorous–resourceful*. Given those traits, the rated conditional probability that the person is also *stubborn–dominating–egotistical* is .77, whereas the rated probability that the person is *withdrawn–silent–helpless* is only .20. Because these sequences represent substantial work, and because they

TABLE 3.2
Relatedness Values and Person Impressions for Sixteen Sets of Traits

Trait adjectives	Relatedness	Impression		
		HL	LH	Primacy
energetic, vigorous, resourceful, stubborn, dominating, egotistical	.77	5.36	4.44	.92
trusting, patient, respectful, withdrawn, silent, helpless	.79	6.16	5.12	1.04
bold, daring, adventurous, reckless, immature, foolhardy	.89	4.00	3.44	.56
trustworthy, dependable, loyal, shy, passive, timid	.86	6.88	6.52	.36
self-disciplined, logical, intelligent, gloomy, cynical, moody	.64	5.20	4.08	1.12
cheerful, humorous, good-natured, noisy, boisterous, childish	.75	5.44	4.72	.72
polite, agreeable, cooperative, mediocre, dull, uninteresting	.72	5.04	4.16	.88
self-confident, sharp-witted, active, irritable, careless, grouchy	.58	4.40	3.68	.72
Mean	.75	5.31	4.52	.79
energetic, vigorous, resourceful, withdrawn, silent, helpless	.20	5.40	4.84	.56
trusting, patient, respectful, stubborn, dominating, egotistical	.40	5.52	4.68	.84
bold, daring, adventurous, shy, passive, timid	.22	5.92	5.80	.12
trustworthy, dependable, loyal, reckless, immature, foolhardy	.26	4.92	4.36	.56
self-disciplined, logical, intelligent, noisy, boisterous, childish	.32	5.36	4.40	.96
cheerful, humorous, good-natured, gloomy, cynical, moody	.25	5.16	4.48	.68
polite, agreeable, cooperative, irritable, careless, grouchy	.26	4.88	4.04	.84
self-confident, sharp-witted, active, mediocre, dull, uninteresting	.25	5.00	4.20	.80
Mean	.27	5.27	4.60	.67

SOURCE: Data from Hendrick and Costantini (1970a). (Copyright © 1970 by the American Psychological Association. Reprinted by permission.)
NOTE: Relatedness values are rated probabilities that a person with the three favorable traits in any set could possess the three unfavorable traits. Impression data from a separate group of 25 subjects who rated likableness on a 1–8 scale of the 16 listed person descriptions in both the listed HL order and the reverse, LH, order.

appear to give reliable primacy effects, they should be useful in future research. Accordingly, they are listed in full, together with the consistency ratings and the impression data from the 25 subjects.

The overall results are clear. Mean primacy for the consistent sequences is .79; mean primacy for the inconsistent sequences is .67. The difference is not significant and is opposite to the prediction from inconsistency discounting. A further comparison can be made between the impression means themselves. This comparison is also opposite to the direction predicted from inconsistency discounting.[a]

These results are convincing although they were apparently contrary to the authors' original expectation. Just as this experiment was being completed, however, the experiment discussed in Section 3.3.4 was published (Anderson, 1968a). These two experiments seemed to disagree in their support for inconsistency discounting.

Hendrick suggested that this disagreement arose because Anderson (1968a) had used a pronouncing requirement. He reasoned that pronouncing could have eliminated attention decrement, thereby producing the recency, and he set about to test this idea.

In their second experiment, therefore, Hendrick and Costantini replicated their first experiment but also included a pronouncing condition in which the subject had to repeat each adjective after it was read. Again, the results were strong and clear. Without pronunciation, mean primacy was .64, very nearly equal for consistent and inconsistent sequences, exactly as in the first experiment. With pronunciation, primacy turned to recency. The mean value of the recency was .38, very nearly equal for the consistent and inconsistent sequences.

In this way, Hendrick corrected the false step related in Section 3.3.4 and set the attention-decrement hypothesis back on solid ground. His results have added theoretical significance because they indicate that inconsistency discounting does not even make a partial contribution to the primacy effect.

Primacy–Recency with Number Averaging. In a related experiment, Hendrick and Costantini (1970b) followed the usual primacy–recency paradigm except that the stimuli were numbers and the subject estimated the average of the numbers on an intuitive basis. Previous work on this number-averaging task (Anderson, 1964d) had found recency when a cumulative average was requested after each number in the sequence. Hendrick and Costantini now showed that primacy was obtained when a response was required only at the end of the sequence. Thus, the pattern of results from number averaging is exactly parallel to that obtained by Stewart for person perception (Section 3.3.3).

Neither inconsistency discounting nor change of meaning can easily account for this primacy effect. There is no inconsistency in having both large and small numbers in the same sequence. Neither is there much reason or opportunity for numbers to change their built-in values. It seems necessary, therefore, to conclude that the weight parameter decreased over trials and that this weight decrease resulted not from inconsistency discounting, not from change of meaning, but from attention decrement.

Nature of Attention Decrement. Establishing the attention-decrement hypothesis marks the end of one line of investigation but the beginning of another. Elimination of the interactive explanations constitutes progress, but attention decrement is a partially defined concept that itself requires explanation. Up to now, little has been said on this topic, and, indeed, little is known.

A memory crowding interpretation of attention decrement was explored by Hendrick, Costantini, McGarry, and McBride (1973). The idea was that the decrement might occur because the processing center was crowded by the initial adjectives and did not have adequate time for the later adjectives. On this hypothesis, the primacy effect should be a direct function of the time interval between successive adjectives.

This idea was tested in three experiments, each of which used two or three interadjective time intervals that ranged from rapid-fire to as long as seemed tolerable. In Experiment 3, for example, the adjectives were given at the rate of one each 5.8 seconds in the slowest condition.

The outcome was simple. Rate of presentation had little systematic effect. A significant difference did appear in the first experiment, but it failed to hold up in the two following experiments. The usual large primacy effect was observed in every condition.

This result is striking. In the fastest condition, the adjectives were read almost as fast as possible. In the slowest condition, the subject had 5 seconds to cogitate each adjective. But although the primacy effect is quite sensitive to certain experimental manipulations, presentation rate had little effect. The implications of this result remain to be determined. Hendrick *et al.* (1973) took the result as evidence against the attention-decrement hypothesis, but that rests on a specific conception of the nature of attention in terms of memory crowding. It is equally plausible that the decrement is not a matter of processing time, but results from a crystallization of impression as a function of amount of information. In this respect, the results appear to be consistent with the hypothesis of two memory systems, one for the impression and another for the adjectives, which is taken up in Section 4.2.

3.4 Organizing Processes

Stimulus interactions are interesting because they represent active organizing processes. They are more easily assumed than proved, however, as has been illustrated by the failures of the plausible interaction hypotheses in the two preceding sections. This section demonstrates the operation of one organizing process and makes some suggestions for further work.

3.4.1 A STUDY OF INCONSISTENCY DISCOUNTING

At the time of the present experiment (Anderson & Jacobson, 1965), two results from previous work stood out. First, considerable evidence had accumulated against the change-of-meaning interpretation of primacy (Section 3.3). Second, provisional evidence had been obtained for the parallelism property. Both results argued against the operation of stimulus interaction, but this conclusion was open to two objections.

The first objection was that previous work, in order to avoid possible selection bias, had selected and combined adjectives by random choice. Interaction among a small proportion of the words could therefore easily be missed. The second objection was that the standard instructions stressed that each word was accurate and equally important. The intent of these instructions, to minimize effects of redundancy and inconsistency, may have been realized only too well. Under more naturalistic instructions, such interactions might become more prominent. Accordingly, the following experiment was designed to study effects of inconsistency under various instruction conditions.

Design. Person descriptions were chosen to embody two kinds of nominal inconsistency, affective and semantic. The description *honest, gloomy, considerate* contains only an affective inconsistency between the two favorable and the one unfavorable adjective. *Honest, gloomy, deceitful* contains similar affective inconsistency as well as semantic inconsistency between *honest* and *deceitful.* [a]

Each description contains two adjectives of equal value and a single adjective of opposite value. The two equal adjectives were expected to form an organizing field for the impression; inconsistency effects would therefore appear as decreased influence of the single adjective. The magnitude of these effects was also expected to depend on the following three sets of instructions, which play an important role in the theoretical interpretation:

- The *standard instructions* attributed each adjective in the person description to a different knowledgeable acquaintance. Subjects were told that all words were accurate, equally important, and should be given equal attention even though that might sometimes seem difficult.

- The *naturalistic instructions*, which were the main focus of interest, differed from the standard instructions by stating that different acquaintances might not be equally good judges of personality. Some words might therefore deserve more attention than others.
- Two *total discounting instructions* were used. These stated that one unspecified adjective in the set did not belong, and that it should be ignored in forming the impression. In one of these conditions, subjects indicated which adjective they thought did not belong, whereas in the other they were told just to ignore it.

Theoretical Discounting Prediction. The logic of the experiment is illustrated in the 2 × 2 design at the left in Figure 3.6. Each row consists of a pair of adjectives and each column consists of a single adjective. This 2 × 2 design yields four descriptions containing three adjectives each. For a numerical illustration of discounting predictions, the initial impression was ignored and the values of positive and negative adjectives were set at 5 and 15, respectively.

The solid lines in the left panel of Figure 3.6 are predictions from the equal-weight averaging model. These two curves obey the parallelism theorem.

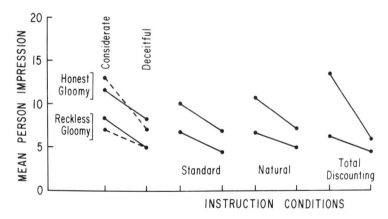

FIGURE 3.6. *Inconsistency discounting. Left panel shows parallelism predicted from equal weighting model (solid curves) and pattern of deviations from parallelism predicted by discounting model (dashed curves). Remaining panels show discounting observed under three sets of instructions. (After Anderson & Jacobson, 1965.)*

The dashed lines show how these predictions change if inconsistent information is discounted. Consider the description at the upper left corner, *honest, gloomy, considerate.* This contains an affective inconsistency, and any discounting should favor the positive pair, *honest–considerate,* at the expense of the single negative, *gloomy.* But if *gloomy* gets lower weight by discounting, then the response should become more favorable, as shown in the graph.

For *honest, gloomy, deceitful,* however, the response should decrease. In this description, discounting should lower the weight of the single positive, thereby raising the relative weight of the negative pair. The same holds for *reckless, gloomy, considerate* in the lower row. For *reckless, gloomy, deceitful,* of course, no discounting is expected since there is no inconsistency.

This design thus forces any discounting tendencies in specified directions. The plotted values were obtained under the assumption that discounting reduced the weight by half. The net result is to produce an interaction or nonparallelism of a specified shape. It is this pattern of nonparallelism that is to be sought in the empirical results.

Empirical Discounting. The predicted results are visible in the three panels of data at the right in Figure 3.6. When standard instructions were used, the two curves were nearly parallel, although there was a small deviation of the predicted form. Even with 64 subjects, however, this was marginally significant only in the second replication of the design.

When total discounting instructions were used, marked nonparallelism occurred and showed the predicted form. Equivalent results were obtained in both versions of these instructions; indicating which adjective was to be eliminated from consideration did not increase the discounting effect. The lower curve is not horizontal because subjects occasionally eliminated an L from HLL descriptions. Although not of primary interest, these results are important as a measure of the maximum discounting that is naturally available in this situation.

When naturalistic instructions were used, distinct nonparallelism was obtained. There is a mild flattening of the lower curve and a more noticeable steepening of the upper curve, a pattern that agrees with the predicted pattern in the leftmost panel.

The amount of discounting occurring with naturalistic instructions seems relatively small compared to that occurring with total discounting. In these 2×2 designs, the discounting effect appears in the two-way interaction and its magnitude can be expressed as a single number. These magnitudes stand approximately in the ratios 1:3:9 for the three instruction conditions of Figure 3.6. The naturalistic instructions thus produce only about one-third of the potential discounting effect. Despite the presence of extreme antonymic inconsistencies such as *honest–deceitful,* discounting effects are only moderate.

Affective and Semantic Inconsistency. An unexpected outcome of this experiment was that semantic inconsistency had relatively little effect. More precisely, descriptions that contained both affective and semantic inconsistency, such as *honest, gloomy, deceitful,* yielded no greater discounting than descriptions that contained only affective inconsistency, such as *reckless, gloomy, considerate.* This disagrees with expectation enough to raise concern that some peculiarity of the adjectives was responsible. Several samples of adjectives were used, however, and this same pattern was obtained across all instruction conditions. This result has been supported by Kaplan (1973), who used somewhat similar design. Kaplan's study has added interest because the direction of discounting was controlled by personality predispositions of the subjects (see Section 4.3.1).[b]

Theoretical Implications. The present experiment provides a clear demonstration of inconsistency interactions in the personality adjective task. This result has more than usual interest in view of the difficulties that have troubled previous attempts to study interaction processes. Some opportunities for further work are discussed in the next section. At the time of this experiment, however, the small size of the inconsistency effects reaffirmed the working hypothesis that the immediate problems of information integration lay elsewhere.

Also important in determining the course of subsequent work was the unexpected lack of semantic effects. One implication is that a general evaluative process plays a dominant role in human judgment, an implication that supports the present research concentration on this judgment dimension.

A related implication concerns the nature of the integration process. Since semantic relations had little effect, it would seem that they play a limited role in the integration. This implication agrees with the study by Anderson and Hubert (1963) showing that person perception involves a second memory system (Section 4.2). Both studies suggest that integration does not occur as an aggregation of features in semantic memory, but instead involves a superstructure of judgmental operations.

A general implication of these results is that inconsistency plays a limited role in human judgment. This is clear in the lack of effect of semantic inconsistency. But even affective inconsistency had only modest effects. These results suggest that people may often be relatively unconcerned about seeming inconsistencies in given information.

This outcome is not peculiar to the personality adjective task, for it has been found in attitude research as well. Using similar design, Himmelfarb and Anderson (1975) attributed three opinion statements on war and peace to a person and asked subjects to judge the person's true attitude. No evidence of inconsistency effects was obtained, even when two extremely militaristic and one extremely pacifistic statement were attributed to the same person.

In hindsight, people's capacity for tolerating inconsistency in the informational field seems eminently reasonable. People and situations in daily life are seldom all of a piece. A person can easily have tendencies to be both friendly and hostile, or harbor conflicting impulses toward spouse or department chairman. Everyday decisions are often made in a welter of conflicting information. Although the pure mind may abhor inconsistency, the practical mind learns early that events often lack order and reason.

3.4.2 FURTHER WORK ON ORGANIZING PROCESSES

This section presents comments and suggestions for further work on organizing processes. Although oriented to the problem of inconsistency, most of these comments apply more generally. The underlying theme is the use of judgment tasks to bring organizing processes under experimental analysis.

A Methodological Problem. The nonparallelism observed in the naturalistic condition of Figure 3.6 could result from other causes than inconsistency discounting. For example, the graph has the same shape as the negativity effect of Section 4.4.2. Also, the nonparallelism could arise merely from nonlinearity in the response scale (Section 1.2.4). Had the naturalistic condition been used alone, the discounting interpretation would have been dubious.

The other two instruction conditions provided validational support for the discounting interpretation in the naturalistic condition. The standard instructions were not of great interest in themselves; indeed they might seem inappropriate for the descriptions that contained semantic contradictions. Nevertheless, this condition provides a baseline that allows for negativity effects and response nonlinearity that may be present. The difference between this baseline and the observed effect in the naturalistic condition is therefore interpretable as inconsistency discounting.

Similarly, the total discounting instructions are not of primary interest because they told subjects to discount. However, they served the double function of validating the pattern of nonparallelism as an index of discounting and of showing how much discounting was available in this situation. Without these bracketing instructions, interpretation of the interaction for the naturalistic condition would have been uncertain. The bracketing instructions rule out alternative interpretations and provide direct support for the hypothesized process of inconsistency discounting.

The method of bracketing instructions should be generally useful because it can be applied to single descriptions. Discounting can thus be assessed in a single judgment under a condition of interest by comparing it with judgments made under no discounting and total discounting instructions.

In general, psychological interactions are difficult to study because they usually, if not always, require reference to a no-interaction standard. To say that two stimuli interact implies that their resultant is different from what it would have been had there been no interaction. Without a no-interaction standard, which can be elusive under the hypothesis that interaction is truly present, there may be little basis for interpretation. Much published work on inconsistency suffers from failure to provide a no-interaction standard and rests merely on assumptions about the operative process. Arbitrary no-interaction standards are often assumed, but that is treacherous, as illustrated in the discussion of implicit additivity in Section 1.6.6. The study of cognitive interaction requires development of new tools and methods.

Base Field Design. One approach to interaction analysis may be obtainable with the following *base field design*. The subject receives a base field of information that is to be treated as given fact, together with one of a number of ancillary fields not so constrained. The ancillary fields are chosen to embody hypothesized inconsistency, redundancy, or other interactive relations to the base field. The problem is to verify these hypothesized interactions and to measure them.

For this purpose, the ancillary fields are constructed from factorial design with stimuli chosen to avoid interaction between factors of the ancillary fields themselves. The base field is then combined with each field in the ancillary design, and these compound fields are presented to the subject. As long as the components of the ancillary fields do not interact among themselves, functional measurement may be employed to evaluate their parameters—regardless of how they interact with the base field. That is the purpose of this design, for the hypothesized interactions should become manifest in these parameter values. Discounting will appear as reduction in weight, and change of meaning as shift in scale value.

There are two main conditions for using base field design, aside from technical scaling problems. First, the ancillary factors must obey an integration model to allow functional scaling. This condition, which is assessed in the test of goodness of fit, is not considered to be a serious limitation on use of this method. Indeed, this condition should be reasonably easy to satisfy in many situations because the integration of the ancillary fields need not have substantive interest in itself.

The second condition is the need for a no-interaction standard to assess interaction. A neutral base field that did not interact with any ancillary field would be desirable. The functional parameters obtained from this neutral base field design would provide tbe desired standard for assessing the hypothesized interactions. This condition is less central than the first but may be more difficult to satisfy.

Technical scaling problems might be avoided by analyzing the raw response. That is essentially what was done in the foregoing experiment on inconsistency discounting. Although this approach is less informative, its simplicity recommends it.

An obvious alternative to evaluate interactions is with direct report, by instructing subjects to judge the meaning or importance of a given stimulus component in their integrated judgment. Unfortunately, these self-estimates may not be valid, as already illustrated in the studies of the positive context effect. An advantage of the base field design is that the functional parameter values provide a validity criterion for self-estimates and thereby a means to develop methods for obtaining valid self-estimates.

Organizing Fields. The base field of the previous subsection may be considered to be an organizer. Because the base field is given and fixed, information in the ancillary field must be organized to cohere with the base field. Inconsistency discounting represents one class of organizing processes, but other classes also have interest. Base field design is intended to facilitate experimental analysis by placing these processes under control of the base field.

Organizing processes will operate even though a base field is not explicitly specified. An example appears in the total discounting condition of the foregoing experiment, in which subjects were instructed to ignore one adjective. Since this adjective was not specified, subjects had a more complex organizing problem that included construction of a base field. The base field comprises the other two adjectives and its construction was generally in line with the a priori organization in the experimental design. An additional complexity arises in the naturalistic condition because the ancillary adjective is only partially discounted and may react on the primary organizing field.

Field construction may be used as an experimental task to provide views of organizing processes in action. For example, one generalization of the total discounting condition would present the subject with n adjectives under instructions to select m that describe the person. A useful procedure for field construction would begin with the single most important adjective and add one adjective at a time that went best with those already selected. These choice data could be supplemented by ratings of how well any adjective went with any given subfield and of the unity of the subfield as a whole. Such response measures would provide indexes of field organization.

Discounting Deviant Opinion. It seems clear that the organizing principles for person descriptions will be complex (Section 4.5.4). Accordingly, it may be helpful to consider simpler tasks such as those illustrated in the following experiment. Anderson (1968c) presented subjects with sets of numbers that represented various persons' estimates of the number of beans in a jar. Each set consisted of two compact groups of numbers, with group size and intergroup distance varied parametrically. The subject's task was to estimate the actual number of beans in the jar. The task mirrors many practical decision situations. There is a best or correct response, but the only clues are fallible opinions that exhibit considerable conflict among themselves.

To resolve such conflict of opinion, subjects presumably attempt to construct a base field, if they do not already have one, and discount elements that are deviant from the base field. The use of two compact groups of opinions in this experiment was intended to force the subject to choose the larger group as the base field and to discount the smaller, deviant group. That is what occurred. The amount of discounting varied directly with the size of the main group, directly with the distance between the main and deviant groups, and inversely with the size of the deviant group.

A notable advantage of this task is that the scale values of the stimulus numbers can be set equal to the numbers themselves, which greatly simplifies the analysis. For the given case of two compact groups, in particular, the weight parameter in the averaging model can be calculated from the response to measure discounting for a single set.[a]

Rationalization. *Rationalization* refers to the use of some verbal formula to explain why something should not be taken at face value. Despite its popularity in common parlance, the study of rationalization has remained largely in the armchair. One or two suggestions for experimental analysis are presented here.

As a concrete illustration, consider a description of a person that contains some seemingly inconsistent action. In addition to forming an impression of the person, the subject judges a given list of rationalizations of the inconsistent action by rating the goodness of each, by selecting one or several as a preferred explanation, or both. Use of a prepared list of rationalizations has various advantages. It avoids complications from subjects' inability to articulate rationalizations. It facilitates joint use of more than one rationalization. In addition, numerical judgments of rationalization goodness are helpful for analyzing partial reliance on any one rationalization.

Most important, this approach allows use of integration designs. The description of the person would be constructed in factorial design with stimulus factors chosen to be relevant to the rationalization under study. This rationalization response could then be analyzed in the same way as the likableness response. By studying the integrated action of two sources of information, it becomes possible to determine the functional role of each.

Discounting, which is one way of handling inconsistency, may be accompanied by more than one type of verbal rationalization. The two simplest types are *ignoral,* which may refer to ignoring the inconsistent element or possibly to ignoring the inconsistency; and *denial,* which may similarly refer to denying the truth of the inconsistent element or possibly to denying the existence of any inconsistency. These two rationalization types have the advantage of being almost universally applicable. A more sophisticated form of discounting that could arise in person perception would attribute the action to external forces in the environment, thereby decreasing its relevance to the judgment of the person. These rationalizations would presumably correspond to a reduced weight parameter in the integration. This weight parameter provides an objective measure against which to compare the verbal rationalization.

Other experimental situations can be treated in the same way. Person perception would have added interest if it could be embedded in some known person, especially the self. Attitudes toward United States presidents (Section 1.3.2) would allow the use of massive amounts of real information about real people whose actions are subject to diverse political and historical attributions about causal motivation.

Two questions about the role of rationalization deserve mention. The first is whether rationalization represents distortion of given information. This interpretation is common, but it rests on an implicit normative assumption that the evidence value resides in the objective stimuli so that there is one normal way to evaluate it. In the present view, however, persons with different background knowledge and belief will naturally evaluate the same piece of information in different ways. It is inappropriate to say that a person who assigns a low weight to the information is distorting it merely because most people would assign a higher weight. Values are personal. One person's values cannot be taken as the no-interaction standard to demonstrate distortion or interaction in another person. It might be claimed that the person is mistaken, that further thought would lead to a different evaluation, but insufficient thought is not distortion. It might also be claimed that the person ought to have different background knowledge and beliefs, but that also is not distortion. The term *distortion* seems ill chosen, therefore, because it implies the operation of a special process that does the distortion, and thereby obscures the true problem with a false explanation.[b]

The second question concerns the causal role of rationalization. It might represent either the actual valuation process or an attempt at verbal justification of more basic processes of valuation. In the latter case, the rationalization may be entirely epiphenomenal or it may serve other functions, such as ego defense. It is also possible that rationalization is more or less unrelated to the integration. Thus, an ostensibly inconsistent element might be neatly rationalized away at a verbal level yet have considerable influence in the integration.

An integration-theoretical approach should contribute to the study of rationalization because it can determine how the person actually does handle information in integration tasks. That provides a standpoint from which rationalizations can be scrutinized to determine their role in cognitive functioning.

Ego Defense. Blame avoidance is a basic characteristic of human behavior. It appears early in life in the form of children's excuses. Its persistence and vigor may be seen in any adult's reactions to criticism. Related to blame avoidance are other processes that arise in reaction to task failure and social belittlement and appear to have primary roles in everyday ego defense.

Blame, failure, and belittlement may be viewed as aversive stimuli. Normally such stimuli would be integrated into the self-concept, thereby decreasing self-esteem. The person employs various ego defenses to avoid or mitigate the aversive integration. These defense processes are not considered the unconscious mechanisms of psychoanalytic theory, but everyday information processing. Although it is interesting to speculate on the origins and social development of such ego defense processes, the present discussion is mainly concerned with the need for experimental analysis of the processes.

Rationalizations provide one strategy for ego defense. An approach to the study of rationalizations may be possible with the attribution formula, Action = Person ⊛ Environment, where ⊛ is a generalized integration operation (Anderson, 1978b). Thus, the action is given, to be explained by attribution to self or environment. Causal attribution to the environment is a standard formula for ego defense. Although specific verbalizations will involve numerous situational factors, they may generally be sophistications of primitive denials of the form, "I didn't do it," or, "It's not my fault." Even when causal attribution to the self cannot be denied, blame can still be mitigated with such formulas as, "I didn't mean to do it," "I wasn't myself when I did it," or "I've reformed."

Some rationalizations can be related to moral algebra. In the averaging rule, Blame = Intent + Damage (Anderson, 1978b; Leon, 1976, 1980), rationalization would reduce the subjective value of intent and thereby of blame. Such mitigation can operate both externally, to ward off punishment, and internally, to defend the ego. The latter may derive from the former as part of the development of the self-concept.

A primary need for the study of everyday ego defense processes is a set of flexible experimental tasks. One class of tasks could be based on third-party judgments in which subjects predict a person's reliance on various defense processes on the basis of given information about that person. This task could parallel the personality adjective task in form, with stimulus information and judgment dimension selected to bear on the defense process under study. Possible extensions could employ a given list of judgment dimensions, as noted in the discussion of rationalization, or discrete choice responses, as discussed in Section 3.5.3.[c]

Third-party judgments of defense processes can be embedded in actual persons. Examples would include parent–child or spouse–spouse judgments in family counseling (Section 4.5.2). Indeed, first-party judgments could be obtained in the same way, the more readily because defenses that seem transparently empty to an observer are often credible and effective for the person using them.

3.5 Related Topics on Meaning Constancy

Most research on meaning constancy has been limited to one experimental task, namely, adjective–adjective combinations in person perception. A few related studies are summarized here. The first of these, on adjective–noun combinations, helps define the issue of semantic integration.

3.5.1 ADJECTIVE-NOUN COMBINATIONS

The trait *romantic* seems desirable in an actor but not in a plumber. Evidently the value of *romantic* depends on the occupational context, which implies that an adding-type model could not account for such adjective–noun combinations. That is not surprising; adjectives and nouns are qualitatively different and so the adjective–adjective integration rules of the preceding sections would not seem generally appropriate. Although relatively little studied, adjective–noun combinations present a number of interesting problems, both for psycholinguistics and for information integration (Anderson, 1971c, pp. 83–84; Anderson & Lopes, 1974; Figure 4.17, this volume). Two aspects of these problems are noted here.[a,b]

Valuation. The basic judgmental operation for adjective–noun phrases is one of valuation. The noun sets up some dimension or context of judgment, within which the adjective is to be evaluated. Central to this valuation is the distance between the adjective and a comparison point, which often represents some optimal or representative value (Section 4.5.4; see also Anderson, 1974b, pp. 18ff, 1974d, pp. 279ff). The closer the adjective is to the comparison, the more polarized the judgment will tend to be.

The simplest form of valuation is one-dimensional. That seems to be typical for inanimate objects evaluated on suitability for some purpose, such as *dull knife* or *switchblade.* The former is judged unfavorably with respect to the normative ideal of sharp knife, whereas the latter may be judged favorably or unfavorably relative to personal aims.

Often, however, valuation is multidimensional. Social roles and occupations, in particular, appear to involve three different underlying dimensions. First is noun-defined proficiency. Some adjectives, such as *capable* and *reliable,* are general proficiency quantifiers, but others are more noun-specific. Thus, *persuasive* and *mechanical* would imply respectively more and less proficiency for the occupation of lawyer. This dimension of valuation obviously rests on the relationships between the adjective and noun (Anderson & Lopes, 1974):

> The weight of any adjective will change from noun to noun, as has been illustrated above, and presumably the same holds for its scale value. This is certainly interaction of a kind, but it is important to recognize that it

occurs in the valuation operation rather than in an integration operation. The noun merely defines the dimension of judgment, and the adjective represents a value on this dimension [p. 73].

Closely associated with the proficiency dimension is the social value dimension. A *capable lawyer* and a *capable thief* may be equally good at their occupations, but the two occupations have different social values. Social value of the adjective–noun phrase thus involves two preliminary valuations, and it may be conjectured that they are combined by a multiplying-type rule.

The third dimension is that of likableness. This judgment has a different nature, as can be seen in the fact that the occupation noun carries information of the same quality as the adjective. By virtue of occupation, a lawyer is expected to have certain personality traits that are as relevant to likableness as any stimulus adjective. This judgment dimension, in contrast to the first two, does require integration of adjective and noun values. Evidence that this integration follows an averaging rule is given by Anderson and Lopes (1974, Figure 1).

An associated complexity is that the noun may have dual effects, referring both to a person and to an occupational role. This may be illustrated by the contrast between *happy mother*, who is no doubt a better mother for being a happy person, and *irresponsible mother*, who neglects her maternal duties and may be more generally an undependable person.

As these considerations show, judgments of adjective–noun phrases on social desirability may have a complicated structure. The character of Falstaff in Shakespeare's Henry IV plays provides an example (see Wilson, 1943). Falstaff is personally likable to a degree that upstages the royal characters, Prince Hal and King Henry. He is an exceptionally proficient con artist and entirely reprehensible on almost every scale of social value. Shakespeare's integration problem is not less complicated, for he seeks to maintain an unresolved but not obtrusive tension between the two aspects of Falstaff's character so that the playgoer may rejoice with Falstaff's successes and justly concur in his final casting off when Prince Hal says, "I know thee not, old man."

Meaning Constancy and Meaning Change. A proper integration problem arises when several adjectives are used to modify the noun, for their several values must be integrated to obtain an overall judgment. This integration problem has been studied mainly with likableness judgments, but the averaging model is expected to hold on other response dimensions. Furthermore, the results presented in this chapter suggest that the meaning and value of each adjective will remain constant within the given context regardless of what other adjectives it may be combined with. It should be emphasized, however, that meaning constancy has hardly been studied beyond the likableness dimension.

A different aspect of meaning constancy arises in the example of *irresponsible mother*, who is worse than, say, *irresponsible clerk*, or even *irresponsible aunt*. Different behaviors are implied by each phrase, and, no doubt, the value, not to mention the weight of *irresponsible*, varies across contexts. However, it seems plausible that *irresponsible* has the same semantic meaning in all these cases, and that the different behaviors reflect a semantic integration of adjective and noun. Each noun defines a class of typical behavior, and the adjective refers to a common quality of these behaviors.

It should be added that some adjective–noun phrases present problems to be solved that seem to require change of meaning. Inconsistency and redundancy interactions represent two such classes of implicit problem-solving tasks. *Honest thief* poses an inconsistency that could be resolved by various rationalizations; for example, that the person was not basically a thief, which could be represented as a shift in noun meaning away from its modal value. The ostensible redundancy of *dishonest thief* poses a similar problem, implicit in the communicational context that *dishonest* is intended to convey nonredundant information. About such phrases, however, little is known.

3.5.2 ADJECTIVES AS ADVERBS

Some adjectives may perform an adverbial function, acting in part to modify certain adjectives with which they may be combined. Ability adjectives, such as *capable* and *resourceful*, may have such an adverbial role. *Resourcefulness* clearly has positive value in a good person, but might have negative value in a bad person.

Two attempts in the writer's laboratory have been no more than half-successful in demonstrating such adverbial functioning. One design factor was the critical adjective, quantified at 2, 5, and 8 on a 1–9 scale (e.g., from 1 = "Not at all resourceful" to 9 = "Extremely resourceful"). The second design factor was the value, positive or negative, of the other context adjective in the person description (e.g., *kind* or *cruel*).

In the positive context, of course, the person impression becomes more favorable as the scale value of the critical adjective increases. That will occur whether the critical adjective has the standard adjectival effect or the hypothesized adverbial effect. The important question concerns the negative context. If the critical adjective has only the adjectival effect, the impression should increase parallel with the positive context. If it has the adverbial effect, however, the impression should decrease as the scale value of the critical adjective by itself becomes more favorable.

In general, the data for the negative context have shown a flat curve, which leaves the matter uncertain. An obvious speculation is that the adverbial effect cancels the adjectival effect. The adjectival effect can hardly be doubted, and so the assumed adverbial effect seems the only reason why the impression does not increase in the negative context as it does in the positive context. To explain a null result by appeal to two cancelling effects is not satisfying, however, and more direct demonstration of the presumed adverbial effect would be desirable.

The foregoing design, with its clever use of quantification for the critical adjective, was introduced by Becker (1971). Becker gave a different, post hoc interpretation in terms of semantic redundancy of the critical adjective and the context. However, he obtained the critical decrease for only one of eight adjectives (*methodical*) and that could well have been by chance.

Two comments on this issue should be made. First, the postulated interaction need not represent change in the meaning of the word. *Resourceful* might have the same meaning in a kind person and in a cruel one, just as *very* has the same meaning in either an honest or a dishonest person. Its implications could still be different, by acting as a quantifier of the context (see Figure 1.16).[a] Second, adjectives such as *resourceful* may influence the response by two different paths, directly on likableness as a person and indirectly, mediated by judgments of social value. Accordingly, manipulation of the social role of the person might reveal the hypothesized adverbial effect. Beyond that lies the more challenging problem of developing quantitative models for two-process effects of single stimuli.

3.5.3 MEANING AS ASSOCIATION

The associationist tradition has fostered a view that defines the meaning of a word in terms of its associations, and a similar idea was used in Asch's synonym experiment, described in Section 3.1.2. The problem of predicting associations to a set of words or to one word within a set may be approached with the integration decision model discussed elsewhere (Anderson 1974a, pp. 246ff, in press, Section 3.10), which allows for discrete response. Each potential association may be considered a response dimension on which each stimulus word has some scale value and weight, which represent associational strength between the two words in question. The overt response is determined by the integrated strengths of the several words of the set: If this value exceeds some threshold, it elicits the association. It follows, of course, that the same word will elicit different associations in different sets. No meaning changes or other interactions among the words are needed or required.

Results reported by Ostrom and Essex (1972) on associations to pairs of words will serve to illustrate some additional implications of this association integration model. Their most interesting result was that 60% of the associations to a pair of words had not been elicited by either word singly. This result was taken to favor the hypothesis that some associations to the pair may be created, not predictable from the associations to the separate words.

However, novel associations can be readily produced within the present model. This requires only that association strength be below threshold for either word separately, but above threshold when the two are integrated. Similarly, the finding of fewer associations to pairs than to singles could reflect responses that were above threshold for one member of the pair, but sufficiently below threshold for the other member of the pair that their integrated strength was also below threshold. Finally, associations common to both members of the pair would tend to occur earlier than the novel associations owing to higher integrated response strength.

Despite its traditional and continuing interest, free association is not readily amenable to exact analysis, especially for testing whether associations to a set of words involve interaction among those words. An alternative approach would be to present a list of potential associations to the subject, with sets of stimulus words varied in factorial design. A choice decision task could then be obtained by asking the subject to choose associated words to each set under some specified criterion. Or, by asking the subject to judge degree of associatedness, a numerical response could be obtained that would be more informative than all-or-none choice data. Present methods could be applied to test the integration model as well as the hypotheses of the two preceding paragraphs.

3.5.4 A CONDITIONING HYPOTHESIS

Theoretical Rationale. Adjectives combined in a person description may change one another's meanings by some conditioning process. Mere pairing of the words could cause each to take on some of the meanings of the others. This kind of meaning change is different from the mutual semantic interactions considered previously, and it is desirable to indicate why this possibility might deserve consideration.

In the experimental procedure (Anderson & Clavadetscher, 1976), subjects judged likableness of a person described by two adjectives, of value H and M, and, on a later trial, judged likableness of a person described by the M alone. The question is whether the initial pairing causes M to take on some meaning of the H.

The potential for such a pairing effect is indicated by the positive context effect considered in Section 3.2. If the subject was asked about the value of the M in the HM description, a substantial effect of the H would appear. A substantial pairing effect is thus available, but the question is whether it lasts beyond the momentary pairing.

Theoretical reason for expecting the pairing effect to last comes from traditional learning theory. Since evaluative responses are elicited from both adjectives in the pairing, the doctrine of association by contiguity implies that each will take on part of the value of the other. Similar reasoning follows from Pavlovian theory of classical conditioning. The implicit affective response (unconditioned response, or UCR) to the H adjective (unconditioned stimulus, or UCS) becomes conditioned to the M adjective (conditioned stimulus, or CS). This conditioning effect should manifest itself in later responses to the M alone. A similar classical conditioning assumption has been employed in attitude theory by Fishbein (1967) and in interpersonal attraction by Byrne (1971; Clore & Byrne, 1974).

Experimental Analysis. The same basic task was used in five experiments, of which Experiment 2 is typical. In the conditioning phase, subjects judged likableness of persons described by one or two adjectives, M^+ or M^- in value, either alone or paired with H or L adjectives. To assess effect of number of conditioning trials, M adjectives were presented one, two, or four times, each time paired with a different H (or L) adjective. In the test phase, subjects judged likableness of persons described by the single M^+ and M^- adjectives. These two phases were repeated a second time with the same stimuli in order to allow additional opportunity for conditioning.[a]

The main results from Experiment 2 are in Figure 3.7. The open-circle curves plot mean response to the M^+ descriptions in the test phase as a function of the paired conditioning adjective (H or L) and the number of conditioning trials listed on the horizontal. The two leftmost points are essentially equal, as they should be since these adjectives received zero pairings and were treated alike in the conditioning phase. A substantial pairing effect is shown by the remaining points; the M^+(H) curve lies approximately 15 points above the M^+(L) curve. Thus, the likableness value of the M^+ adjective in the test phase depends on whether it was paired with H or L adjectives in the conditioning phase. The two filled-circle curves, for the M^- adjectives, show similar pairing effects. Overall, the pairing effect is highly significant, $F(1, 19) = 32.53$. In Experiment 1,

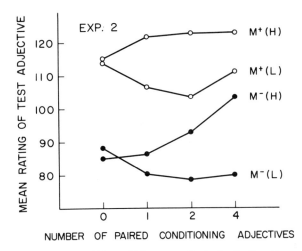

FIGURE 3.7. *Rated likableness of M⁺ and M⁻ personality-trait adjective after zero, one, two, or four prior pairings with adjectives of H or L value. (From Anderson & Clavadetscher, 1976. [Copyright © 1976 by the American Psychological Association. Reprinted by permission.])*

which used a somewhat different pairing procedure, the effect was not significant, but a preliminary experiment more similar to Experiment 2 had obtained a significant pairing effect. Thus, the conditioning phase does appear to produce substantial changes in adjective value.

However, this pairing effect does not appear to be a true conditioning effect because it does not depend on the number of pairing trials. This appears most clearly in the top curve in Figure 3.7, which is flat over the last 3 points. Although the M⁻(H) curve shows a suggestive rise, the main effect of trials did not approach significance, $F(2, 38) = 1.21$. Moreover there was essentially no increase in the pairing effect between first and second presentations, which is also contrary to the conditioning hypothesis. Accordingly, the conditioning hypothesis does not seem viable.

Since a maximal effect is achieved with a single pairing trial, it may be some kind of familiarization effect. This interpretation was tested in two further experiments by including a preliminary familiarization phase. In Experiment 3, each test M appeared four times in the familiarization phase, twice with H and twice with L adjectives. In Experiment 4, each test M appeared once by itself in the familiarization phase. This familiarization phase was followed by an initial test phase, a conditioning phase, and a final test phase. Assessment of the familiarization effect itself was omitted in order to allow the increased sensitivity of a within-word test for any conditioning effect.

The outcome supported the interpretation of the first two experiments. Whether familiarization was produced by pairing with other adjectives or by a single solitary presentation, no trace of subsequent conditioning was obtained.

Classical Conditioning of Attitudes. The present results provide no support for the classical conditioning assumption that formed the basis of Fishbein's (1967) theory of attitudes and Byrne's (1971) theory of interpersonal attraction. Both approaches were heavily influenced by Hullian learning theory and by various attempts to extend classical conditioning ideas to language. The primary theoretical assumption for both Fishbein and Byrne was one of classical conditioning, but this has been taken for granted and neither of the associated research programs seems to have made direct tests of this assumption. Accordingly, the present results raise a question about the conceptual foundation of both formulations.[b]

Change of Meaning. The present experimental task provides reasonably clear evidence for true changes in meaning. However, these changes seem to be familiarization effects, which is quite different from the usual conception. Indeed, the implication of Experiment 4 that familiarization can be achieved merely by presenting the solitary adjectives clearly eliminates any interpretation in terms of interaction. For many purposes, therefore, the effect would be an artifact to be eliminated by preliminary familiarization.

Use of the term *familiarization* narrows the field of explanation but should not obscure the explanatory puzzle that the effect presents. It is still necessary to explain why familiarization should be able to change value. One interpretation is in terms of the integration-theoretical view of valuation as a constructive process (Anderson & Clavadetscher, 1976):

> In this view, values are not essential aspects of the adjective itself, but instead are constructed by a valuation operation as needed for the judgmental task at hand. Thus, the first presentation would require a valuation operation and that could be influenced by the immediate context of the first presentation, in particular by any other adjectives then present [p. 17].

3.6 Implications for Cognitive Analysis

Although the studies summarized in this chapter are concerned primarily with the meaning-constancy hypothesis in the personality adjective task, a few general implications of these results deserve consideration. The first two sections take up some implications for semantic representation and cognitive processing. The next two sections deal with problems that are jointly conceptual and methodological, especially with the role of conscious report in psychological research. The final section comments on the meaning-constancy investigations as a case history of psychological research.

3.6.1 IMPLICATIONS FOR PSYCHOLINGUISTICS

Semantic Memory and Judgment. The meaning-constancy hypothesis suggests that judgment and decision involve a superstructure of operations that are distinct from semantic memory, which plays a more passive role. The main finding of the experiments summarized in this chapter is that adjectives have a fixed meaning that they maintain from one combination to another. Despite the consistency constraint imposed by the condition that the several adjectives describe a single person or object, each adjective seems insensitive to its companions and there is little interaction among them.

In sharp contrast is the extreme sensitivity of each adjective to the specified dimension of response. Innumerable response dimensions may be specified, and the weight and scale value of each adjective changes with every response dimension. People have swift facility in these inferences, which show subtle dependence on context and background.

This contrast between the insensitivity of the adjectives to one another and their sensitivity to the dimension of response indicates that valuation and integration are distinct processes. This independence of valuation and integration suggests the following scheme for information processing. Each adjective has a fixed, unitary existence in a semantic memory. The task defines a response dimension and sets up a corresponding valuation operation. Valuation represents an inference from the contents of semantic memory that yields the weight and value parameters relative to the operative dimension of response. These parameters are computed and transferred to an integration center where they are integrated to produce the overall judgment. The valuation operation is sensitive to the task requirements, but it operates independently on each separate adjective so that there is no interaction among them.

A different view frequently appears in theories of semantic memory, a view that also seems to underlie much of the work that has sought for change of meaning. In this view, the integration takes place in semantic memory itself. Each adjective is represented as a set of features, and the integrated resultant is a combination, sometimes set-theoretical, sometimes configural, of these several sets of features. The actual judgment may be produced by a valuation operation acting on this integrated resultant. Such a representation is natural for formulations that emphasize semantic interaction.

Accordingly, the success of the meaning-constancy hypothesis raises doubt about such semantic memory conceptualization. Even when stimulus interactions are observed, they may affect weight rather than scale value or meaning. Collateral support for the present judgmental view is provided by the results of Section 4.2, which indicate the operation of separate memory systems for the words and for the impression.

It hardly needs emphasis that valuation–integration independence cannot hold in general. With long-term attitudes, for example, valuation operates on a cognitive structure developed through previous integration. Even in the adjective task, stimulus interaction implies that valuation and integration are related (Section 3.4).

Nevertheless, valuation and integration appear to be independent in some important cases. That is a fortunate fact. It underlies the success of cognitive algebra, which is helpful for cognitive analysis. More generally, it suggests that the main cognitive apparatus resides not in semantic memory, but in a superstructure of dynamic operations that subserve ongoing activity.

Functional View of Language. The present approach embodies a functional view of language in which meanings lie in the communicator rather than in the communication. The communicator is attempting to convey a meaning, or referent, and tries to delimit it by using a number of words. The intended meaning is not the referent of any one word, but of all the words, not to mention associated nonverbal cues and context-background information. It is inappropriate to say that the words change meaning, for the meaning results from semantic integration. A chemical analogy may help clarify the issue (Anderson, 1971c): "Word mixture does not imply meaning change any more than a mixture of sweet and sour solutions implies chemical change. The changed associations are no better guide in the one case than the changed taste is in the other [p. 84]."

This functional view may help explain why the meaning of a word often seems to depend on context. Typically, a focal word or phrase points to the main area of meaning, which is further delimited with qualifying words and phrases. These qualifiers may appear to change the meaning of the focal word—if the meaning is treated as the referent of the focal word. In the present view, however, the meaning is in the word user, and the qualifiers help delimit that meaning without necessarily affecting the meaning of the focal word.

As an alternative formulation, it might seem reasonable to say that the meaning of any word is defined only by the entire context. However, that is a theoretical interpretation that requires experimental justification. It is certainly correct to say that the meaning of the word user is defined only by the entire context. To go beyond that to consider the meaning of a single word within context requires a theory of integration, as Section 3.2 has illustrated.

Of course, the valuation operation for a single word will depend on context-background factors that help clarify the communicator's intention or purpose. This process is similar to that governing the dimension of judgment, and a given word can certainly have different implications in different contexts. It does not follow, even so, that its meaning is different (e.g., Sections 1.1.3 and 3.5.1). For a fixed context, moreover, integration of context and background information will exert direct effects on the constructed meaning that should not be attributed to changes in interpretation of the focal word.

It is entirely possible, of course, that words do change meaning as a function of context. The integration approach makes this question decidable, at least in certain tasks. For the personality adjective task, the evidence reviewed in this chapter provides little support for the change-of-meaning hypothesis.

3.6.2 HALO THEORY

The halo concept has long been popular as an explanatory concept. Most applications have taken the halo concept more or less for granted, however, and it has not received much experimental study. In fact, supporting evidence is hard to find. The purpose of this section is to comment on the nature of the halo concept and to make some suggestions for further experimental analysis.

The Halo Concept. The concept of "halo" originated as an attempt to explain certain seeming inaccuracies in rating evaluations (see e.g., Guilford, 1954; Tagiuri, 1969). A supervisor's ratings of various aspects of workers' job performance, for example, seemed to deviate from more objective measures—in the direction of the supervisor's personal likes and dislikes. As one of the pioneer investigators put it, the general impression colors particular judgments (Thorndike, 1920, p. 25).

The essential idea of the halo concept is that such general impressions exist and act as causal determinants of particular judgments. Traditional evidence for the halo concept has typically been in the form of trait intercorrelations. Ratings of particular traits of persons showed much higher correlations than the facts seemed to warrant. Physical appearance, for example, showed relatively high correlations with ratings on leadership, intelligence, honesty, and so forth. Such results could readily be accounted for with the halo assumption of a general impression as a common determinant of diverse particular judgments. The intercorrelations would then represent this common factor.

Two general objections may be raised to such evidence from trait intercorrelations. The first questions the evidence itself; that is easy to do because objective measures of accuracy are usually not available for judgments of persons. Without objective measures of leadership and honesty, for example, there is no definite basis for claiming that the correlations between their judged values and physical appearance are too high.[a]

The second objection is that evidence of inaccurate judgments is not adequate to demonstrate a halo process. Although raters' judgments may be systematically inaccurate, their judgments may reflect their proper beliefs. Their ratings of leadership, honesty, or job performance may reflect causal determinants of their general impressions, not vice versa. If so, the halo interpretation is not correct. This second objection is not tied to the question of accuracy. No matter which way the first objection may be resolved, the second remains.[b]

Halo as Information Integration. The problem of the halo process is fundamental to person perception and more generally to any theory of information integration. The results cited in this chapter provide clear evidence of the halo process, as well as some indication that this process obeys the averaging formulation. This work has provided both an empirical foundation and a conceptual framework for the halo concept.

In the present approach, however, the halo effect is viewed less as an "error" than as normal information processing. The traditional approach is basically normative, viewing the judgments relative to an objective standard of accuracy so that any inaccuracy seems to require psychological explanation. That is a typical failing of normative approaches (Section 1.8.1). Inaccuracy is no doubt undesirable, but what requires psychological explanation is the behavior that occurs, not its deviation from normative standard.

From an integration view, the halo process makes good sense. The general impression of another person contains predictive information about that person's behavior in various situations. This information is integrated along with other available information to obtain a judgment. The available information may be inaccurate or misevaluated, to be sure, but the integration strategy is not unreasonable for an organism attempting to function in an uncertain environment. Halo effects may thus reflect the operation of general integration processes (Section 4.1.8).

Further Research on Halo Processes. The research summarized in this chapter has focused on one aspect of the halo issue, namely, the interpretation of the positive context effect in the personality adjective task. The results support the interpretation as a true halo, but their generality remains untested. Moreover, study of the halo process itself has been slighted owing to concern with meaning constancy. Extensions of this research to other domains and other problems is desirable, therefore, and should be fruitful in view of the large magnitude of the positive context effect.

A straightforward but important line of inquiry concerns the use of entities other than persons. For example, judgments of social groups open up new questions about effects of integration constraints on the positive context effect, as discussed in Section 4.1, and also have substantive interest for group dynamics. Other entities, including physical objects and events and social episodes, may be considered in similar manner.

Another line of study concerns the structure of the halo process. Halo has traditionally been treated as an evaluative, "good–bad" factor, but halo processes may be present on any specific dimension. The overall impression of the person would be treated as one piece of information to be evaluated along any specified dimension. This valuation operation may yield multiple halo components, those specific to any given dimension, plus a generalized good–bad component that mediates effects along any given dimension. The personality adjective task allows systematic manipulation of the given stimuli, the dimension of the person impression, as well as the dimension of the component judgment, and so may provide enough leverage for structural halo analysis.

3.6.3 ASCH'S GESTALT POSITION

Asch's (1946) theoretical view rested on one simple, attractive thesis: The several adjectives in a person description undergo mutual, dynamic interaction. The impression forms a whole in which the role and meaning of each adjective depend on its interrelations with all the others. Asch attempted to support this thesis with several lines of evidence. When this evidence is examined, it appears to be noncontroversial but to have little relevance to Asch's thesis. Inasmuch as no previous critique has been given of Asch's position as a whole, the main points will be summarized here.

Method. Several different response modes were used by Asch, including the checklist method already considered in Section 3.1.2. In that example, *warm* Person A and *cold* Person B elicited considerably different response frequencies on the 18 checklist pairs. As already noted, this result merely shows that *warm* and *cold* produce different impressions, an obvious fact that has no bearing on the question at issue.[a]

The interpretation of these same checklist data created another problem, essentially a confusion between stimulus and response. Claiming that the adjectives in the person description interacted with one another, Asch (1946) stated that "If a man is intelligent, this has an effect on the way in which we perceive his playfulness, happiness, friendliness [p. 264]." But, in this case, *intelligent* was a piece of stimulus information that was relevant to each of the three cited response dimensions. That has nothing to do with interaction among stimulus words in a description because it holds for one-word descriptions. Asch's statement is true, but merely shows that stimuli affect responses.

A more plausible line of evidence comes from other experiments in which the same word seemed to produce different effects in different lists. An example is provided by the synonym production task already considered in Section 3.1.2, in which subjects were instructed to give synonyms to *calm* in two different descriptions. Since this same word elicited different synonyms in the two cases, Asch concluded that its meaning was different in the two cases. The flaw in this reasoning became apparent when it was observed that some given synonyms were not synonyms of *calm*, but of other words in the lists. The given synonyms were different for the two lists because the other words in the two lists were different. When properly unravelled, most of the evidence presented by Asch has just this form, that different words produce different effects.

A more subtle form of the same difficulty appears in other experiments, in which subjects made direct comparative judgments that *unaggressive* was different in an *active–helpful–unaggressive* person and in a *weak–sensitive–unaggressive* person [pp. 281–282]. Here again, the obvious alternative hypothesis is that this phenomenological difference represents a direct effect of the other words not mediated by any interaction with *unaggressive*. Indeed, this result can readily be interpreted as a manifestation of the positive context effect considered in Section 3.2.

Primacy. One other piece of evidence was presented by Asch, namely, the primacy effect considered in Section 3.3. This is proper evidence, not subject to the foregoing objections. However, the change-of-meaning interpretation of primacy turned out to be incorrect.

Central Traits. As part of his thesis, Asch claimed that some traits were *central*, others *peripheral*. The central traits were considered to dominate the impression and to control the meanings of the peripheral traits in each description. This concept of central trait rests squarely on the assumption that the traits undergo changes of meaning. Since the evidence indicates that the traits do not change meaning, Asch's concept of central trait does not require separate discussion.

However, the concept of central trait has been used in two other ways, both essentially different from that of Asch. One usage equates centrality with importance. Obviously, some traits are more important than others, but that does not imply that they affect the meanings of the other traits. Asch (1946) was explicit on this point: "It is inadequate to say that a central trait is more important [p. 268]."

The other usage equates centrality with breadth of implication. Thus, *warm* is relevant to judgments about numerous other aspects of personality, whereas *polite* has a narrower range of implication. However, this is a fact about the implications of a single given trait and has no relevance to the basic question of whether two given traits interact.

This latter usage may require further comment. In one of his initial experiments, Asch showed that inclusion of *warm* and *cold* in two otherwise identical descriptions produced greater differences in checklist responses than did *polite* and *blunt*. However, Asch's conclusion that *warm* and *cold* were central traits—in the sense that they interacted with and controlled the meanings of the other given traits—does not follow. The same result will be obtained without any such interaction merely because *warm* and *cold* have greater breadth of implication than *polite* and *blunt.*

Greater breadth of implication could be easily demonstrated by getting checklist responses to *warm* and *polite* as single traits; there cannot be any interaction or meaning change in a one-word description. In fact, Asch did essentially this in one of his last experiments (1946, p. 277) and noted in passing that the result undercut his earlier conclusion that *warm* was a central trait. Asch's argument reaffirming his earlier conclusion begged the question, but he clearly recognized that breadth of implication was not sufficient to demonstrate centrality.

It is possible to redefine *central trait* apart from the issues of interaction and change of meaning. Such redefinition should be explicit, however, because evidence on relative importance or on breadth of implication has no necessary or likely relation to the issues of interaction and organization.[b]

Conceptual Analysis. The main criticism of Asch's formulation is that it failed to clarify the conceptual issues. His basic thesis, that the adjectives in a description interact with one another in forming the impression, is eminently reasonable. Merely to verify the seeming pervasiveness of such interactions, as Asch aimed to do, would have been an important contribution. Asch's evidence did not do that because it rested on inadequate conceptual analysis, which did not make a clear distinction between evaluation of single stimuli and integration of several stimuli or even between stimulus and response. Most of Asch's evidence is not relevant, therefore, as has already been shown in the preceding discussion of method. When the conceptual structure of the problem is made clear, Asch's evidence is seen largely to reach conclusions that are obvious and uninformative.

Such shortcoming in conceptual analysis should be distinguished from a more empirical kind of failure that is common in research and may be illustrated by the primacy effect. Primacy requires explanation, and Asch's change-of-meaning explanation was reasonable. It has turned out to be incorrect, but not because of inadequate conceptual analysis, for its rationale remains conceptually valid. In contrast, nearly all Asch's other evidence loses its interest under closer scrutiny.

No one would question that our perceptions of other persons have organization and structure. The same is no doubt true in the personality adjective task. The problem is to develop method and theory that can contribute to the analysis of such organization and structure. Although Asch called attention to an important problem, his phenomenological approach failed to clarify the theoretical issues and his experimental work revealed little about either trait integration or structure in the impression.

3.6.4 CONSCIOUS REPORT AND PSYCHOLOGICAL KNOWLEDGE

The change-of-meaning hypothesis has rested primarily on the evidence of conscious report. Subjects and experimenters alike give detailed, confident introspections about intricate interactions among the adjectives in a person description. Asch noted that his own conclusions were based mainly on subjects' written sketches, with checklists used more as a convenience for summary and exposition. Any investigator who studies a few such sketches finds it hard to avoid the belief that the change-of-meaning hypothesis must somehow be true.

But the change-of-meaning hypothesis is not true after all. When subjected to experimental scrutiny, the change-of-meaning hypothesis appears to be wrong in just that situation in which conscious report gave it the clearest and strongest support. Evidently, conscious report is not a reliable guide to psychological truth.

The limitations of conscious report have long been recognized in psychology. The once-dominant introspectionist school, which identified psychology with the study of consciousness, was replaced long ago by the behaviorist reaction, which denied that consciousness had access to the causes of behavior and so declared consciousness to be irrelevant. The work on meaning constancy discussed in this chapter emphasizes this distrust of consciousness by showing the apparent change of meaning to be a cognitive illusion.

A similar view has been adopted by Nisbett and Wilson (1977a,b), who carried it to an extreme. They present further evidence for the halo hypothesis, as well as a variety of other evidence on inadequacies of conscious report. However, the view reached by Nisbett and Wilson differs from the present position in one basic respect: They take the extreme view that people can almost never give valid reports about the effect of stimuli on their responses; what accuracy they exhibit is coincidental, resulting from implicit causal theories about plausible effects of stimuli on responses. In contrast, the present view recognizes the unique value of consciousness as a source of psychological knowledge.

There is no doubt that conscious report is sometimes seriously misleading and invalid—that is a primary methodological lesson from the research on the meaning-constancy hypothesis.[a] But it does not follow that conscious report is always invalid.

The issue may be focused on the question of personal knowledge of stimulus values. Nisbett and Wilson claim that subjects' assessments of the effects of stimuli on their own judgments cannot be more accurate than assessments that could be made by external observers. In particular, therefore, another's estimates would be as good as the subject's own estimates of stimulus weights and values.

Experimental tests may be obtained in integration tasks similar to the task used by Nisbett and Bellows (1977). A pair of subjects, A and B, would be asked for overall response and also for self-estimates of the effective values of the separate stimuli. A's self-estimates may be used to generate predicted responses for both A and B, and similarly for B. The claim is that predictions of A's overall responses are no more accurate when based on A's self-estimates than when based on B's self-estimates; and similarly for B. Alternatively, the self-estimates may be compared directly to functional measures of stimulus effects derived from the overall response. A statistical test may be obtained by comparing accuracy for the two sets of predictions across a number of subject pairs.

Relevant data appear in Figure 1.24, in which the predicted dating preferences for each female are based on her self-estimates. But these self-estimates are rather different for the three females, both in shape of the probability curve and in rank order of male's attractiveness on the horizontal. Using one female's estimates to predict another's dating preference will apparently lose substantial accuracy. This outcome, if verified statistically, would show that subjects do have personal knowledge of the effects of stimuli on their responses.[b]

No one would claim that people can become aware of the complete integration process—as the work on meaning constancy illustrates. But even the halo interpretation of the positive context effect concedes some ability of subjects to recognize the effects of each stimulus adjective on their response. Other pertinent examples come from the use of conscious knowledge in defining cognitive units (Section 1.8.2) and from studies of intuitive physics (Section 4.5.3). To reject the evidence of consciousness on the ground that it is fallible is to reject a valuable tool of cognitive analysis (see also Mandler, 1975).

3.6.5 MEANING CONSTANCY AS A CASE HISTORY

The research on the two hypotheses of meaning constancy and meaning change provides an interesting example of scientific inquiry. This research has aimed at understanding three main phenomena: the primacy effect, the positive context effect, and the evidence of conscious introspection. All three phenomena have an obvious and attractive explanation in terms of the change-of-meaning hypothesis. The introspective evidence, in particular, is so persuasive that further proof might seem unnecessary.

But experimental analysis, including important contributions by Clyde Hendrick, Martin Kaplan, Thomas Ostrom, and Robert Wyer, led to deeper understanding. In the present interpretation, this research shows that meanings do not change; they stay constant. Primacy—and recency—result from attentional processes, not from change of meaning. The positive context effect represents a halo process, not change of meaning. The evidence of conscious introspection is a cognitive illusion.

An apology is due those investigators whose views and work have been too briefly noted here. Not all agree with the present interpretation and indeed an earlier draft of this chapter reached a more tentative position (Anderson, 1974e, p. 62). Not much evidence has been added since then, but the review for this chapter led to a firmer conclusion. Of special significance are those tasks that are sensitive to partial contributions of change of meaning to the positive context effect. This effect is large; if any substantial part of it reflected change of meaning, that should have become clear in those tasks. There were two or three uncertain signs of such contributions but, even if they were verified, they would be much too small to account for the phenomenon.

It deserves comment that the body of research on meaning constancy has been efficient. This is a difficult judgment to make, of course, because each experiment needs to be related to its place in the growing structure of knowledge. Good ideas do not always pan out. Experiments that seem cogent at one state of knowledge can lose much of their force in the light of later developments, even when they helped that development along. Initial work in any area, when the issues are not well defined, the methodology still rudimentary, often has an awkwardness that in retrospect may be difficult to understand. Nevertheless, most of the experiments on meaning constancy and meaning change retain their significance. That represents a high efficiency index, far higher than in the area of mathematical models of person perception, for example, which has suffered from use of weak inference (Section 1.2.8).[a]

These investigations also illustrate the conceptual role of mathematical models. Although the common image of model analysis centers on algebraic formulas and predictive precision, conceptual aspects are not less important. Most experiments in this chapter were concerned with qualitative predictions of direction rather than exact predictions of amount. Explicit model analysis became necessary in some instances, but the main function of the models was in conceptual analysis. In the primacy question, the model framework helped lead to the alternative attentional interpretation as well as to ways of testing both interpretations. Clarification of the conceptual distinction between the two interpretations of the positive context effect was also facilitated by the model framework. In neither case were specific formulas of primary concern; the concepts that they represented were primary.

The usefulness of an informational approach also deserves comment. Simple information principles were able to provide a unified account of a ramified body of results and to do so with considerable precision. Similar outcomes have resulted in other studies of cognitive interaction in this research program. Despite their intuitive attractiveness, assumptions about interactive processes have typically been found wanting, whereas simpler information principles have done reasonably well. Seeming tendencies toward consistency, balance, and so forth can often be seen as a consequence of more basic information principles. This outcome does not mean that cognitive interaction does not exist (Section 3.4), but it does mean that assumptions about cognitive interaction cannot be taken for granted.

The meaning constancy investigations also show that inductive theory (Section 1.8.1) is not merely empirical generalization. The positive context effect is an empirical result that is silent about its own meaning. Meaning constancy and meaning change are theoretical concepts originating from general background knowledge and gaining precision in the model conceptualization. Similarly, the surface pattern of primacy and recency in the adjective task did not lead to understanding. The concept of attention decrement, which provided a unified account of primacy–recency results in the adjective task, came from elsewhere than the data under study. Rejection of the deductive view of psychological science does not lessen the importance of theory, but rather emphasizes the need for close relations between theory and experiment.

Science has truly been called "the endless frontier," for each end is a new beginning. Although the original question of meaning constancy seems to be settled, its study has opened up important new questions. The concept of attention decrement provides an explanation for primacy–recency but this concept is itself in need of explanation. Similarly, the halo concept, which was also introduced incidentally to the main inquiry, now presents many interesting problems in its own right (Sections 3.6.2, 4.1.7). Other results have suggested a judgmental approach to psycholinguistics and semantic representation. Last, but perhaps most interesting, the question of what link in the chain of information processing lies open to view by the conscious mind may be determinable with an integrational-theoretical approach.

Notes

3.2.1a. The diagrams of Figure 3.2 provide a visual indication of the difference between the two interpretations of the positive context effect, but they are simplified in two respects. The first is that judgments of single traits may not yield their scale value per se, but may include the initial impression (Section 1.6.2). This property of the averaging model could become relevant in quantitative tests of the meaning-constancy interpretation. The second simplification is that the operations of valuation and integration need not, and generally will not, be distinct when the stimuli interact (e.g., Anderson, 1971a, p. 199). The separation of these two operations in the right panel is an expositional convenience that does not affect the point at issue.

3.2.2a. Figure 3.1 is averaged over several replications of stimulus adjectives, which could average out semantic interactions within the separate replications. However, the relevant replication effects were not significant in the statistical analyses (Anderson, 1971c, p. 80). This does not eliminate the possibility of small but real deviations from parallelism in the separate replications, but such deviations clearly could not account for the large positive context effect that is visible in Figure 3.1.

3.2.3a. Mention should also be made of a study by McKillip, Barrett, and DiMiceli (1978), who found that descriptions with low ambiguous traits yielded more extreme impression responses and also more extreme component ratings than descriptions with high ambiguous traits. For various reasons, however, these data are difficult to interpret. One problem is the evident appearance of substantial regression artifacts stemming from the selection on both value and ambiguity. Thus, adjectives selected to be equal and neutral in normative value yielded substantially different responses, both for the person impressions and for the context ratings. The more extreme impression responses may therefore merely represent regression artifacts in the values of the single adjectives (see Anderson, in press, Section 7.6).

This regression problem may not be serious for the component ratings, if, as seems to have been the case, the components were balanced across contexts. But although the differential ambiguity effect seemed substantial, it was marginally significant in a test that used inappropriate pooling of error terms.

3.2.4a. In an experiment based on that of Hamilton and Zanna (1974), Kaplan (1975a) obtained likableness judgments of a component trait combined with context traits equated in value but either semantically related or semantically unrelated to the component. Kaplan found no significant differences between the positive context effects in the two conditions and concluded that it was mediated by a purely evaluative halo effect. This result was not replicated by Zanna and Hamilton (1977) and presumably represents an inappropriate acceptance of the null hypothesis. It perhaps should be added that Zanna and Hamilton incorrectly cite Anderson (1974e) to support their assertion that a halo formulation could not account for their results; the page they reference explicitly states that "The judgmental [halo] interpretation could allow for an effect of semantic relatedness [p. 50]."

3.2.4b. Schümer (1973) has written that there was a typographical error in which the two entries in the M^+ column of his Table 1 were transposed vertically, and similarly for the H column. In both columns, therefore, the larger value should be in the bottom row so both effects are in the predicted direction although small and nonsignificant.

3.2.4c. Ostrom's (1971) result provides a qualitative test of the halo averaging formulation since the prediction is not sensitive to the exact numerical values of the parameters. The model analysis could be pursued further to obtain exact tests by varying set size and composition.

3.3.2a. For both sequences of adjectives, Asch reports checklist responses in which subjects checked which of two polar opposites, such as generous–ungenerous, better described the person. The mean percentage of favorable choices may be calculated from Asch's Table 7 to be 53.5 and 38.8 for Persons A and B described by the adjectives listed in the text. For the second sequence of adjectives, however, the two corresponding values were almost identical, 58.2 and 57.5, respectively. Despite this near-zero effect, Asch (1946) asserted that "there is still a definite tendency for A to produce a more favorable impression with greater frequency [p. 272]."

3.3.2b. Also of interest are the likableness ratings of persons A and B from Asch's (1946) experiment (see text). When these were given as the first description of the experiment, the mean ratings were .87 and −.94 for A and B, respectively, yielding primacy of 1.81, which was statistically significant with 32 subjects. When these two descriptions were given at the end of the experiment, the effect was only one-half as large and was nonsignificant even with 64 subjects. This test has less power than that described in the text, of course, because it is based on a single score.

3.3.2c. The finding of equal mean primacy was complicated by an interaction with sex, with males and females about equal on the gradual sequences but with females showing greater primacy and males less primacy on the abrupt-change sequences.

3.3.2d. In a study of generality of the primacy effect, substantial primacy was also obtained in judgments of meals (Anderson & Norman, 1964), a result that also raised a question about the interactive explanation. A food does not have a variety of meanings in the sense that a trait adjective seems to have. Moreover, as children know, there is no inconsistency in having both liked and disliked foods in the same meal.

This result, like several of those cited in the text, is not interpreted as being critical for the interactive explanations. Rather, such results help define what an interactive explanation would have to be like if one wished to pursue it. The interaction hypotheses lack explanatory power because they are constrained by the data, not vice versa. A weak explanatory hypothesis may, of course, point toward a true explanation.

The cited results would suggest an interpretation based on affective or evaluative interaction rather than semantic interaction. In fact, there is a well-known position of this kind, namely, the perceptual contrast effect. However, it produces recency rather than primacy (see also Anderson, 1971b).

3.3.3a. Detailed tabulations in Anderson and Hubert (1963) give recall probability as a joint function of ordinal position in recall and ordinal position in list (Tables 2–4) and frequency of doublet recall as a joint function of ordinal positions of first and second doublet members (Tables 5–6). In addition, assessment of recall decrement over trials suggested that proactive inhibition was induced by experience with recall itself, not by experience with the impression task without recall.

3.3.3b. Cumulative responding procedure requires experimental care. Stewart observed in his pilot work a strong tendency for subjects to respond on the basis of just the latest adjective. This problem was resolved by appropriate changes in procedure, and individual analyses showed that almost all the subjects exhibited individual curves similar in shape to those of Figure 3.5 (Stewart, 1965, p. 164).

A similar tendency to respond only in terms of the latest piece of information seems to be present in a similarity–attraction study of Byrne, Lamberth, Palmer, and London (1969, Table 2). Leach (1974) reports a similar recency effect under standard conditions with the introvert–extrovert behavior paragraphs used by Luchins (1957) (as did Luchins and Luchins [1970] under a similar condition). Leach also suggested that the recency effect obtained by Stewart with continuous responding was an artifact of procedure, failing to note that Stewart had explicitly considered this matter.

3.3.3c. A clever direct test of the change-of-meaning interpretation of primacy was made by Tesser (1968), who incorporated component judgments within the serial task. After an initial reading of the six adjectives, each of the 126 subjects rated the person. The adjectives were then read again, and subjects rated each as a trait of that person. To explain the obtained primacy effect, change of meaning requires the component ratings of the single traits to be less in the second half of the sequence than in the first half. No such trend was found. This lack of trend agrees with the halo model because the impression I developed after the first reading is the same for each trait judgment in the second reading. This experiment was carefully designed, although its power is limited by the weak primacy that was obtained.

3.3.4a. It was also a mistake. The accumulated evidence for attention decrement was sufficiently strong, even at that time, that some pecularity of procedure should have been suspected. However, the phenomenological validity of the interactive view made it easy to accept.

3.3.5a. In a subsequent experiment with these same descriptions, Hendrick (1972) presented all six adjectives together and asked for a preliminary judgment of how well the last three would characterize a person described by the first three. Following this, subjects rated likableness on the assumption that the person was actually characterized by all six traits. Likableness was lower than when no preliminary judgment was required for the inconsistent sequences but equal for the consistent sequences. In Hendrick's interpretation, the preliminary judgment made the inconsistency salient, but subjects treated it as a negative characteristic rather than discounting it. This interpretation is supported by the reported lack of effect of order (H–L and L–H) of the six adjectives; discounting would presumably reduce the effect of the last three adjectives, thereby producing an order effect.

3.4.1a. The term *affective inconsistency* may be misleading, for the operation of semantic relations cannot be ruled out merely because these adjective descriptions did not contain semantic contradiction. The essential result, however, is that discounting was no greater with than it was without semantic contradiction.

A rigorous definition of *inconsistency* will not be attempted here. Properly speaking, however, there is a basic distinction between nominal inconsistency, as defined by the investigator, and effective inconsistency, as defined by subjects' reactions. To say that subjects are not very sensitive to inconsistency thus rests on some assumption that the nominal manipulation has validity beyond the investigator's intuitions. The justification for this conclusion in the naturalistic condition of the discounting experiment depends heavily on comparison with the total discounting condition.

3.4.1b. A few additional results from the discounting experiment (Anderson & Jacobson, 1965) should also be noted. The complete design was actually a 2^3 factorial so constructed that all discounting effects would theoretically be localized in the two-way interaction shown in Figure 3.6. All other interactions were nonsignificant, a result that supported the theoretical interpretation although it is limited by the low power of these particular tests in the design.

Figure 3.6 also suggests greater discounting in HHL than in HLL sets. This effect is prominent under total discounting instructions, in which the greatest difference from the standard condition is for the HHL set *honest, gloomy, considerate* in which *gloomy* is discounted. This effect was general across the eight stimulus replications.

When subjects explicitly indicated which adjective did not belong, their eliminations were predominantly, although not always, in the expected direction. In the HLL sets that contained antonyms, one antonym was always eliminated, but only in 22 of 32 cases was this the H adjective. In general, the odd adjective was eliminated more often in HHL than in HLL sets. This result and that described in the previous paragraph could mean that HH forms a stronger base field than LL or it could reflect a bias toward favorable evaluations. Study of this elimination task in conjunction with Kaplan's disposition measure (Section 4.3.1) could be interesting.

3.4.2a. That the psychological scale values of the number stimuli were equal to the objective number values seems reasonable but need not be true. Evidence to support the assumption that the functional values of the numbers are a linear function of the numbers themselves was given in Anderson (1964d). Further work on this task is given by Levin (1975, 1976).

3.4.2b. Various processes for resolving inconsistency, including forms of denial and ignoral, have been discussed by numerous writers (see Abelson *et al.* 1968). It is hoped that the present approach will contribute to the experimental analysis of these and other organizing processes.

3.4.2c. The equation Blame = Intent + Damage may help explain the finding that actors tend to attribute their actions to the environment, whereas observers tend to attribute those actions to the actor (see Jones, 1979). These attributions may, in part, reflect generalized tendencies to avoid blame in the former case and to assign blame in the latter case. This interpretation can be extended to predict a reversal of this attributional error for actions with positive outcomes because the actor will tend to take credit for such action.

It should be added that comparisons of actor and observer attributions involve certain methodological problems. In particular, actor and observer have different background knowledge and that may set up communicational constraints that will produce different responses even when their views on the causal factors are identical (Section 4.5.4).

3.5.1a. Previous work on the adjective–noun problem in person perception, which was introduced by Osgood, Suci, and Tannenbaum (1957), is discussed by Anderson and

Lopes (1974), who make the general criticism that neither the conceptual framework nor the response measures used in previous formulations took cognizance of the two distinct value components, namely, personal likableness and social value. The adding-type models used in previous work were therefore inadequate to represent the basic noun-dependent valuation of the adjective.

More recent articles by Higgins and Rholes (1976) and by Kirk and Burton (1977) have applied a similar criticism to integration theory, apparently having failed to recognize the theoretical role of the valuation operation discussed in this section. The basic premise of both articles, namely, that evaluation of adjectives may depend intimately on noun context, agrees with integration theory (Anderson 1968b, p. 732, 1971c, p. 83).

The experiment by Higgins and Rholes is interesting because it employed inanimate nouns that may have a simpler judgmental structure than social roles. The study by Kirk and Burton is interesting for its use of similarity judgments and multidimensional scaling to analyze temporal development in the structure of social roles.

3.5.1b. This discussion of adjective–noun combinations also serves to re-emphasize the constructive nature of the valuation process (Section 1.8.2). Meanings do not reside in words per se, but in the inferences drawn by the perceiver. Ostrom (1977) seems to imply that this position is unique to meaning-change interpretations when he asserts that "No amount of data will lead the meaning shift theorist to abandon the assumption that the cognitive representation we have of other persons contains more than the informational stimulus items (e.g., traits) describing the target person, and that the implicational associates contained in the cognitive representation are crucial in determining impression responses to the target person [p. 498]." There is no need to think of abandoning this assumption. It is an integral part of integration theory, which possesses the notable advantage of providing ways to measure inferences and implications (see also Sections 1.6.5, 1.8.3).

3.5.2a. Relative quantifiers such as *tall* and *expensive* seem amenable to similar analysis. The sentence, "An expensive shirt is not as expensive as an expensive car" uses the term *expensive* in two ways, as a relative quantifier and as an absolute cost. The relative quantifier may be represented as a proportionate distance from the corresponding reference point, so that *expensive, very expensive, not very expensive,* etc., would have the same proportionate distances from the standard reference for shirt as for car. This formulation leads to a bilinear fan analysis similar to that for adverbial quantifiers (see Figure 1.16).

3.5.4a. A few procedural details about the Anderson and Clavadetscher (1976) conditioning experiment should be noted. Each subject judged a total of 32 single conditioned adjectives in the test phase of Experiments 1 and 2 and a total of 24 single adjectives in both test phases of Experiments 3 and 4. Stimulus balancing was elaborate, both within and between subjects, but not more so than seemed required by the desire for stimulus generality and for experimental sensitivity. Experiments 3 and 4 employed 36 and 48 subjects, respectively, and both found a 2.0 mm difference to be significant, a difference of only 1% on the graphic response scale. This particular effect had no substantive significance for it merely reflected mean differences in value between two sets of test adjectives. However, it testifies to the sensitivity of the experiment as well as to the usefulness of within-word balance.

It may be questioned whether single conditioning presentations could produce a noticeable effect, especially when so many words were presented over the course of the experiment. However, substantial conditioning effects were obtained under analogous conditions in Anderson (1969d) and, of course, in the familiarization result cited in this text.

3.5.4b. In more recent work, Fishbein appears to adopt an information-integration approach in place of his previous classical conditioning approach (e.g., Fishbein & Ajzen, 1975, p. 235).

3.6.2a. Thus, Johnson (1955) refers to his earlier review of evidence on the halo effect with the statement: "Despite the popularity of this effulgent phrase, when a search was made for evidence of the 'halo effect,' none could be found [p. 310]." Johnson defined halo in terms of a difference between subjective and objective trait–trait correlations, so that failure to find halo effects was in part a consequence of failure to find objective measures of trait occurrence.

3.6.2b. Some of the reported correlations in the halo literature have been so large that there is little doubt about their objective inaccuracy. Practical prescriptions for reducing halo effects are thus sensible but may be more effective with better understanding of the structure of halo processes.

3.6.3a. A frequently cited study by Kelley (1950) replicated Asch's result using real persons in place of hypothetical adjective descriptions. Students in regular classes formed different impressions of a strange lecturer depending on whether the words *rather warm* or *rather cold* had been included in a short written description passed out to the students just before the lecture.

The customary interpretation of this result follows Asch and Kelley in thinking that the warm–cold variation caused a transformation or distortion of the lecturer's performance and perceived personality. However, the same outcome would be predicted from an information principle without any assumption of transformation or distortion. The warm–cold variation would be simply another piece of information, perhaps, indeed, the major one in this experiment, on which to judge the lecturer on such dimensions as sociability and popularity. This argument does not imply that the Asch–Kelley interpretation is wrong, only that it does not follow from such data.

3.6.3b. Proper central traits can readily be obtained using the base field method of Section 3.4.2. By specifying one trait as given, it becomes the organizer for the field construction. It is thus simple to demonstrate the existence of interaction and organization; the real problem is to find ways to analyze them. Even with these specified central traits, there is the problem of whether the organizing process affects only weights or also values and meanings.

3.6.4a. Nisbett and Wilson (1977b, p. 251) treat Asch's change-of-meaning view and the present meaning-constancy view as equivalent interpretations of the positive context effect. This fails to recognize the point to the associated research and the polar opposition between the meaning-constancy view and the evidence of conscious report, which remains the main support for the change-of-meaning view. Thus, the primary objection of Schneider, Hastorf, and Ellsworth (1979, p. 270) to the integration models is that they do not seem phenomenologically valid.

3.6.4b. A similar argument has been made by Smith and Miller (1978) and White (1980) (see also Anderson, in press, Section 6.2.6). The test suggested here might be objected to on the ground that B's self-estimates may be less accurate in predicting A's overall response than would be obtained by asking B to act as an external observer and estimate A's self-estimates. That may well be true in fact; but the operative hypothesis of Nisbett and Wilson (1977a,b; Nisbett & Bellows, 1977) is that any accuracy in self-estimates derives solely from common knowledge shared with other subjects. On this hypothesis, B could only be more accurate as an external observer by corresponding lesser accuracy in own self-estimates; and similarly for A. Any such difference therefore cancels out in the four sets of predictions from the two sets of self-estimates of the pair of subjects. An even simpler test could be obtained by asking each subject to give a set of self-estimates and in addition a set of estimates for a specific or generalized other. The hypothesis claims that the former estimates will not be more accurate than the latter.

3.6.5a. A similar but more extended discussion of a closely focused research area is given by Cartwright (1973), who considered the research on risky shift. Risky shift is concerned with the question of why groups make riskier decisions than individuals. Because of its concentration on one task and issue, the research on meaning constancy and meaning change may also be viewed as a "line of investigation [Cartwright, 1973, p. 223]" in Cartwright's sense. Cartwright reaches a rather negative judgment about the efficiency of the risky shift research. A comparison of similarities and differences between these two lines of investigation, not to mention others, would be interesting.

4

Special Problems of
Information Integration

Five problems of information integration are taken up in this chapter. The first concerns a form of stimulus interaction—an adding-with-contrast hypothesis—that provided a theoretical alternative to the averaging hypothesis. This adding model failed, but the results raised a question about meaning constancy that required extensive additional work.

The second problem involves the relation between judgment and verbal memory. A serendipitous result indicated that the common assumption of isomorphic relation was inadequate; instead, judgments appeared to involve a memory system different from the verbal materials from which they originated. This result supports the conception of an operational memory, distinct from semantic memory, as the locus of judgmental processing.

The third problem concerns the concept of initial impression or prior attitude. This concept has a key theoretical role but its reality has been questioned for judgments of hypothetical persons because there is no specific prior information about them. The work reviewed on this problem supports the concept of initial impression as an expectancy about people in general and shows how it is related to personality predispositions.

The fourth problem is really a cluster of problems that involve the concept of weight. These include the meaning of the concept and a variety of weighting effects, such as source effects and configural weighting.

The final problem is that of generality of results obtained with the personality adjective task. This is discussed in the final section, which adds some suggestions for further work.

4.1 The Positive Context Effect

4.1.1 ORIGIN OF THE PROBLEM

By the time that the initial experiments of Chapter 2 were completed, the outlook for integration theory was favorable. The parallelism property indicated that some simple process was operative; critical tests had ruled out the adding model in favor of the averaging model; and the averaging model had given a good account of the set-size data. An auxiliary line of experiments had largely ruled out change of meaning as an interpretation of the primacy effect found with serial presentation. Moreover, a simple extension of the basic model was showing promise for quantitative analysis of serial integration. Accordingly, the main direction of work began to shift to the study of stimulus interaction and to extending the integration-theoretical approach to other areas.

However, one theoretical objection remained unsettled. The adding model could account for the averaging-adding tests of Section 2.3 by assuming a contrast effect. In Table 2.2, for example, contrast could cause the M^+M^+ traits to take on a slightly negative cast when combined with HH traits. The lower response to HHM^+M^+ than to HH would thus reflect addition of a negative value, not averaging of a positive value.[a]

This adding-with-contrast assumption could provide a very good account of known and probable facts. That the response to HHM^+M^+ is only slightly lower than the response to HH would be taken to mean that contrast made M^+M^+ only slightly negative. Increasing the number of H's would increase the contrast, and hence the added M^+'s would produce a larger decrement. On the other hand, contrast need not go so far as to make the M^+'s actually negative, and so they could also produce an increment, as is sometimes observed. Most important, the adding interpretation for the set-size effect would hold because no contrast would result from adding M^+M^+ to M^+M^+. The contrast assumption would thus serve the adding model in as good stead as the initial impression serves the averaging model, and might seem more attractive on grounds of parsimony.

The contrast assumption itself was not novel. On the contrary, the concepts of contrast and assimilation were then popular in social psychology and had even been made principal explanatory concepts in some theoretical formulations. Although there was reason to believe that contrast effects reported in social judgment were largely artifacts, the adding model with contrast required explicit consideration.

This section begins with an experiment designed to test the contrast assumption. As had been expected, the data showed no contrast—an unexpected positive context effect was obtained instead. As a consequence, the issue of meaning constancy, which had seemed settled, arose again in a more troublesome form that has been discussed in Chapter 3.

4.1.2 TEST FOR CONTRAST EFFECTS

Conceptual Status of Contrast and Assimilation. For clarity, a preliminary remark is required on the usage of *contrast* and *assimilation*. These concepts originate in traditional perception, as in brightness contrast and hue contrast, where their existence is clear. They refer to changes in the perception of a focal stimulus as a direct effect of contextual stimuli (e.g., Section 1.3.7; Clavadetscher, 1977; Clavadetscher & Anderson, 1977; Massaro & Anderson, 1970,1971). In social psychology the concepts of assimilation and contrast have been popular because they promised an explanatory frame at the more basic level of perception and psychophysics.

It is well known, however, that apparent contrast and assimilation can be produced by various artifacts. Response language can produce one such artifact (see, e.g., Anderson, in press, Section 1.1.5); another appears in the halo effect discussed later. In these cases the context has its effect on the response, not on the perception of the focal stimulus.

Unfortunately, contrast and assimilation have come to be used in two ways: nominally, to refer to an observed effect; and causally, as an interpretation or explanation of that effect. This double usage creates confusion, especially as the two are often loosely intermingled.

Present usage will take contrast and assimilation as explanatory theoretical concepts, not merely as names for observed effects. This agrees with the usage of Helson (1964) and of Sherif and Hovland (1961), who have been most responsible for the popularity of these concepts in social psychology. The observable effects themselves will be termed *negative* and *positive context effects*, respectively, neutral terms that do not carry surplus theoretical meaning (Anderson, 1970a).

Theory and Design. The obvious way to test for contrast is to ask the subject, much as is done in perceptual studies. Accordingly, a procedure of component judgments was employed in the several experiments of this section, all of which had similar design.

In the first experiment, subjects received cards listing three trait adjectives that they read aloud under one of two conditions. In the Person condition, subjects were told that the three adjectives described a person. They first rated likableness of the person, and then rated one specified adjective "corresponding to how much you like that individual trait of the person." In the Word condition, each card was presented as three unrelated adjectives that might be used to describe people. Subjects rated the same adjective as in the Person condition, "corresponding to how much you like the traits implied by the adjective."

The one adjective that was rated will be called the *component*, or *test*, *adjective*, and the other two adjectives on the card will be called the *context adjectives*. The design was a 2 × 4, Test × Context factorial, with M^- and M^+ as test adjectives and with LL, M^-M^-, M^+M^+, and HH as context adjectives. From the subjects' view, however, test and context adjectives were indistinguishable until the experimenter pointed to the one adjective to be rated.

Before looking at the data, the theoretical scene at the time of this experiment should be recalled. The outcome predicted by the social judgment theories of Helson and of Sherif and Hovland was a contrast effect. The rating of the component test trait should be displaced away from the two context traits. This contrast effect should appear in the Word condition, reflecting perceptual influence by the contiguous contextual words.

No Contrast. No contrast was obtained. That rules out the hypothesis of an adding-with-contrast model, which was the original concern of this experiment.

The data of the Word condition, which bear directly on the contrast assumption, are shown in the left panel of Figure 4.1. If contrast is present, then the rating of the M^+ test adjective should be more favorable in the less favorable contexts. When graphed, the M^+ curve should slope upward. Contrast should also cause the M^- curve to slope upward.

In fact, the curves are essentially flat. The effect of context did not approach significance for either curve. This null result seems to hold generally. An earlier experiment of similar design (noted in Anderson & Lampel, 1965, p. 434) had obtained quite similar results. Subsequent experiments by other investigators have also included Word conditions, and none has obtained an effect of context. In short, the adding-with-contrast model lacks empirical support.[a]

Positive Context Effect. The Person condition, in the right panel of Figure 4.1, shows an extremely strong context effect. When the three

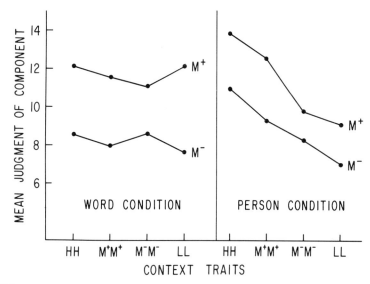

FIGURE 4.1. *Positive context effect. Subjects see three personality trait adjectives and rate likableness of one of them, namely, the M^+ or M^- trait listed as curve parameter. When the three adjectives describe a person (Person condition, right panel), rating of component is shifted toward the value of the other two (context) adjectives listed on horizontal axis. When the three adjectives are unrelated (Word condition, left panel), rating of component is unaffected by value of context traits. (After Anderson & Lampel, 1965.)*

traits describe a person, the M^+ trait is rated about 14 when combined with the favorable HH traits, about 9 when combined with the unfavorable LL traits. This is a positive context effect since the rating of the component trait is pulled toward the values of the context traits. The M^- trait shows a parallel pattern, as can be seen in Figure 4.1. This finding also has been verified in many different laboratories.

The positive context effect might be called an *assimilation effect* but that term carries explanatory overtones. Since the theoretical interpretation of the effect is problematical, the more neutral terminology seems preferable.

Comments on Experimental Procedure. A number of experimental details deserve comment. The words were in random order on each card and the subject was required to read them aloud. Only after the subject finished reading all three words did the experimenter point to the one word that was to be rated. These precautions were used to ensure exposure to all the words on the card, a necessary condition for testing for perceptual contrast.

Special care was taken to avoid the response language artifacts that had plagued earlier work on contrast. Practice trials were given to familiarize the subject with the rating scale, the procedure, and with the range of stimuli to be judged. The person was rated on a scale with seven verbal labels, whereas the component trait was rated on a 1–20 numerical scale. These two scales were made different to minimize possible interference between the two response measures. The more important scale used 20 steps as it was thought that a greater number of steps would also help minimize response language artifacts.

Each subject judged only one trait in each description in order to avoid possible bias from responding to other traits in the same set. Different adjectives were used in every set so that no subject saw the same adjective twice. The main purpose of these precautions, later found to be unnecessary, was to avoid possible response transfer effects.

For generality, four different stimulus replications were used, with adjectives chosen randomly from larger lists at each value. Stimulus replication was not significant in any analysis, which attests to the generality of the results. Work in other laboratories also supports the present result, for the Word condition in particular (Griffitt & Jackson, 1970; Takahashi, 1971b; Wyer & Dermer, 1968; Wyer & Watson, 1969).[b]

4.1.3 INTERPRETATIONS OF THE POSITIVE CONTEXT EFFECT

It seems clear that the positive context effect cannot be interpreted as perceptual assimilation. The data from the Word condition show no interaction among words simply experienced together. The perceptual assimilation–contrast concepts that have been basic to various theories of social judgment (e.g. Helson, 1964; Parducci, 1974; Sherif & Hovland, 1961) do not seem to hold. If the positive context effect does represent true stimulus interaction, then it must be cognitive interaction, not perceptual interaction.

Indeed, the simplest and most obvious explanation of the positive context effect is in terms of change of meaning. At face value, the subject's rating of the component trait is a direct index of its value or meaning within the given context. Consistency assumptions could explain the effect. For example, the subject could select shades of meaning of each adjective so as to reduce any felt inconsistencies among them or to increase the overall consistency of the person. Other forms of cognitive interaction could also produce the positive context effect.

A different view is also possible, a view in which the positive context effect does not represent true interaction. Once the adjectives have been integrated into an overall impression, they may no longer possess separate, individual meanings. The subject may therefore be unable to respond to the part as such once it has been integrated into the whole. The judgment of the component part may not be to that part alone; instead, it may reflect action of the other parts, or of the whole, directly on the response.

Each view of the positive context effect points to a different locus of processing. Change of meaning assumes that the adjective changes before or while it is integrated; this changed value is the one that is effective in the impression. The alternative, judgmental view assumes that each adjective has a single fixed meaning that is effective in the impression, regardless of context. In this view, the positive context effect is produced by a process subsequent to the formation of the person impression.

These two alternative interpretations of the positive context effect have been compared in Chapter 3 on the meaning-constancy hypothesis. The remainder of this section considers the positive context effect as an empirical phenomenon of intrinsic interest.

4.1.4 THEORY OF THE POSITIVE CONTEXT EFFECT

Halo Interpretation. One interpretation of the positive context effect is that it represents a generalized halo process. In a standard example of halo, teachers rate docile children as more intelligent than obstreperous children. Their judgment of intelligence is seemingly influenced by their overall liking for the child. The positive context effect of Figure 4.1 differs from the usual concept of halo effect in that the component to be judged is an explicit part of the whole, which is an important feature for experimental analysis.

The traditional halo concept seems reasonable and has been largely taken for granted. The essential idea is that a general impression exists and acts as a causal determinant of judgments of particular attributes. But it is also possible that the same stimuli that produce the general impression act directly on the judgments of attributes, with the general impression having no causal role. Indeed, the causal chain may go in the opposite direction, with judgments of attributes being causal determinants of the general impression. The behaviors that determine the teacher's judgment of the child's intelligence may also be determinants of the general impression, either directly or as mediated by the judgment of intelligence. The traditional correlational analyses do not go very far in delineating such causal networks.

This problem arose in the issue of meaning constancy considered in Chapter 3. The evidence cited there does support the concept of a halo process (Sections 3.2 and 3.6.2). The rest of this section presents other work on this problem.

Averaging Model for the Positive Context Effect. According to the halo interpretation, the judgment of the component is a composite of two factors: the component itself and the overall impression. That gives a qualitative account of the results shown in Figure 4.1. For the Word condition there is no context effect because there is no overall impression and the response is determined entirely by the specified adjective. For the Person condition, however, the overall impression is integrated with the component itself. This integration would produce the directional effects observed in Figure 4.1; see also Figure 3.1.

Quantitatively, it is an attractive hypothesis that this integration obeys an averaging rule. Thus, a model for the positive context effect can be written

$$s^* = ws + (1 - w)I, \tag{1}$$

where s^* is the in-context rating of the component, s is its context-free value, w is the relative weight, and I is the overall impression.

It should be recognized that two quite different integrations are involved: first, in forming the overall impression; second, in judging the component. The component is assumed to have the same scale value in both integrations as long as the same response dimension is involved in both judgments. The component may, however, have different weight parameters in the two integrations. In Eq. (1), therefore, w would depend on the amount of information that went into the overall impression, as well as on the relation between the impression and the component judgment dimension.

Testing Goodness of Fit. Testing the context model is complicated by the fact that s is one determinant of I. Consequently, s and I cannot be manipulated as independent stimulus factors to obtain the usual parallelism test. However, the parallelism property does hold if both integrations are linear. Accordingly, if component and context are varied in factorial design, the component judgments should fall on parallel curves.

Just such a test is contained in the right panel of Figure 4.1. The curves are roughly parallel and the interaction did not approach significance. That supports the halo–averaging model for the positive context effect.

Meaning Change. The context model of Eq. (1) was derived on the assumption that the component had the same value s in all contexts. However, this context model is not necessarily inconsistent with the interpretation of the positive context effect as change of meaning. It could be postulated that Eq. (1) quantifies the meaning change so that s^* is the effective value in forming the person impression. Although Eq. (1) was derived from a judgmental view, its success does not rule out the change-of-meaning hypothesis. Hence the problem of meaning change still requires investigation (see Section 3.2).

4.1.5 TESTS OF THE CONTEXT MODEL

A Problem of Power. The experiments discussed in this section were designed to get more exact tests of the context model. The main problem was methodological: to get enough power to provide sensitive tests of the model.

That this was a problem can be seen in the right panel of Figure 4.1. The two curves are not overly parallel, yet the lack of significant interaction means that the observed nonparallelism could reasonably be attributed to prevailing variability. Since a real effect of the size observed could well be psychologically important, this experiment seemed to lack power to provide a satisfactory quantitative test.

Not many ways to increase power were open because the experiment was already on a within-subject basis. Increasing the number of subjects or replicating within subjects did. not seem likely to meet the need. However, it was possible to shift from between-adjective to within-adjective design. A within-adjective design in which the same component adjective appeared in every context would decrease error variability. It would also, of course, entail possible carryover or transfer effects from one context condition to the next. It was to avoid the possibility of such carryover effects that the experiment of Figure 4.1 used between-adjective design in which no adjective appeared more than once to a given subject.

Fortunately, work that had been completed just prior to this experiment suggested that carryover effects would not be a problem. That work had used both between-adjective and within-adjective designs and had obtained the same pattern of results from both (Anderson, 1965b, Figure 1, in press, Figure 1.5). At the same time, the within-adjective design had yielded substantial increases in power. Accordingly, it was decided to try within-adjective design to study the positive context effect.

Four Experiments on the Context Model. Four similar experiments were conducted, all based on a 4 × 4, Component × Context design. The component factor consisted of four single adjectives (H, M^+, M^-, and L); the context factor consisted of four pairs of adjectives (HH, M^+M^+, M^-M^-, and LL). Three of these experiments used within-adjective balance so that each component adjective appeared four times for each subject, once with each context. Similarly, each context pair appeared four times, once with each component.

General procedure was the same in all four experiments. Each description was typed on a card that was handed to the subject who read the words slowly aloud. After a 10-second wait for the subject to consolidate the impression, the experimenter said "Rate," whereupon the subject rated the person on a scale with nine verbal categories that ranged from "extremely dislikable" to "extremely likable." The experimenter then pointed to one adjective that the subject rated on a 1–20 numerical scale "corresponding to how much you like that individual trait of the person." In some conditions the subject also rated the other two adjectives in the same way.

Figure 4.2 shows the judgments of the component test trait from the last of these experiments. The main question concerns the parallelism prediction. The four curves are reasonably parallel, which agrees with the context model. Actually, the interaction term in the analysis of variance was just significant. That is perhaps a testimony to the sensitivity of the experiment since the deviations from parallelism are small, only .18 in mean magnitude on a 20-point scale. At the time, this interaction was not considered meaningful. However, there is a slight convergence of the curves as context value becomes more negative. Such convergence would be expected if the weight of the impression in Eq. (1) is greater when it is based on negative adjectives.

Between the two experiments of Figures 4.1 and 4.2, three unpub-

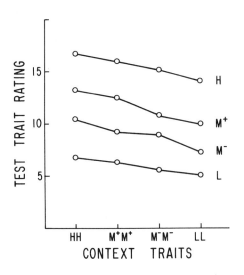

FIGURE 4.2. *Parallelism analysis of positive context effect. Rated likableness of component test trait (curve parameter) plotted as a function of two context traits (horizontal axis) in person description. Parallelism supports averaging-halo model of positive context effect. (From Anderson, 1966a. [Copyright © 1966 by the Psychonomic Society. Reprinted by permission.])*

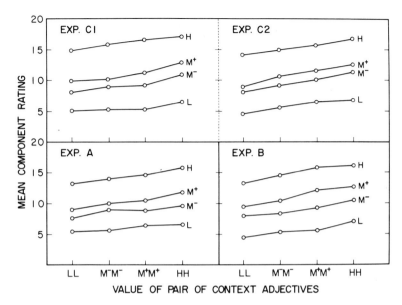

FIGURE 4.3. *Parallelism analysis of positive context effect. Rated likableness of component trait (curve parameter) plotted as a function of two context traits (horizontal axis) in person description. Near-parallelism supports averaging-halo model of positive context effect. (From Anderson, 1965c.)*

lished experiments were run to study effects of certain procedural details. Only graphical analyses were performed on these data, and these analyses will be reported here. The data on the component judgments are summarized in the four panels of Figure 4.3.

Theoretical interest in Figure 4.3 again centers on the parallelism prediction, and it seems reasonably well satisfied in all four panels. Nevertheless, there is some sign of convergence as context becomes more negative. This tendency may reflect greater weighting of negative adjectives as just noted (see also Sections 4.1.6 and 4.4.2).

Much of the concern in these three unpublished experiments was methodological. Experiments A and B used complete within-adjective balance. In Experiment A, only the one component adjective was rated; in Experiment B, all three adjectives were rated. Experiment C reverted to the procedure in which each subject saw each adjective only once, but with a more complex latin-square balance to remove a part of the between-adjective variability in the component judgments. In Condition C1, only the one component was rated; in Condition C2, all three adjectives were rated. The pattern of results is similar in all four panels. It would seem, therefore, that these procedural details are not important, and that the context model is fairly robust.[a]

Person Impressions. The judgments of the person are shown in Figure 4.4. Parallelism seems fairly well satisfied for Experiments A and B, but the two plots for Experiment C are less regular, possibly because of the less complete adjective balance. The person judgments for the experiment of Figure 4.2 (see Anderson, 1966a) had roughly the same shape as appears in Panel B of Figure 4.4 though with a significant interaction.

The context model implies that each graph of component judgments should be linearly related to the corresponding graph of person judgments. That should remain true even when the person judgments themselves show interactions, as long as they are equally weighted in Eq. (1). Hence corresponding panels of Figures 4.3 and 4.4 should have similar shape. That is more or less true for the vertical spacing of each set of curves, but it is hard to say whether or not the same holds for the observed deviations from parallelism. Unfortunately, these person judgments may suffer because, for the reason given in Section 4.1.2, they were collected on a scale of nine verbal categories. Indeed, it is perhaps surprising that this scale, converted directly to numbers, did as

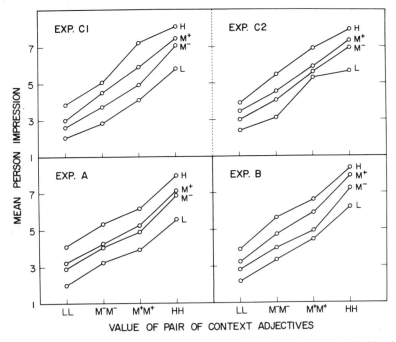

FIGURE 4.4. *Parallelism analysis of person impressions. Likableness of person described by three trait adjectives plotted as factorial graph, with one trait as curve parameter, two traits on horizontal axis. (From Anderson, 1965c.)*

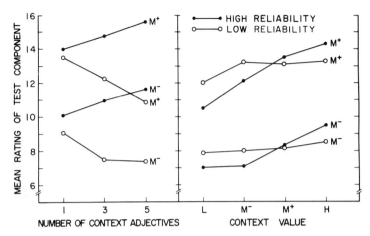

FIGURE 4.5. *Four tests of averaging model for positive context effect; see text. (After Kaplan, 1971a.)*

well as it did. Further study of this question might give equal procedural attention to both response measures and perhaps deliberately seek non-parallelism by variation in set size, as in the experiment of Figure 4.5.

4.1.6 RELATED WORK ON THE POSITIVE CONTEXT EFFECT

The positive context effect lends itself to a variety of interesting questions. Many of these are directly related to the issue of meaning constancy and have been discussed in Chapter 3. A few others of more general interest will be considered here. These experiments can all be related theoretically to Eq. (1): Any manipulation that affects the value of w or I will also affect the positive context effect.

Work by Kaplan. An ingenious and important article by Kaplan (1971a) tested four predictions from Eq. (1) in two experiments that employed procedures similar to those already described. The first prediction was that the positive context effect should be a direct function of the number of context adjectives. As the number of context adjectives is increased, the impression I will become more extreme by virtue of the set-size effect. Also, the relative weight $(1 - w)$ of I in Eq. (1) might increase because larger sets would produce a stronger or weightier impression. Both factors would act in the same direction to increase the positive context effect.

The data on the first prediction are in the left panel of Figure 4.5. The top curve shows that a given M^+ adjective is rated about 14 when the context is a single H adjective, but near 16 when the context includes five

H adjectives. The other M^+ curve shows corresponding decreases for negative contexts varying from one to five L adjectives. Parallel results were obtained for judgments of the M^- component in the two lower curves. Thus, the first prediction is well supported.

Kaplan's second prediction related to a reliability manipulation. Subjects were told, in effect, that the probability was either high or low that the given traits were an accurate description of the person. The person impression should be more extreme with the more reliable information because the initial impression then has lower relative weight. Since I is more extreme, the component judgment should also be more extreme by Eq. (1).

The data on the second prediction are in the right panel of Figure 4.5. The upward slope of these curves shows the usual positive context effect. This effect is much greater for the more reliable information, as predicted. Indeed, each pair of crossed curves appears to follow the linear fan form of the multiplying model, $(1 - w) \times I$, represented in the second term on the right of Eq. (1).

The third and fourth predictions tested by Kaplan concerned parallelism. These were also successful, as may be seen in both panels of Figure 4.5. In the left panel, the two upward sloping curves represent a Component $\times$ Context design and so should be parallel, as indeed they are. The two downward sloping curves should also be parallel. Some deviations from parallelism are visible in this case, but the interaction test was not significant. In the right panel of Figure 4.5, the two curves for each level of reliability should also be parallel, as is approximately true. This parallelism prediction depends on the auxiliary assumptions noted in Section 4.1.4. In particular, the assumption that the person impressions themselves obeyed a linear model was supported by the analysis of those data in Kaplan's article (see also Section 4.3.1).

Work by Wyer. Reports by Wyer and Dermer (1968) and Wyer and Watson (1969) have explored an interesting variation of the component judgment task. The idea was to manipulate the integration constraint so as to vary the interrelatedness among the adjectives. The two extremes of relatedness were the Person condition, in which the adjectives described a person, and the Word condition, in which the adjectives were unrelated personality traits. For example, the subject could be instructed that the adjectives described a person, a disorganized person, a group of people, or merely told to judge the average value of the adjectives. The experiments were designed around the meaning-constancy issue, but present attention will be restricted mainly to the averaging model for the positive context effect.

Wyer's data are variable, but the main points seem to be as follows: In four experiments, no context effect was obtained for the Word condition, in which the adjectives were merely unrelated personality traits. This lack of context effect agrees with the experiments of Section 4.1.2. In four experiments, Wyer included a comparison condition in which subjects were instructed that the adjectives were not intended to describe a single person, that it would be difficult to think of them as describing a single person, but that they were nevertheless to judge how well they would like the traits "considered in combination." These instructions produced substantial positive context effects, as large or larger than when the adjectives described a person in three of the four experiments.

This result was interpreted by Wyer to mean that the positive context effect could appear even in a set of unrelated words. An alternative interpretation may be suggested; namely, that these instructions acted contrary to the way intended. In fact, a substantial proportion of subjects said in subsequent questioning that they had considered the adjectives as describing a single person (Wyer & Dermer, 1968, p. 13). Post hoc analysis found no difference between the 9 subjects who did and the 15 subjects who did not say that they considered the adjectives as describing a single person. It is uncertain reasoning to conclude from this that the positive context effect is independent of whether the adjectives describe a person. Since the instructions evidently failed for 9 subjects, they might also have failed for the other 15.

In one experiment, in which the adjectives described members of a group, the positive context effect seemed to be as large as when the adjectives described a person. This result seems contrary to the integration model, which would expect a reduced effect on the grounds that a group is a less unified entity than a single person. The group instruction should set up less integration constraint, therefore, and hence a smaller value of $(1 - w)$ in Eq. (1). Although Takahashi's (1971a) data support the halo-averaging model on this point, Wyer (1974b) has replicated his finding.[a]

Finally, it should be noted that nearly all of Wyer's experiments find significant statistical interaction between component and context adjectives, both for the component judgments and for the overall impression. The interaction in the component judgments is not necessarily a problem for the context model because it predicts no interaction only when there is no interaction in the person impression. Unfortunately, only F ratios were reported, and it is unknown whether the interactions in the component judgment showed any interpretable pattern.

The interaction in the person impressions has more immediate as well as more general concern. Wyer does not discuss the shape of these interactions, merely stating that they seem inconsistent with linear models. The one graph that is presented (Wyer & Dermer, 1968, Figure 2) shows a substantial interaction of the form that would be expected in an averaging model if negative stimuli have greater weight. This problem deserves further study.

Work by Takahashi. Takahashi (1971a) presents a valuable parametric study in which three levels of interrelatedness were combined factorially with four levels of context adjectives, Japanese equivalents of H, M^+, M^-, and L. High relatedness was obtained by saying that the four adjectives described a person; medium relatedness by saying that each adjective described one student of a group from the same college who were much alike in character; low relatedness by saying that each adjective described one of four students from four different colleges who were not alike in character. Different subjects were used in each of these three instruction conditions.

Qualitatively, the results were as expected from the model. The positive context was large for the person condition, much less for the two group conditions, which themselves were not greatly different. Surprisingly, there seemed perhaps to be an effect even in the low-relatedness condition. This would be important theoretically since it would imply that even a slight integration constraint can produce a positive context effect.

Quantitatively, Takahashi's experiment may be analyzed in terms of Eq. (1). The value of the context adjectives will affect I, and the relatedness instruction will affect $(1 - w)$. Relatednesss did not affect I itself, as shown by the group impression data. For a given component adjective, therefore, the model has a simple multiplying form: $(1 - w) \times I$. Accordingly, ratings of a component H in the Instruction Condition $\times$ Context design should follow a simple linear fan in which the slopes of the lines correspond to the values of $(1 - w)$. Ratings of a component M^+ should also follow the bilinear form, but in this case the curves should cross over, much as they do in the right panel of Figure 4.5. Analogous results should hold for the negative component adjectives. This complex pattern of theoretical predictions appears quite clearly in the four successive panels of Takahashi's Figure 1.

4.1.7 COMPONENT JUDGMENTS AS AN EXPERIMENTAL TASK

Past work on the positive context effect has been narrow in scope because of dominating concern with the meaning constancy hypothesis. That has obscured the broad interest of the general task of judging components of an integrated whole. Some possibilities have been explored in the work described in the preceding section, but these and other problems deserve further study in their own right."

One problem that has already received attention is the effect of the integration constraint. The positive context effect is obtained only when the stimuli are interrelated in some way, as by the task instructions. Furthermore, the magnitude of the context effect depends on the nature and extent of the integration constraint. More systematic study of this problem would be useful, as in Takahashi's (1971a) parametric experiment.

Another problem concerns response dimensions. Most experiments have used a likableness dimension for both the person judgment and for the component judgment. Instead, both might be judged on cheerfulness, say, or on punctuality. Or the person might be judged on sociableness and the component on intelligence. Systematic variation of judgment dimensions could be helpful in mapping the structure of the person impression.

In particular, it is important to know whether and to what extent the structure of the impression depends on the assigned response dimension. If the component judgments on any dimension depend on which dimension was assigned for the person judgment, then it would seem that the impression was itself dependent on the assigned response dimension, not simply on the information presented about the person. Such results would be consistent with the conception of the valuation operation discussed in Chapter 1 and with the two-memory hypothesis of Section 4.2, whereas they would raise complications for a set-theoretical representation.

A third problem, somewhat different in nature, concerns the use of other integration tasks. An interesting example is judgment of meals, which could form a useful complement to judgment of persons because of the lack of semantic interrelationship among foods. The halo interpretation would imply that meals should also produce a positive context effect, a result that would be awkward for change-of-meaning semantics. Other possible integration tasks include judgments of commodity bundles and job situations. Judgments about groups of people (Sections 4.1.6, 3.6.2) would have special interest because of the experimental flexibility of this task.

As a final application, there is a variant of the halo interpretation in which the context adjectives act directly on the judgment of the component, not indirectly through the overall person impression. In this case, the component judgment is a direct weighted average of the adjective values. A way to distinguish between these two processing models can be obtained by expressing the I term in Eq. (1) as an average of the adjective values. The expression for the component judgment s^* is then also an average of the adjective values and requires the weight of each context adjective to equal $(1 - w)$ times its weight in the person impression. The direct action model does not place this constraint on the weights.

This weight constraint could provide a discriminative test in terms of component ratings along dimensions other than likableness. If each context adjective acts directly on the judgment of the component, then its effective weight will depend on its own semantic relation to the judgment dimensions. In general, therefore, different context adjectiveness would have unrelated weights. Under the true halo interpretation, in contrast, these weights would be proportional across judgment dimensions. This reasoning provides another illustration of the potential simple judgment tasks have to reveal cognitive structure.

4.1.8 GENERAL INTEGRATION PROCESS

The positive context effect may be interpreted as one manifestation of a very general integration process. The response to any focal stimulus will be influenced by contextual stimuli because of a general tendency of the organism to integrate all stimuli in the prevailing field. This integrational tendency is considered to be a basic biological capacity related to the tendency to form units or wholes (Section 1.8.2).

The weight of each stimulus in the response will be determined by attention and relevance factors. However, these weighting processes will often be crude, at least from a rational point of view. This point is well illustrated by the traditional concept of halo effect noted at the beginning of Section 4.1.4. That the teacher's judgment of intelligence is influenced by docility is ordinarily considered to be an error of judgment, but that may be an error itself. The teacher's judgment may reflect important and relevant characteristics of the child, although not exactly the characteristics that the intelligence tester had in mind. More generally, the halo effect may represent a sensible biological strategy of utilizing available information in what is often an uncertain predictive task.

The following experiments appear to involve a general integration process. Most were conducted from other perspectives, and the present interpretation is somewhat speculative. It is thought, however, that the concept of a general integration process may provide a unifying theme to much of this work.

Social Stereotypes. Social stereotypes represent a tendency to attribute some characteristic to individuals by virtue of their membership in some social group. Common stereotypes involve ethnicity, sex, age, and occupation. Because of their preoccupation with ethnic prejudice, psychologists have generally considered stereotypes to be more or less irrational (see, e.g., Brigham, 1971). In recent years, however, there has been a growing feeling that stereotypes represent normal information processing.

Such an informational view of stereotypes follows naturally from the integration-theoretical approach. Membership in a group constitutes a piece of information about the person; that information is integrated with other information in making judgments about the person. What is surprising is the often substantial effect of group information that has slight relevance. This very fact, however, supports the interpretation in terms of the general integration process.

A nice illustration of this matter is provided by the studies of Wyer and Watson and of Takahashi, already discussed. In these experiments, the group had only the barest hypothetical existence, yet substantial positive context effects were obtained. Similarly, Sigall and Landy (1973) found that judgments about a man were influenced by the physical attractiveness of his woman companion. However, no effect was obtained when the woman was physically present but not associated with the man, similar to Condition W in Figure 4.1. Also relevant is a study by Anderson, Lindner, and Lopes (1973), who found greater positive context effects for a member designated as leader, a result that was interpreted in terms of the weight parameter of the averaging model. All these studies had other primary concerns, however, and more systematic informational analyses of group belongingness would be desirable (see also Eiser, 1980, pp. 64ff).

Physical Attractiveness. The injunction to "look your best" on job interviews, dates, and other social situations testifies to the operation of a positive context effect based on physical attractiveness. Physical appearance is one among several pieces of information that may be integrated into a diversity of judgments about the person. For example, Dion, Berscheid, and Walster (1972) found that persons described by more attractive photographs were generally rated more favorably on a variety of judgment dimensions (e.g., sociable, sophisticated, sexually warm, sincere, kind, and altruistic), and were expected to enjoy better jobs and marriages.

One theoretical problem concerns the nature of the valuation operations that transform physical appearance into its implications for various dimensions of judgment. Physical appearance has direct relevance for dating, for example, and may present direct cues, such as a smile, for judgments such as sociableness. However, there seems no obvious causal link between physical appearance and altruism.

One theoretical interpretation is that physical appearance is transformed into a generalized good–bad value that is then integrated into the judgment on any given response dimension. This interpretation seems consistent with the hypothesis of Berscheid and Walster (1974) that "what is beautiful is good." This view seems too simple, however, because physical appearance could be expected to have opposite scale values for different dimensions of judgment (Section 1.1.3). Semantic inferences would also be required to allow different relevance weights for different dimensions.

Notably lacking in studies of physical attractiveness and of stereotypes in general is the use of tasks that require information integration. A photograph is a meager basis from which to make inferences about such personality traits as altruism, and so there is a natural concern that subjects' responses in such tasks may not mean very much. If another piece of information that had solid relevance to the judgment were to be included, then a meaningful relative importance could be determined for the photograph information. More generally, the study of integration rules could provide a basis for mapping the implicit personality theory of stereotypes (Section 1.8.3).

Cognitive Reorganization. An interesting manifestation of the general integration process occurs when new information is given that bears on some previous experience. The added information should be integrated with the original information, possibly causing reorganization of the memory representation.

In one experiment of this type, by Snyder and Uranowitz (1978), subjects read a realistic case history about Betty K. Subsequently, they were told either that Betty K. had a lesbian or a heterosexual life style, and this added information was followed by a multiple-choice test of factual material from the case history. Answers to certain questions, about Betty K.'s dating habits in high school, for example, were reliably influenced by the added information about sexual life style. This result is interesting because the added information is actuarially irrelevant to the question (Snyder & Uranowitz, 1978, Footnote 5). What was integrated into the response was the stereotype implications of the added information about sexual life style.

The simplest interpretation is that the information about Betty K.'s sexual life style was merely added to the memory store, a passive accumulation without any true reorganization. Since the stereotypic implications of this information would seem relevant to certain multiple-choice questions, they would be integrated as determinants of the choice response. The problem is thus one of recognition memory, to which an integration-decision model (Anderson, 1974a, p. 246; in press, Section 3.10) would be applicable, with the various pieces of information affecting the memory strengths of the choice alternatives. Even the simplest information integration analysis can thus account for the "reconstruction of past events [1978, p. 942]" claimed by Snyder and Uranowitz.

A theoretically more appropriate analysis would have to allow for the two memory systems discussed in Section 4.2. The added information would be integrated into the person impression, although not necessarily into the verbal memory for the specific facts. Response to the factual questions would be determined by both memory systems, essentially as in the preceding analysis, with the person impression producing the observed positive context effect.

Snyder and Uranowitz adopted a different interpretation, claiming that the added information caused reorganization of the memory for the factual material per se. This interpretation is difficult to establish, of course, because the two preceding interpretations could also account for the data. Indeed, their interpretation is conceptually similar to Asch's (1946) change-of-meaning hypothesis discussed in Chapter 3, and it faces essentially the same difficulties. Snyder and Uranowitz recognized the problem, but their control condition does not seem satisfactory.[a]

The problem of memory structure is too complex to be pursued here, but it deserves mention that serial integration typically involves memory reorganization. In the cumulative integration mode, at least, the memory representation is continually updated and restored (Section 2.5.2). Although the answering process for a multiple-choice question in the cited experiment need not involve memory reorganization, the answer itself could be restored, and that would constitute memory reorganization. However, this possibility was not tested in the Snyder and Uranowitz experiment. To do so would require somewhat different procedure (see, e.g., Loftus, Miller, & Burns, 1978).

4.2 Two Memory Systems: Verbal Memory and Judgment Memory

Person perception depends on memory, of course, and this dependence was explored in an early study in this research program (Anderson & Hubert, 1963). It was found that impression memory had no simple relation to the remembered adjectives. This result, which had considerable influence on subsequent experiment and theory, is taken up in this section.

4.2.1 TWO HYPOTHESES ABOUT MEMORY AND JUDGMENT

Verbal Memory Hypothesis. The verbal memory hypothesis says that attitudinal judgments are based on the contents of verbal memory. In the personality adjective task, therefore, the impression at any time would be directly related to the adjectives that were recallable at that time.

The verbal memory hypothesis has obvious attractions. It promises an explanatory framework for judgment in terms of readily observable facts. It avoids the error of taking the physical stimulus at face value, for the effective stimulus would be what the subject remembers. That allows for inattention, lack of comprehension, and forgetting, which is only sensible. Above all, the verbal memory hypothesis proffers the hope that judgment theory can be placed on a foundation provided by the enormous mass of research on verbal learning over the last century.

For these and other reasons, the verbal memory hypothesis became widely accepted in social judgment, especially in attitude theory (see McGuire, 1969), where it was often taken for granted. Various findings of low correlations between attitude and recall were troublesome, but these individual difference data could be explained in various ways. There had to be some relation between judgment and memory, moreover, and no alternative hypothesis suggested itself.

Words and Meanings. An alternative to the verbal memory hypothesis states that impressions and attitudes are stored in a memory system different from the verbal materials from which they originate. This two-memory hypothesis, as it may be called, rests on an assumption that ideas are psychologically separable from words. The word may function only as a carrier of the meaning. Once the meaning has been extracted, the word is a verbal husk, no longer needed, that may be stored in a verbal memory or simply forgotten. According to this two-memory hypothesis, there need be no particular relation between what is remembered and the person's impressions and attitudes.

This two-memory hypothesis arose serendipitously in the experiment described in Section 4.2.2, which studied verbal recall in conjunction with primacy–recency in the personality adjective task. If the verbal memory hypothesis was correct, then primacy–recency effects in the impression of the person would mirror those in the recall of the adjectives. This implication was not supported.

4.2.2 EVIDENCE FOR THE TWO-MEMORY HYPOTHESIS

In these experiments, subjects heard sequences of six or eight adjectives that described a person, then rated the person on likableness. In certain conditions, they were also asked for casual recall of the adjectives.

Recall Data. Typical recall data are shown in Figure 4.6, which plots percentage of recall at each serial position for sequences of eight adjectives. The first word is recalled about 45% of the time, whereas the second and third words are recalled only about 30% of the time. Thereafter recall increases steadily, reaching levels of 50%, 70%, and 90% over the last three serial positions.

Both curves, from separate experiments, show the same bowed shape with a small initial downswing and a large terminal upswing. Recall thus shows a small initial primacy and a large terminal recency. Similar curves have been reported in numerous experiments on verbal learning.

What is different about the recall curves of Figure 4.6 is that subjects also made an attitudinal judgment about the likableness of the person described by the adjectives. This judgment requires an integration of the information in the eight adjectives. If the verbal memory hypothesis is correct, the eight adjectives will have unequal effects on the judgment because they are not equally remembered. The last four adjectives should predominate in determining how likable the person is. Thus, the impression should show a recency effect.

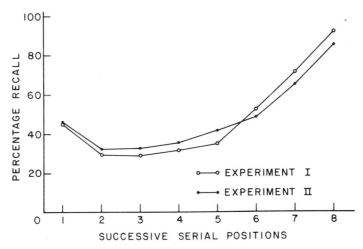

FIGURE 4.6. *Bowed serial curves of recall in personality adjective task. (After Anderson & Hubert, 1963.)*

Likableness Judgments. This recency prediction from the verbal memory hypothesis could not be tested if all adjectives had equal value. With scale value constant across serial position, different weighting patterns would yield equal judgments of likableness. However, this experiment included High–Low sequences in which the first four adjectives were favorable and the last four were unfavorable. If the last four adjectives in this High–Low sequence dominate the person impression, it should be somewhat unfavorable, a recency effect. If the same eight adjectives are given in the opposite, Low–High order, similar recency should be obtained, so the person impression should be somewhat favorable. With this choice of sequences, therefore, the person impressions should exhibit recency that mirrors the recency in the verbal recall.

The impression data disagreed with this prediction. A slight impression recency was obtained in one condition in Experiment I, but it was much smaller than expected from the verbal memory hypothesis. Moreover, substantial impression primacy was obtained in other conditions in which recall was not requested (see Table 3.1).

A possible objection to this interpretation is that the initial primacy component in the recall curves of Figure 4.6 might exert disproportionate effects on the person impression. Moreover, the terminal recency in the recall could be spuriously high owing to short-term memory effects.

Accordingly, the experiment was replicated with modified sequences in which the first two adjectives were always medium, followed by six adjectives in High–Low or Low–High order. The two medium adjectives filter out the initial primacy component in the recall so it cannot contribute to either primacy or recency in the person impression. The corresponding recall curve for Experiment II in Figure 4.6 shows pure recency over the last six recall positions, whereas the person judgments actually showed slight primacy.

Unwarned Recall. One sequence had special interest because the request for recall was made without warning in the middle of Experiment II. Recall was very low, only two words out of eight, but it also showed pure recency over the last six serial positions. In contrast, the person judgments showed primacy. This opposition of recency in recall and primacy in judgment implies that the verbal memory hypothesis is inadequate.

It was concluded, therefore, that impression memory is distinct from the verbal memory for the stimulus materials that gave rise to the impression. The verbal materials served as carriers of ideas and meanings, but these ideas and meanings were psychologically distinct from the words and had their own form of memory storage. This means that impressions and attitudes can persist long after the experiences from which they were formed have been forgotten.[a]

4.2.3 COMPLETE SERIAL CURVES: JUDGMENT AND RECALL

The recall curves of Figure 4.6 give a revealing picture of verbal memory. Similar curves for person impressions, to show how much the information at each serial position contributes to the likableness judgment, would be desirable. Such curves would provide stronger tests of the verbal memory hypothesis and more detailed information about person memory.

But the likableness judgment is an integrated resultant of all the adjectives, a single response that comes only at the end. Whereas the recall curve can be had for the asking, the serial impression curve requires some method for dissecting the single impression response into constituent parts that represent the contributions of the adjectives at each serial position. Such methods have been presented in Section 2.5 and some applications are given in the following subsections.

Serial Impression Curves. The first serial curves that were obtained for person impressions showed a straight-line primacy effect (Anderson, 1965b). Examples are shown in Figure 2.9 of Section 2.5.3, which also describes the method. These curves show that each successive adjective exerted a linearly decreasing influence on the person judgment at the end of the sequence. Those adjectives nearest in time to the person judgment had least influence. This linear primacy is clearly inconsistent with predictions from the verbal memory hypothesis.

Similar straight-line primacy was obtained in a subsequent experiment with improved methods of curve analysis (see Table 2.7). The verbal memory hypothesis requires these serial impression curves to have a bowed shape similar to the shape of the recall curves in Figure 4.6. Both experiments therefore provide strong evidence against the verbal memory hypothesis. However, it would still be desirable to obtain both recall and impression curves from the same experiment, especially with time intervals between the stimulus adjectives and the response to allow for short-term memory decay (Anderson, 1974e, pp. 14-15). The following two experiments provide more extensive information on these questions.

Comparison of Impression and Recall Curves. Dreben, Fiske, and Hastie (1979) presented sentences that described behaviors of a hypothetical person (e.g., Alan bought groceries for an elderly lady next door who was ill). In the two conditions of present interest, subjects rated the person on likableness after receiving a sequence of sentences. In one of these conditions they also recalled the contents of the sentences.

Both conditions yielded similar serial curves for the likableness judgment, as is shown in the left and center panels of Figure 4.7. These curves portray the effect of the sentence at each serial position on the overall judgment of likableness using the method of Section 2.5. Both curves are similar and show overall primacy, with the first two serial positions having greater effect than the last two.

The main theoretical implication rests on comparison of these serial impression curves with the recall curve in the right panel. The shapes are quite different, contrary to the verbal memory hypothesis.

Effect of Manipulating Recall. The logic employed by Riskey (1979) was to manipulate degree of recall and look for corresponding changes in attitude that would be required by the verbal memory hypothesis. Subjects heard a sequence of eight personality-trait adjectives that described a hypothetical person. They then counted backward by three's for 0, 15, or 30 seconds, following which they rated the person on likableness and also, on an occasional unwarned trial, recalled the adjectives. In the experimental condition of present interest, recall was requested only infrequently for a few descriptions, and subjects were told they should not make any special effort to remember the adjectives.

The recall data are shown in the left panel of Figure 4.8. The dif-

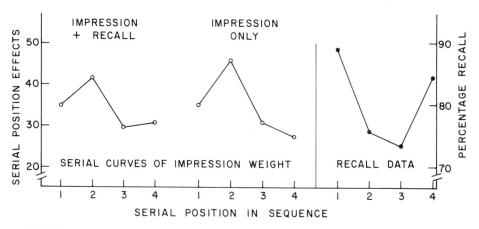

FIGURE 4.7. *Test of verbal memory hypothesis. Left and center panels show importance of stimulus sentences at each of four serial positions on judgment of person at end of sequence. Right panel shows recall of same sentences. Difference in curve shape argues against verbal memory hypothesis. (After Dreben, Fiske, & Hastie, 1979.)*

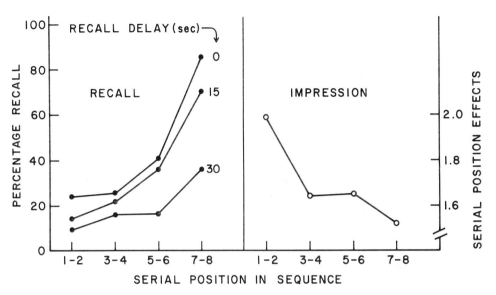

FIGURE 4.8. *Test of verbal memory hypothesis. Right panel shows importance of stimulus adjectives at each of four serial positions on judgment of person at end of sequence. Left panel shows recall of same adjectives under three memory interference conditions. Uniform primacy in person impression and uniform recency in recall argue against verbal memory hypothesis. (After Riskey, 1979.)*

ferences among these curves show that counting backward decreased recall markedly, in line with standard results in verbal learning. However, substantial recency remains even with 30 seconds delay. If the verbal memory hypothesis is correct, then these differences in recall will produce corresponding differences in impression. No such differences were found. The serial curves for the likableness judgment were very similar in all conditions.

Supplementary evidence against the verbal memory hypothesis is obtained by comparing the recall curves in the left panel with the impression curve in the right panel. The decline in the impression curve shows uniform primacy, whereas the recall curves show uniform recency. In another of Riskey's conditions, the use of recall on every trial eliminated the impression primacy with minor effects on the shape of the recall curves. Both results corroborate those of Anderson and Hubert (1963).

4.2.4 RELATED WORK ON THE TWO-MEMORY HYPOTHESIS

Beginning with a somewhat different experimental task, Greenwald (1968) independently arrived at a theoretical view that argues against the verbal memory hypothesis. Subjects first read a message containing 6 short arguments favoring and 6 opposing foreign aid. They were then asked to memorize the 6 underlined arguments, following which they were tested on their attitude on foreign aid and, unexpectedly, on their memory for all 12 arguments. The 6 underlined arguments were those favoring foreign aid in one experimental condition and those opposing foreign aid in the other condition. Subjects recalled an average of 3.0 underlined arguments but only .3 nonunderlined arguments. If the attitude was based on this recall, it would have to be quite different in the two experimental conditions. In fact, no significant difference was reported.

Greenwald's procedure is interesting and potentially important because it avoids the need for serial curves. However, the conclusion rests on acceptance of a null hypothesis that seems difficult to believe; given the asymmetry of attention to the pro and con arguments, some difference in attitude seems inevitable. Unfortunately, this experiment is not reported in detail and no followup seems to have been done. The objection just raised might be avoided by using fewer underlined than nonunderlined arguments. That could tip the attitude toward the nonunderlined position while maintaining the opposite direction in recall.

Other work on this question has been presented by Chalmers (1971) and Rywick and Schaye (1974), who have favored a verbal memory approach, and by Brink (1974) and Crano (1977), who have supported the present, two-memory view. Earlier studies have been reviewed by McGuire (1969). This work is not covered here, but it is fitting to conclude with Crano (1977) that "It seems apparent now . . . that the simple isomorphism of retention and attitude, so long an article of faith of the classical attitude theorists, simply does not exist [p. 94]."

A general implication of these results is that impressions and attitudes have a life of their own, independent of the particular experiences out of which they were formed. The S–R concepts of reinforcement and association have limited relevance to learning of attitudes, at least as these concepts have traditionally been applied. Learning and memory are certainly basic to impressions and attitudes, but it seems desirable to investigate approaches keyed to problems of judgment theory (Anderson, 1973c, pp. 6–7; Sections 3.6.1 and 4.5.4).

4.3 Initial Impression and Prior Attitude

Prior attitudes influence present judgments no less than present information. The averaging model assumes that such prior attitudes, denoted by s_0 or I_0, are averaged in with the external stimuli. In the personality adjective task, prior attitude is usually called the "initial impression," and it plays a key theoretical role in accounting for effects of set size (Section 2.4), source reliability (Section 4.4.3), and other variables.

However, the concept of initial impression may seem odd in the case of judgments of hypothetical persons about whom the subject knows nothing and who, in truth, do not exist. Some intuitive justification can be given by viewing initial impression as generalized expectancy. More solid justification arises from the demonstrated ability of the averaging model with initial impression to unify a complex aggregate of data.

Still, it would be desirable to have direct substantive referents for the initial impression. One such referent is provided by the following studies, which relate initial impression to personality predispositions.

4.3.1 KAPLAN'S WORK ON PREDISPOSITIONS

The work of Martin Kaplan occupies a special place in information integration theory, for it includes early, basic contributions regarding several problems. One series of experiments, summarized in Chapter 3, includes two of the best published papers on the meaning-constancy interpretation of the positive context effect. Collateral papers have studied set-size and redundancy effects (e.g., Kaplan, 1971e, 1972a). Other work has applied an integration-theoretical approach to interpersonal attraction and, more recently (Kaplan, 1977a,b; Kaplan & Schersching, 1980), to group decision and legal judgment.

Kaplan attacked still another problem, that of personality predispositions, in the series of papers considered in this section. The immediate concern of this summary is to relate Kaplan's work to the concept of initial impression. However, the general problem at issue concerns the role of response predispositions in person perception. Accordingly, this work also bears on general theory of interpersonal attraction as well as on the problem of individual differences in person perception.

Response dispositions are tendencies to evaluate another person in certain ways, independent of specific information about that person. Some people have generally favorable expectations about other people in advance of any specific information, whereas others have generally unfavorable expectations. Such dispositions have long been of concern in personality theory but have been studied primarily from a correlational view. Kaplan's work provides a more experimental approach.

Theoretical Analysis. Kaplan's basic assumption was that response disposition is one determinant of initial impression. Accordingly, disposition should be considered as an initial piece of information, to be averaged in with external stimulus information presented by the experimenter. By the set-size equation of Section 2.4, the response to a set of n stimuli of equal weight w and equal value s can be written

$$R(n) = [nw/(nw + w_0)]s + [w_0/(nw + w_0)]I_0. \tag{2}$$

It is assumed that different persons have the same values of w and s but differ in their initial I_0 values.[a]

Five predictions follow from this formulation:

1. Subjects with positive dispositions will rate other persons more favorably than subjects with negative dispositions.
2. For a fixed number of stimuli, disposition and stimulus value are integrated by an adding-type rule. Thus, the data should exhibit the parallelism property.
3. Disposition effects will decrease as amount of information increases.
4. Disposition effects will decrease as w increases. Hence increases in status or expertise of the source of information, for example, should decrease the disposition effect.
5. If the stimuli have equal weight but unequal value, then the response will be a linear function of $\bar{s}$ for fixed n, where $\bar{s}$ is the average value of the stimuli. Equation (2) remains true, therefore, with s replaced by $\bar{s}$.

Prediction 1 is the main working assumption in Kaplan's articles. The other predictions weave this qualitative conception of individual differences into the quantitative theory.

Prediction 2 follows directly from Eq. (2), which is additive in s and I_0. Predictions 3 and 4 reflect the averaging requirement that the relative weights sum to unity. The second term in brackets on the right, which is the relative weight of disposition, depends inversely both on n and w. Prediction 5 shows how a linear law for interpersonal attraction follows from integration theory.

Measurement of Disposition. Response dispositions can be measured with the Kaplan Checklist, given in Table 4.1. This list contains 36 adjectives, 12 each of high, medium, and low likableness value. Subjects check 12 words that they would be likely to use in describing people in general. Their disposition score is the number of high words minus the number of low words. The usual cutoff scores are +6 and −4 for selecting subgroups that are high and low in disposition.

Originally, a cumbersome continued association procedure was used

TABLE 4.1
Kaplan Adjective Checklist

Dignified	Bad[n]	Friendly[p]
Inconsiderate[n]	Subtle	Cruel[n]
Mean[n]	Restless	Angry[n]
Blunt	Considerate[p]	Skeptical
Thrifty	Careful	Honest[p]
Good[p]	Gentle[p]	Sophisticated
Cold[n]	Sad[n]	Nervous[n]
Cautious	Outspoken	Kind[p]
Loving[p]	Sweet[p]	Stupid[n]
Sincere[p]	Dumb[n]	Obnoxious[n]
Intelligent[p]	Nice[p]	Happy[p]
Loud[n]	Materialistic	Moderate

SOURCE: After Kaplan (1976). (Copyright © 1976 Duke University Press, Durham, North Carolina.)
NOTE: [p] and [n] denote positive and negative disposition traits, respectively. Subjects are instructed to check the 12 words they would use most frequently in describing *people in general.*

in which subjects had 6 minutes to list traits that described people in general. Disposition score was the number of high minus low traits among the first 12 listed. The checklist consists of traits that were most frequent in the continued association data and appears to give results comparable to the association method (Kaplan, 1976). The checklist is much easier to use, of course, and seems preferable on both theoretical and methodological grounds.

The rationale for trying out this measure of disposition is straightforward and needs no discussion. Its substantive validity was examined in the following experiments.

Additivity of Disposition and Stimulus Information. The parallelism prediction may be tested by combining disposition and external stimulus information in factorial design. One illustrative experiment is shown in the left panel of Figure 4.9. Open and closed circles represent subjects with positive and negative dispositions, respectively. Each subject judged the likableness of hypothetical persons described by adjectives of L, M^-, M^+, and H value, as listed on the horizontal axis. Positive disposition subjects gave more favorable judgments, which supports Prediction 1. The two curves are approximately parallel, which supports Prediction 2, that external stimulus information is averaged in with prior disposition. Similar support for Predictions 1 and 2 may be seen in the similar experiments of the center and right panels of Figure 4.9.

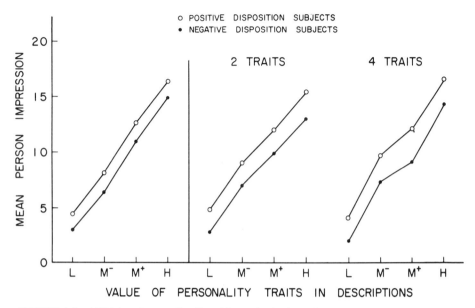

FIGURE 4.9. *Additivity of personality disposition and stimulus information. Mean judgment of likableness plotted as a function of subject personality disposition and stimulus trait value. Observed parallelism supports additivity prediction. (Data in left panel after Kaplan, 1971b, Figure 1. Data in right panel after Kaplan, 1971d, Figure 1.)*

Effects of Set Size. By Prediction 3, disposition effects should decrease as amount of external stimulus information increases. Figure 4.10 shows one experimental test (Kaplan, 1972b). Subjects selected for positive and negative disposition rated likableness of persons described by two, four, or six trait adjectives of positive or negative value. Prediction 3 is supported since the curves for the two disposition groups converge as number of adjectives increases.

In an interesting experimental variation, Kaplan (1971c) studied judgments of sociableness instead of the customary dimension of likableness. Each person description contained two kinds of adjectives, those that connoted sociableness and those that connoted changeableness, a dimension uncorrelated with sociableness. Subjects were classed as high or low in sociableness disposition according to whether the first 12 words given in a person association test contained more than 5 or less than 2 sociable words.

The theoretical assumption was that subjects who were high or low in sociableness disposition would be correspondingly high or low in expectancy for others to be sociable. In terms of the model, this expectancy affects the value of the initial impression parameter for judgments on the sociableness dimension.

The data, shown in Figure 4.11, bear on three theoretical predictions. The left panel shows judgments for sets of four traits, varied from 0 to 1 in their proportion of sociable traits. As the proportion of sociable traits increases, judged sociableness naturally also increases, in a near-linear way. The parallelism of the two curves provides further support for Prediction 2, that internal disposition and external information are integrated by an adding-type rule.

The right panel of Figure 4.11 shows data for sets of eight traits judged by the same subjects as in the left panel. Here the difference between the two disposition groups has all but disappeared, giving further support to Prediction 3. This disappearance of the disposition effect as set size increases has special importance because it supports

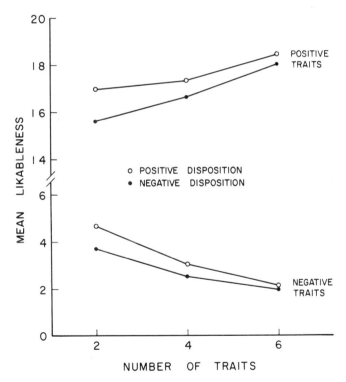

FIGURE 4.10. *External stimulus information decreases effect of personality disposition. Judgment of likableness plotted as a function of subject disposition and of number and value of descriptive traits. (After Kaplan, 1972b.)*

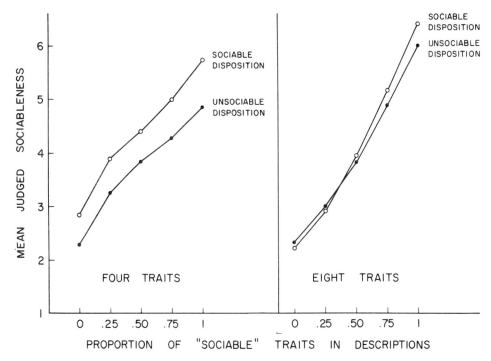

FIGURE 4.11. *Sociability predispositions affect impressions of other persons. Subjects selected as sociable or unsociable in disposition rate sociableness of persons described by four or eight trait adjectives with five proportions of sociable and changeable traits. Parallelism of two curves in left panel implies that disposition is averaged with external stimulus information. Near-identity of two curves in right panel implies that sufficient external information averages out the initial predisposition effect. (After Kaplan, 1971c.)*

Kaplan's assumption that both disposition groups have the same scale values for the external stimuli and differ only in the value of I_0. Thus, the disposition effects in the left panel cannot be interpreted to mean merely that subjects with positive dispositions assign higher values to the traits. Furthermore, both disposition groups were equivalent on judgments of changeableness, which shows that they used the rating scale in similar ways.[b]

The contrast between left and right panels of Figure 4.11 is a central feature in Kaplan's interpretation. Disposition has traditionally been considered to be a response bias, but a simple response bias would be expected to appear as a constant difference, the same in both panels. Even a more complex bias interpretation might find it hard to explain why bias should vanish as amount of information increases.

It should be noted, however, that not all of Kaplan's experiments have supported Prediction 3: Compare the center and right panels of Figure 4.9, in which the disposition effect is as large with four traits as with two. Kaplan has attempted to argue that this represents a redundancy effect, but the evidence is uncertain and the problem needs further study.

Linearity. The curves of Figure 4.11 are approximately linear, a result that agrees with the averaging model. To apply the model in this case, the sociable traits are assumed to have one value on the dimension of judgment, that is, sociableness, and the changeable traits another. Then the average value $\bar{s}$ of the traits will be proportional to the proportion of sociable traits plotted on the horizontal axis of the figure. The linearity of the curves then follows from Prediction 5.[c]

That the curves of Figure 4.11 have greater slope in the right than in the left panel also follows from the model. The theoretical expression for the slope is the first term in brackets on the right side of Eq. (2). This slope term is an increasing function of the number of stimuli and hence is larger for eight traits than for four.

Source Effects. By Prediction 4, an increase in the weight parameter, w, will decrease the effect of subjects' dispositions. Kaplan tested this prediction in two experiments by varying the characteristics of the source of the given information. In Kaplan (1971b), each given trait was ascribed to a source characterized by an occupation either high or low in social status. In Kaplan (1971d), each trait was characterized by the source's certainty, high or low, that the trait really described the person. In each experiment, source weight and subject disposition were varied in 2×2 design. Theoretically, the difference between high and low disposition subjects should be smaller when source status or certainty is higher. This prediction was verified in both experiments.

In these experiments, the predicted effect manifests itself as the interaction term of the 2×2 design, which has a simple algebraic form. Let w_H and w_L be the two source weights, and let I_0^+ and I_0^- be the initial impressions of the two disposition groups. The expression for the interaction can be shown to be

$$[(w_H - w_L)(I_0^+ - I_0^-)nw_0]/[(nw_H + w_0)(nw_L + w_0)].$$

Theoretically, this expression is not very large because it is limited by the difference between the two levels of source weight and by the difference between the two levels of initial impression. The statistical reliability of Kaplan's result depended on getting substantial main effects of both variables.

Dispositional Discounting. Personality dispositions can play an active role in information integration. When external information is inconsistent some of it may be discounted, and Kaplan hypothesized that the direction of discounting would be influenced by disposition. Subjects with negative dispositions would discount positive information, and vice versa.

To test this reasoning, Kaplan (1973) used descriptions that contained equally many favorable and unfavorable traits. Two instruction conditions were used, similar to the equal-weighting and naturalistic instructions described in Section 3.4.1. The naturalistic instructions were intended to facilitate any disposition-induced discounting, whereas the equal-weighting instructions were intended to inhibit discounting.

The results were in the predicted direction, though of marginal reliability. The person impressions in the left panel of Figure 4.12 show the usual large difference between the two disposition groups. More pertinent, the curves for each disposition group slope in the predicted direction. Subjects with positive dispositions were expected to discount negative traits in the naturalistic, but not in the equal-weight, condition. Such discounting would produce more favorable responses, so the curve for positive disposition should slope upward. Subjects with negative disposition should show the opposite effect and slope. Both effects appear in the graph, but the interaction term, which tests for difference in slope between the two curves, fell short of significance.

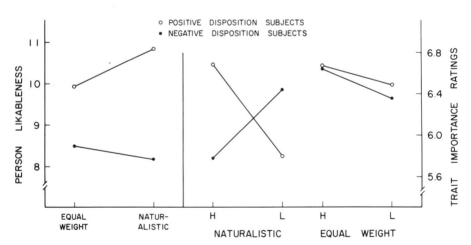

FIGURE 4.12. *Test of hypothesis that discounting of information is controlled by personality disposition. Left panel plots judged likableness as a function of subject disposition and discounting instructions. Right panel plots rated importance of each separate trait in the person descriptions as a function of subject disposition, discounting instructions, and trait value. (After Kaplan, 1973.)*

Kaplan also obtained component ratings of the importance of each adjective within each person description. The center panel of Figure 4.12 shows these importance ratings for the naturalistic instructions. These importance ratings follow the expected pattern, with positive disposition subjects assigning significantly greater importance to positive than to negative traits and vice versa for negative disposition subjects. The pattern of data for the equal-weight condition in the right panel shows no discounting effect, also in accord with expectation.[d]

The concept of dispositional discounting is important for social and clinical judgment. Once people begin to entertain some particular opinion or interpretation they may begin to discount information that disagrees with that opinion or interpretation. Initial attitudes may thus tend to perpetuate themselves against the weight of the evidence. In extreme cases, as in social stereotypes, they may even become self-maintaining within the person's belief system. Because of this social relevance, the ideas and methods of Kaplan's study deserve further investigation.[e]

Situational Determinants of I_0. Although disposition is considered to reflect enduring personality characteristics, it may also be affected by momentary situational factors. To test this hypothesis, Kaplan and Major (1973) exposed subjects to a simulated radio broadcast during a waiting period at the beginning of the experiment. In the positive (negative) condition, the radio commentator portrayed man in a favorable (unfavorable) light, and in the neutral, control condition, a song was played. Following this, Kaplan and Major used a person perception task developed by Byrne (1969) in which the subject received information about a hypothetical person's attitudes on either three or six social issues. This information was chosen so that the ratio of attitudes on which the person agreed with the subject's attitudes was 1:3 or 2:3.

Two response measures were used. The first was a measure of interpersonal attraction that has been standard in Byrne's work. The other was the judgment of likableness that has been standard in applications of integration theory to person perception.

Figure 4.13 shows that the subject's judgments of the person are markedly affected by the situational mood information. Both the attraction response (upper panel) and the likableness response (lower panel) decrease as the situational information goes from positive to neutral to negative. This supports the hypothesis that transient situational factors, here represented by the radio broadcast, can affect momentary dispositional state. Similar results have been reported for emotions by Schiffenbauer (1974) and for self-esteem by Baron (1974).

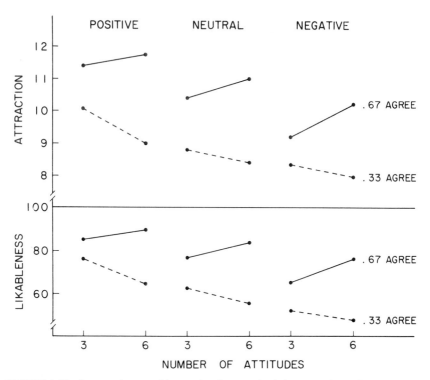

FIGURE 4.13. *Comparative test of integration theory and reinforcement theory for interpersonal attraction. Upper and lower layers plot judgments of interpersonal attraction and likableness, respectively, of persons described by their attitudes on certain social issues. (After Kaplan & Major, 1973.)*

This result has considerable interest for personality theory because it shows that dispositions toward people in general will apply to specific others. More generally, this result suggests that integration theory can provide a unified treatment of both enduring disposition and temporary mood.

Theory of Interpersonal Attraction. In the present view, attraction toward another person results from integration of information about that person. A different view has been advocated by Byrne (1969) and Clore and Byrne (1974), who have attempted to set interpersonal attraction within a framework of classical conditioning that has been popular in social psychology. The two theoretical approaches were compared by Kaplan and Anderson (1973a,b), who criticized the classical conditioning approach on two grounds (see also Section 3.5.4).

The first criticism was that the conceptual framework of classical conditioning seemed inappropriate. In this framework, the personality trait adjectives are treated as "unconditioned stimuli." They evoke affective responses that become conditioned to the "conditioned stimulus" represented by the hypothetical person. Similar attempts to transpose the framework of classical conditioning to social judgment have been made by others, reflecting the influence of Pavlov and of Hull. However, semantic relations such as are involved in reactions to redundancy and inconsistency, for example, are awkward to conceptualize as conditioned reflexes. The same applies to the dependence of subjective stimulus values on the dimension of judgment.

The second criticism was that the model employed in Byrne's theory disagreed with established facts. This model states that attraction is a linear function of stimulus proportion and so implies that the curves in the left and right panels of Figure 4.11 should have equal slope. More generally, it was shown that neither this model nor any likely extension could handle the set-size effect.

In reply to these criticisms, Byrne, Clore, Griffitt, Lamberth, and Mitchell (1973a,b) argued that results obtained with the personality adjective task do not pertain to their theory of interpersonal attraction, which was instead developed around descriptions of hypothetical strangers characterized by their attitudes on social issues. But this latter task was used in the experiment of Figure 4.13, and the data disagree with Byrne's model, which requires all 12 curves to be horizontal. The nonzero slope is just the usual set-size effect and is readily accounted for by integration theory. Indeed, the integration model can even account for the increase in slope of the solid curves from right to left in Figure 4.13 and the corresponding decrease in slope of the dashed curves.

In general, the attitude similarity task and the personality adjective task appear similar and should obey similar laws. Byrne and his associates have obtained a variety of interesting results on interpersonal attraction, including some in the neglected area of motivation. However, none of these results appears to present any difficulty to integration theory.

Comment. Kaplan's work is impressive for its systematic exploration of the concept of initial impression as a personality variable. Although uncertainties remain that require further study, this group of papers is notable for its experimental and theoretical coherence. Three points deserve comment. The most direct consequence of this work is the solidity and breadth that it adds to the concept of initial impression. The Kaplan papers indicate that individual differences in dispositions toward others appear as corresponding differences in the I_0 term of integration theory.

Kaplan's analysis shows that this correspondence is not merely a correlation between two personality tests, but reflects deeper structural properties of the way that information is integrated.

At the same time, Kaplan's work illustrates how integration theory can provide further leverage on the analysis of individual differences. Test-and-correlation methodology, which has dominated the traditional study of personality, is not very analytical. The correlations are typically weak relations between two global variables, taken over a group of individuals. Such data say little about the reaction of any one individual to any one stimulus situation. Within integration theory, at least some of these individual differences can be represented in the parameters of the information integration model. By focusing on these parameters, it becomes possible to study the individual at the individual level across a variety of situations. As Kaplan (1975b) concludes in his review, "Much of what we know about individual differences in social judgment can be included in an information-integration approach to judgment [p. 162]."

The third aspect, related to the second, is illustrated by Kaplan's discounting experiment. There it was seen that personality dispositions can play a central role in the way that people process information. This is an old idea that appears in the body of work on social stereotypes, but it gains new force from the possession of new analytical tools.

4.3.2. WORK BY OSTROM: PRIOR OPINION IN JURY TRIALS

Prior opinion and personality dispositions are important in jury trials. This can be seen in selection of jurors and in other aspects of trial procedure, such as concern with pretrial publicity. Nevertheless, jurors do have prior opinions, and legal procedures to allow for them may be off the mark in the absence of systematic research. An ingenious and important study by Ostrom, Werner, and Saks (1978) applied information integration theory to this problem: "How can a juror be objective yet still incorporate a prior opinion? An analysis of the w_0 and s_0 parameters in the averaging model suggests four conceptions of fair-mindedness [p. 437]."

The four conceptions of fair-mindedness may be summarized as follows, with s_0 replaced by I_0 and with minor changes from the discussion of Ostrom, Werner, and Saks.

First, jurors may attempt to avoid prejudgment and assume innocence and guilt to be equally likely. There is some evidence that people describe their prior disposition in this manner. With response measured as probability of guilt, this conception of fair-mindedness would correspond to $I_0 = .5$.

A second conception of fair-mindedness is given by Bayesian decision theory. To be as objective as possible, jurors should base prior opinion on the outcome of previous trials. More convictions than acquittals result from jury trials. Hence I_0 should be larger than .5. If the jurors are unaware of the actuarial data, then they would have to use their own belief, which could approach 1.0.

Legal theory in the United States rejects both of the preceding conceptions. Instead, jurors are instructed to presume innocence. That would correspond to $I_0 = 0$.

In the final conception of fair-mindedness, jurors would completely ignore their initial dispositions. This seems attractive because it does not ask jurors to attempt to change their disposition or to hold two dispositions at once. In model terms this corresponds to $w_0 = 0$.

Integration theory thus provides a theoretical vocabulary to represent these four conceptions of fair-mindedness. Furthermore, Ostrom *et al.* developed a simple method that allows a largely qualitative test among the four conceptions by varying set size and evidence value in a two-factor design. With evidence of low but nonzero scale value, for example, the set size curve should slope down if $I_0 \geq .5$; it should slope up if $I_0 = 0$; and it should be flat if $w_0 = 0$. The actual design used in this study has special value because it allowed for estimation of the parameter values.

The model analyses showed that w_0 was greater than zero and that I_0 was near zero. Under the conditions of this experiment, at least, the subjects did presume innocence.[a]

Despite their ability to adopt a presumption of innocence, the subjects did have prior dispositions and these became manifest in other parameter values. Disposition was assessed by agreement with a few general questions such as, "Too many criminals are set free by the courts," and subjects were classed as pro- or antidefendant by a median split. Both disposition groups appeared to have equivalent values for I_0 and for the scale value of the more incriminating evidence. However, antidefendant disposition appeared in two other parameters: higher guilt value for the less incriminating evidence and higher weight for each single piece of evidence. Both parameters lead to higher probability of guilt judgments by subjects with antidefendant dispositions. Quite similar results were obtained from college students and from a random sample of exjurors.

4.3.3 DISPOSITION AND VALUATION

The work of Kaplan, Ostrom, and their associates emphasizes the need to consider individual differences in information integration. It deserves notice that these studies all use the same basic integration model, an illustration of process generality (Section 1.8.2). Individual differences appear in the valuation operation and thereby in the model parameters.

Kaplan assumes that the external stimuli have the same scale value for all persons (see Note 4.3.1a). In his jury trial experiments, Kaplan's assumption implies that all persons will reach the same ultimate judgment. As evidence accumulates, the relative weight of initial disposition tends toward zero in the averaging model. As soon as enough evidence is available, it will prevail. This rational view is much like that in the normative Bayesian approach.

But this rational view depends on a doubtful assumption; namely, that external stimuli have the same value for different persons. The rational view may hold empirically in certain situations. It is attractive for legal situations, which usually have a yes–no orientation and in which, correspondingly, the scale values ought to be 0 or 1. However, the parameter estimates cited from Ostrom *et al.* (1978) indicate that the assumption of equal values need not be correct. This assumption cannot be correct in general, moreover, because stimulus values result from a constructive process that depends on prior knowledge and personal values. Two persons might thus place opposite scale values on the same piece of evidence. An increase in amount of objective information may not increase opinion agreement, therefore, but may increase disagreement.

In general, weights and scale values will be unique to each person. That does not deny a degree of commonality in matters of fact or even in matters of taste. However, even simple physical facts are apprehended not directly but through the individual's system of cognition. It is not necessary for two persons to like the same food; neither is it necessary for them to construct the same incriminatory value from a given piece of evidence.

This discussion brings out a needed distinction between the initial impression and the more general concept of "initial state." This distinction parallels that of Section 1.8.2 between attitude and attitudinal judgment. The initial state represents a knowledge structure, and the I_0 term represents the valuation of that knowledge structure with respect to the judgment task at hand. This valuation is a constructive process, as noted in Section 1.8.2, and so depends on and reflects the individual's cognitive organization.

Schemas, prototypes, and similar concepts are thus in the same spirit as the present concepts of knowledge structure and of valuation as a constructive process. All these concepts represent attempts to come to grips with the analysis of cognitive organization. These concepts are not explanations, however, but rather that which needs to be explained.

The primary need is for methods of analysis. In a modest but effective way, integration theory has provided a base for cognitive analysis, as illustrated in the work of Kaplan and Ostrom and in other sections of this chapter (see also Sections 1.8, 3.4, and 3.6).

4.4 Weight

4.4.1 CONCEPT OF WEIGHT

Inductive Definition. The concept of "importance" or "weight" provides an interesting illustration of the inductive mode of scientific definition. The need for such a concept begins in commonsense thinking, but the concept develops its proper definition and full meaning only within a theoretical framework. Definition and meaning emerge gradually as part of the scientific process, so that the concept contains and represents accumulating knowledge.

Intuitively, the need for a concept of weight seems clear. It seems natural and meaningful to ask, for example, whether negative information is more important than positive information (Section 4.4.2). In communication theory, a concept of weight also appears in the distinction between a message and its source. A given message will have greater influence when attributed to a more reliable source, and it seems natural to interpret this in terms of weight, with scale value being determined by the content of the message.

Under closer scrutiny, however, the concept of weight begins to blur into the concept of scale value. Negative information might have greater effect than positive information merely because its scale value has greater magnitude, not because of any difference in weight. Similarly, the argument that source attribution should not affect the scale value of the message rests on a verbal rationale that may lack substantive validity. The concept of scale value might suffice, with a separate concept of weight being unnecessary and unjustified.

To put the concept of weight on a solid basis, therefore, it must be distinguished from the concept of scale value at a more operational level. This is not entirely or even primarily an empirical problem, for it depends on the theoretical model. Indeed, adding models may not allow an identifiable distinction between weight and scale value (Section 1.2.6). For this reason, it has sometimes been argued that the concept of weight is empty or unidentifiable and ought to be merged into scale value. This argument rests on a faith in adding and linear models that has not found empirical support.

One outcome of averaging theory is that it does provide a solid basis to the concept of weight. At the theoretical level, weight and scale value each have a well-defined, conceptual existence within the averaging model. More important, the averaging model has substantial empirical support. The reality and definition of the concepts of weight and scale value are not merely hypothetical, therefore, but have some claim to scientific validity.

The concept of weight is also important in the normative–descriptive distinction considered in Section 1.8.1. Normative approaches require normative weights and so are strained to account for real behavior, which is often nonoptimal. In a study of a weighted sum model in utility theory, Anderson and Shanteau (1970) comment on the different conceptual status of the weights in the descriptive formulation:

> A second, more specific difference arises in the interpretation of the weight parameters. Integration theory applies to value judgments generally, whether or not they involve chance elements. The weight of an informational stimulus represents its importance in the total evaluation, and this may be large or small for a variety of reasons. . . .
>
> In decision making with uncertain information, the weight parameter would reflect the reliability or likelihood of the informational stimulus. But other factors would also affect the weight. A potential loss, e.g., might have greater felt importance than an equivalent gain . . . [pp. 449–450].

Determinants of Weight. The weight parameter will be affected by innumerable experimental manipulations, and no attempt will be made to catalog them here. Most manipulations, however, appear to fall into four categories: relevance, salience, reliability, and quantity of information. The nature of these categories is best explained ostensively:

- *Relevance* refers to the implicational relationship between the stimulus information and the dimension of judgment. A given stimulus can be important in one judgment, unimportant in another. In the experiment illustrated in Figure 4.17, for example, the trait *mechanical* is important in a plumber, unimportant in a lawyer. Similarly, *warmth* would be more relevant to judgments of sociableness than of honesty. The problem of relevance is central in implicit personality theory (Section 1.8.3).

  Both examples cited in the previous paragraph appear to involve similarity comparisons. In the first example, the similarity is between the stimulus adjective and the occupational prototype. The second example involves semantic similarity of stimulus and response adjectives. Other inferences are discussed in Section 4.5.4 and the problem of similarity is considered in Section 5.6.5.

- *Salience* refers mainly to attentional factors. As one example, the dependence of weight on serial position can be interpreted as a salience effect, at least according to the attention hypothesis (Section 2.5). Numerous other attentional factors, including repetition and perceptual emphasis, would also affect salience weighting.
- *Reliability* is a probabilistic concept, referring to the subjective probability that the given information is a valid indicator. Source factors typically operate upon reliability. In person perception, for example, source reliability can be manipulated by specifying how well or how long the source had known the person (Figures 4.5 and 4.15), or the number and variety of occasions on which the source had observed the person (Himmelfarb, 1972). These manipulations can be viewed as determinants of source reliability, that is, of the subjective probability that the source information is correct.
- *Quantity* of information can be defined by experimental operations, at least in simple cases. Thus, the set-size variable studied in Section 2.4 refers to the number of equivalent stimulus items. Analogously, the weight of an extended message will depend upon its length and aggregate content.

It is interesting to speculate that all four categories are reducible to one. Perhaps that is not possible, as there seems to be a clear distinction between reliability, which is a probabilistic concept, and relevance, which does not require any notion of probability. Perhaps, nevertheless, all four categories can be subsumed under a general concept of "informativeness." That concept is immediate for quantity of information; it seems reasonable for the other three categories on the basis that a more relevant, salient, or reliable stimulus is considered to be more informative.[a]

When all is said, however, these terms are still emerging from the common language. They are useful for communicating and for pointing to various experimental manipulations; they cannot be considered as explanatory concepts. The one concept that has progressed beyond the naive level is the concept of weight. If weight can be equated with informativeness, that constitutes a definition of informativeness, not of weight.

4.4.2 NEGATIVITY AND EXTREMITY EFFECTS

The concept of weight becomes important in the interpretation of nonparallelism. One particular deviation from parallelism has appeared repeatedly since the early stages of this research program, and the data from three experiments are shown in panels A, B, and C of Figure 4.14.

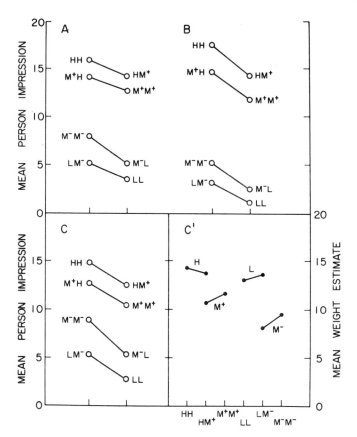

FIGURE 4.14. *Negativity effects in person impressions. Panel A plots person impressions from two 2 × 2 designs. Upper graph, for favorable adjectives, exhibits parallelism. Lower graph, for unfavorable adjectives, exhibits nonparallelism, as though very unfavorable L information had greater weight than mildly unfavorable M⁻ information. Panels B and C show similar data from other experiments. Factorial design is indicated by the letter symbols by each data point; each letter symbol stands for two, three, or two adjectives of indicated value in the three respective panels.*

Panel C' shows self-rated importance of single adjectives in the impression responses of Panel C. On the curve labeled L, for example, the left point represents rated importance of L adjectives in the LL descriptions and the right point represents rated importance of L adjectives in LM⁻ descriptions. These data are discussed in Section 4.4.4. (Panel A after Anderson, 1965a; Panel B after Anderson, 1968a; Panels C and C' after Anderson & Alexander, 1971.)

In each panel, the upper pair of curves represents a 2 × 2 design with positive adjectives; this pair of curves is essentially parallel for each experiment. The lower pair of curves also represents a 2 × 2 design, but with negative adjectives; this pair of curves shows a systematic deviation from parallelism taking the same shape in each experiment. These deviations are not large, but they were statistically significant. Further support for this finding is given in the review by Kanouse and Hanson (1972).[a]

These deviations from parallelism have a simple explanation in terms of averaging theory. It is assumed that the very negative L information has greater importance or weight than the mildly negative M⁻ information. Given this weighting, it is straightforward to show that the response to LM⁻ (and to M⁻L) will lie closer to LL than to M⁻M⁻. In this way, the averaging model can provide a simple account of these data (Anderson, 1965a, p. 399).

The convergence shape in Figure 4.14 could be interpreted as a negativity effect or perhaps as an extremity effect. A negativity effect means that negative information has greater importance than positive information or, more generally, that importance is an inverse function of scale value. An extremity effect means that importance is a direct function of absolute scale value. Within averaging theory, negativity and extremity effects reflect weight parameters that are not constant, but instead are correlated with scale value. These effects are thus special cases of differential weighting.

Negativity and extremity effects are not merely empirical phenomena, but must be treated as theoretical terms. As a consequence, it is not easy to get conclusive evidence for their existence. The various attempts to demonstrate negativity effects by selecting stimuli to be equal and opposite in scale value are subject to the difficulties discussed in Anderson (in press, Section 6.1.4). These difficulties are bypassed with the factorial design test illustrated in Figure 4.14. A related advantage of the factorial design test is that it applies to arbitrary stimuli, not merely those that are equal and opposite in scale value. Hence negativity and related effects can be assessed along the entire stimulus continuum.

It should be recognized that even the nonparallelism in Figure 4.14 is not conclusive evidence of negativity effects. These deviations from parallelism are small and could easily be eliminated by monotone transformation. It could be argued, therefore, that they merely reflect nonlinearity in the response scale; perhaps some kind of floor effect. In that case, of course, it would be inappropriate to speak of a real negativity effect. An adding formulation would also argue for response nonlinearity because adding implies parallelism even with differential weighting.

In fact, all three experiments in Figure 4.14 included a critical test that eliminated the adding formulation and supported averaging. The interpretation of the nonparallelism as real is theoretically consistent with the averaging rule as already noted. Even so, it could still be argued that these small deviations from parallelism in these particular experiments stem from residual nonlinearity in the response scale. That does not seem likely, but it cannot definitely be ruled out.[b]

4.4.3 SOURCE EFFECTS

Source effects are fundamental in information integration; the concept of information seems almost to necessitate the idea of a source of the information. The source may be implicit, as when the experimenter presents the information as factual, or source characteristics such as reliability and expertise may be explicitly specified.

Source–Message Integration. To illustrate the main ideas of the theoretical analysis, suppose that source and message are varied in a two-way design. Assume that source characteristics affect only the weight parameter and that message content affects only the scale value. Under these simplifying assumptions, the theory makes two main predictions. First, a more reliable source should in general produce a more extreme response. Second, integration of source and message should follow either a multiplying rule or a semilinear rule.

The first prediction follows because the message information is to be integrated with the initial impression. Accordingly, the effect of a single piece of information is given by the weighted average

$$R = C_0 + (ws + w_0 s_0)/(w + w_0). \tag{3}$$

As w is increased in Eq. (3), the response will move closer to s, the value of the given information. If s is more extreme than s_0, as would usually be the case, then the response will become more extreme.

The form of the integration rule for source and message depends on the value of s_0. If $s_0 = 0$ and $w_0 \neq 0$, then Eq. (3) reduces to the multiplying rule

$$R = C_0 + [w/(w + w_0)]s. \tag{4}$$

This theoretical multiplying rule is not the simple product, ws, but instead involves the more complex function of w given in brackets. Nevertheless, if w and s are varied independently, the data will plot as a linear fan. If the weight manipulation corresponds to the row factor in the design, then the row means will estimate the weight function given in brackets.

If $s_0 \neq 0$, then Eq. (3) has the form of a semilinear model. The data will plot as a sheaf of straight lines, but they will not in general have a common point of intersection as with Eq. (4). The marginal means of the scale value factor will still provide a linear scale of value. The marginal means of the weight factor will not be simply interpretable, but the slopes of the lines are still given by the expression in brackets of Eq. (4).

Integration across Two Sources. An important experimental study of source reliability was given in a senior honors thesis conducted in the writer's laboratory by Rebecca Wong under able guidance of Michael Birnbaum. Hypothetical persons were described by two acquaintances, each of whom contributed one trait adjective of High, Medium, or Low likableness value. Source reliability was manipulated by specifying how long the acquaintance had known the person: one meeting, 3 months, or 3 years. Each factor was varied separately to form a four-factor design. Further details and a related experiment are presented in R. Wong (1973), L. Wong (1973), and in Birnbaum, Wong, and Wong (1976).

Theoretically, the information about personality characteristics is represented by a scale value and the information about source reliability is represented by the corresponding weight parameter. Extension of the basic averaging model of Eq. (3) to allow for two source–message combinations yields

$$R = C_0 + (w_A s_A + w_B s_B + w_0 s_0)/(w_A + w_B + w_0). \qquad (5)$$

In this expression, w_A, w_B, and w_0 are the weights of Source A, Source B, and the initial impression, respectively; and s_A, s_B, and s_0 are the scale values of the associated information.[a] The Low, Medium, and High adjectives presented by Source A have an average value near zero, so the average of s_A is approximately zero, and similarly for the average value of s_B. The value of s_0 may also be assumed to be approximately zero, as indicated by previous work.

Five predictions from this model can be tested in the data. The first two assume that the adjectives have equal weight, although this assumption is not critical. First, consider how the reliability of a given source is integrated with the trait adjective given by that same source. This prediction becomes clearest if Eq. (5) is rewritten as

$$R = C_0 + [w_A/(w_A + w_B + w_0)]s_A + (w_B s_B + w_0 s_0)/(w_A + w_B + w_0) \qquad (6)$$

and averaged over the values of s_B. Then the second term on the right is near zero, since both s_0 and the average value of s_B are near zero. This term may be neglected, therefore, and the equation reduces to a product of s_A and the function of w_A within the brackets. Accordingly, R should obey a multiplying model; the data should plot as a linear fan with curve slopes given by $w_A/(w_A + w_B + w_0)$.

This first prediction is tested in the left panel of Figure 4.15. The horizontal axis lists the adjectives given by Source A, plotted at their functional scale values. Each data curve represents the reliability of one level of Source A. The linear fan shape and the crossover are in exact accord with theoretical prediction.

Second, consider the integration rule that relates the reliability of one source with the trait adjective given by the other source. The second term on the right of Eq. (6) can be neglected, according to the argument already given. The first term can be considered as a product of s_A and the function of w_B within the brackets. Hence R should again exhibit a simple multiplying form.

This second prediction is tested in the right panel of Figure 4.15. The horizontal scale is the same as in the left panel, but now the reliability values for the three curves are those of Source B. Here again, the crossover and linear fan shape of the three curves support the theoretical model.

It may seem surprising that the reliability of one source should have any relation to the effect of an entirely independent source. That agrees with the averaging model, however, in which the effective weights are interrelated by the requirement that they sum to unity. As the reliability and weight of Source A go up, the relative weight and effect of Source B necessarily go down.

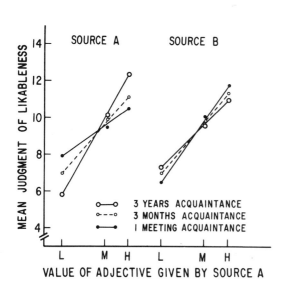

FIGURE 4.15. *Source reliability tests of averaging model. Subjects judge likableness of person described by two trait adjectives, each contributed by a different source. Adjective value and source reliability varied in four-way design. Left panel plots factorial graph for value of adjective from Source A (horizontal axis) and reliability of Source A (curve parameter). Linear fan pattern supports multiplying rule for integration of adjective value and source reliability derived from averaging model. Right panel plots factorial graph for value of adjective from Source A (horizontal axis) and reliability of Source B (curve parameter). Linear fan pattern, inverted relation of curves across the two panels, and lesser vertical spread of curves in right panel are all predicted by the averaging model. (After R. Wong, 1973.)*

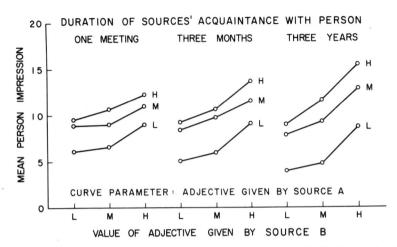

FIGURE 4.16. *Effect of initial impression in averaging model. Subjects judge likableness of persons described by two trait adjectives, each contributed by a different source. Each panel represents 3 × 3 design with value of adjective from Source A as curve parameter, value of adjective from Source B listed on horizontal axis. Both sources have equal reliability, defined by acquaintance durations of one meeting, 3 months, and 3 years in the three respective panels. As source reliability increases, so do curve slope and curve separation, as predicted by averaging model. (After R. Wong, 1973.)*

The third prediction concerns two notable contrasts between the left and right panels of Figure 4.15. The slopes of the curves are a direct function of Source A reliability in the left panel but an inverse function of Source B reliability in the right panel. This agrees with Eq. (6) since the slope function, $w_A/(w_A + w_B + w_0)$, is a direct function of w_A and an inverse function of w_B. In addition, the range of slopes, indexed by the vertical spread of the curves, is greater in the left panel than in the right panel. That also agrees with Eq. (6).[b]

These last two results provide striking support for the averaging formulation; an adding model would predict that all three curves in the right panel of Figure 4.15 would be identical. This test is robust in the same sense as the crossover test of Section 2.3; that is, it requires only a monotone response scale and it applies to a general class of adding models with diminishing returns.

The fourth prediction concerns the role of the initial impression. This can be illustrated by considering the three cases in which both sources have equal reliability. These data are shown in Figure 4.16, in which each 3 × 3 design is based on the same adjectives. As source reliability increases from left to right, the slopes of the curves increase, and so also their vertical spread. This is exactly as predicted.

To establish the fourth prediction, note first that the slopes are given by the weighting coefficient of s_B in Eq. (6). This coefficient is an increasing function of w_B, and hence the slopes increase with source reliability, in agreement with the data. Next, consider the difference in response when s_A is High or Low. From Eq. (5), this difference can be obtained as

$$[w_A/(w_A + w_B + w_0)](s_{High} - s_{Low}),$$

which is an increasing function of w_A. Theoretically, therefore, the vertical spread increases with source reliability, in agreement with the data. Of course, slope and vertical spread are only different views of the same effect.

The initial impression plays a key role in this derivation. If the initial impression did not exist, that is, if $w_0 = 0$, then the model would require the three panels of Figure 4.16 to be identical. As it is, these data provide useful support for the concept of initial impression.

The fifth, and last, prediction concerns parallelism. Figure 4.16 shows a mild divergence interaction, especially in the rightmost panel. The shape of this nonparallelism is that of the negativity effect already noted in Figure 4.14. Of itself, the nonparallelism is not a theoretical problem for the averaging model since it can be explained in terms of differential weighting.

"The Case Against Averaging." A reliability manipulation in a two-source design was used by Wyer (1974a), who rested his "case against averaging" on the outcome. This ingenious experiment requires detailed consideration. It will be shown that neither of the two arguments presented by Wyer is a serious problem for averaging theory. Furthermore, when the logic of Wyer's test is applied to data from the preceding experiment, the results support averaging.

In Wyer's experiment, subjects rated likableness of persons described by two adjectives, one from each of two sources, A and B, who could be either low or high in source credibility. Schematically, the adjectives may be represented by their scale values along the dimension of judgment:

For convenience, it is assumed that both adjectives are favorable, with $s_A > s_B > 0$.

The question is, how will the response change if the reliability of Source B is increased so that its weight increases from w_B to, say, $w_B + x$. Increasing reliability is equivalent in form to adding a piece of favorable information with scale value s_B and weight x. Any adding model must imply that the response will become more favorable, and just such results were obtained. Two earlier experiments with similar results were cited as additional support for the conclusion that the "averaging model is invalid [Wyer, 1974a, p. 305]."

It might indeed seem that the averaging model could not account for the observed increase in response. If the response is an average of s_A and s_B, then it lies between them in the diagram just shown. If the weight of B is increased, the averaging model requires the response to move toward s_B, thereby becoming less favorable. On this argument, the obtained result is contrary to the averaging model.

But something is missing from the previous diagram, namely, the initial impression. The appropriate diagram is as follows:

Theoretically, the response is the average of s_A, s_B, and s_0. This average may lie to the left or right of s_B in the diagram. If all three weights were equal, for example, then the response would lie to the left of s_B. Increasing the weight of B would then cause the response to move toward s_B—thereby becoming more favorable, as was observed.[c] In short, the averaging model with initial impression can account for the results just cited. The theoretical rationale is the same as was used previously to account for the set-size effect.

This fact was recognized by Wyer, and he presented a second argument based on data in his Table 9.4 (p. 304). In this table, the increase in response produced by an increase in reliability of Source B is roughly independent of the values of the two adjectives. This lack of trend is contrary to an averaging model, as Wyer points out. But it is also contrary to the adding model, as can readily be demonstrated.[d] It is inappropriate, therefore, to conclude that these data "support a summative model of information integration over an averaging model [p. 305]."

Indeed, when the logic of Wyer's test is applied to the preceding two-source experiment, it provides a critical, qualitative test that supports averaging and eliminates the summative model. To illustrate, consider the case in which Source A contributes a High adjective, and Source B contributes a Medium adjective. As the weight of Source B is increased, for acquaintance durations of one meeting, 3 months, and 3 years, the mean impression progressively decreased toward neutral, with respective values of 11.67, 11.38, and 11.35. A similar pattern appeared on the negative side, for the case in which Source A contributed a

Low adjective and Source B contributed the same Medium adjective: As the weight of Source B increased, the response increased toward neutral, with respective means of 6.98, 7.30, and 7.63. These results are as expected from the averaging hypothesis. The Medium adjective contributed by Source B has a value near the neutral center point around 10 on the response scale; increasing its weight pulls the response in toward the center. No adding or summation formulation can account for this pattern of data.[e]

Source Effects in Intuitive Statistics. Three theoretical treatments of source effects were compared by Lichtenstein, Earle, and Slovic (1975) in a probability learning task. In initial training, subjects predicted a numerical criterion, given the value of one of two numerical cues, A or B, with respective cue–criterion correlations of .8 and .4. Each cue thus gave probabilistic information of more or less validity about the criterion. Following training, test trials were given on which the values of both cues were presented. The main theoretical question was how the subject integrated the information from the two cues to predict the criterion.

Integration theory provides a straightforward analysis in line with previous treatments of probability learning (e.g., Anderson, 1964a, 1969a; Friedman, Carterette, & Anderson, 1968; Himmelfarb, 1970). On any trial, the two cues constitute two source–message pieces of information. They are averaged with the prior expectancy to determine the prediction on that trial. In this task, the scale value of each cue may be identified with its numerical value; the weight of each cue is determined by its source validity as manipulated in the cue–criterion correlation.

This task has special interest because it imposes an environmental constraint of correct or optimal behavior. This is a prototypic task in Brunswik's (1956) formulation, in which the organism is viewed as striving to maximize achievement in a probabilistic environment. The information in the cues makes it possible to predict more accurately than chance, and there is an optimal way to do so. Indeed, the optimal rule is the ordinary linear regression equation, with regression weights determined by the cue validities. Bayesian decision theory and Brunswik's formulation both agree that this linear equation provides the proper framework for studying the behavior. Both represent a normative approach, in contrast to the descriptive approach of integration theory.

Detailed analysis showed that neither of these two normative formulations could account for important qualitative aspects of the data. Among other problems, a medium value cue made the response less extreme, not more extreme as the regression model requires. That, of course, is the standard result obtained in the adding–averaging tests of Section 2.3.

Lichtenstein *et al.* interpreted their results to support an averaging formulation rather than either of the normative formulations. They did not use the concept of prior expectancy, however, but instead assumed that the value of each cue was "regressed" toward the mean of the distribution, with the response determined by the average of these regressed cues.[f]

In a cogent followup experiment, Birnbaum (1976) analyzed linear, adding, and averaging models by using a two-source design that allowed clear discrimination among these models. The design was analogous to that of Figure 4.15, and a similar two-panel graph of linear fans was obtained. This result rules out the linear and adding models as well as the averaging model with regressed values suggested by Lichtenstein *et al.* Moreover, the weights estimated from the regression analyses were not constant, but varied with particular cue combinations. However, the averaging model from integration theory gave an excellent fit to the data and, as Birnbaum emphasized, provided a unified interpretation of an apparently complex set of results.

These results also bear on the normative–descriptive distinction discussed in Section 1.8.1. Lichtenstein *et al.* (1975) comment that the linear regression model seemed to do well when assessed by means of the customary statistics but that these statistics were misleading: "The traditional measures... were not useful in uncovering these serious discrepancies from the model [p. 85]." The main reason is that the traditional measures originate from the normative orientation and are ill suited to descriptive process analysis. A relevant illustration comes from a study of intuitive statistics in which the most accurate group was least in accord with the averaging model (Anderson, 1968c): "Under such circumstances, emphasis on a normative criterion may retard the development of an adequate theory [p. 392]."

More Complex Source Effects. An important extension of averaging theory to more general problems of source effects has been given by Birnbaum and Stegner (1979), who studied simultaneous effects of expertise, bias, and point of view. Subjects judged the dollar value of used cars, given their "blue book" value and the value estimated by a source person who could be low, medium, or high in automotive expertise. Bias was manipulated by describing the source as a friend of the buyer, a friend of the seller, or an independent. In addition, subjects were instructed to adopt the point of view of either the buyer, the seller, or an independent.

Theoretically, the subject's judgment is an average of the initial impression, the blue book value, and the value estimated by the source. The question is how the source parameters, namely, weight and scale value, depend on source characteristics. In terms of the functional measurement diagram, this is a problem of valuation.

The more expert source should receive greater weight, in line with the previous experiments, but the expected effect of bias is less clear. A buyer will certainly pay more attention to his or her own friend than to a friend of the seller, but this attention might manifest itself in the weight parameter, the scale value parameter, or both. Furthermore, bias might be expected to interact with expertise, with the buyer more distrustful of the seller's friend who has the greater expertise. At the same time, the valuation operation will obviously depend on point of view. Meaningful dissection of these intertwined effects would hardly seem possible without model analysis.

The data provided excellent support for averaging theory in both qualitative and quantitative tests. Source weight was mainly determined by expertise, with weight estimates in a rough $3:2:1$ ratio for high, medium, and low expertise. The bias manipulation had its main effect on scale value; these estimates had essentially the given dollar value for the independent source, but were adjusted up or down by roughly $30 according to whether the source was a buyer's friend or seller's friend. In addition, weights were noticeably higher for the independent source but, surprisingly, were essentially equal for both friend sources, independent of point of view. Point of view had the expected effect on scale value, together with a configural weighting effect by virtue of which total absolute weight was conserved, with weight interchange between the two sources. Accordingly, Birnbaum and Stegner (1979) concluded that "a simple algebra can account for the complex effects of bias in information integration [p. 72]."

Other Work on Source Effects. The first systematic attempt to study source effects in person perception was made by Rosenbaum and Levin (1968, 1969). Their work is interesting because the sources were chosen to be high or low in occupational prestige (e.g., social worker, doctor, laundry man, meter reader). Subjects judged likableness of such hypothetical persons as a Mr. A., who "has been described by a laundry man as *broad-minded, happy,* and *trustworthy.*"

It was assumed that the source manipulation affected the weight parameter, whereas the stimulus adjectives determined the scale value in the averaging model. The main result of this work was that likableness was more extreme, positive or negative, when the trait adjectives were assigned to a source with higher prestige. This result, which later became important in the theory of interpersonal attraction (Kaplan & Anderson, 1973a, p. 308), supported the basic theoretical assumption that the source acts as a multiplier on the value of the message.

An interesting test of Eq. (3) by McKillip (1975) used three sources (friend, acquaintance, competitor) as a between-subject factor, and three adjective descriptions (negative, neutral, positive) as a within-subject factor. Subjects rated credibility of each source–description combination, and also the likableness of the person described. Credibility was uniformly lower for acquaintance than for friend, and the corresponding person impressions showed the crossover expected from Eq. (3). For the competitor source, however, there was an interaction such as McKillip hypothesized: When the description was negative, competitor was rated less credible than acquaintance; when the description was positive, competitor was rated as more credible than friend. These credibility ratings were partly, though not entirely, reflected in the person impression.

McKillip interpreted his results in terms of an initial valuation stage in which source and description interact to determine the w-parameter of the integration model. Related results on attitudes are presented by McKillip and Edwards (1975). This approach deserves further exploration, both for the great interest of interactions and for the potential use of credibility judgments as self-estimated weights.

Other work on source effects in information integration has mainly been on specific questions. For example, Anderson and Jacobson (1965) varied source reliability as a means to study inconsistency discounting. Kaplan (1971a,b) made good use of source effects in his work on personality dispositions, as discussed in Figure 4.5 and Section 4.3.1. Himmelfarb's (1970) study of cue integration in a probability learning task obtained some support for an averaging model in which cue weights were taken as a rational function of their probability of being correct during the reinforced training. Himmelfarb (1972, 1975) and Himmelfarb and Anderson (1975) showed how certain results in social attribution could be interpreted as effects of information reliability on the weight parameter of the averaging model.[g]

4.4.4 OTHER WEIGHTING EFFECTS

Relevance Weighting. Since the valuation operation depends on the dimension of judgment, so too will weight and scale value. The two experiments following illustrate this dependence, which may be called *relevance weighting.* Also of interest is a qualitative test that demonstrates weight effects independent of scale value.

An extreme case of relevance weighting is shown in Figure 4.17, adapted from Anderson and Lopes (1974). Subjects judged occupational proficiency of lawyers and plumbers, each described by two personality traits in various 2 × 3 designs. In each design, the row factor had the same two levels, *extremely reliable* and *averagely reliable,* shown by the curves. Separate designs were employed for each of the adjectives— *persuasive, mechanical,* and *musical*—which were chosen for their relevance properties. The three levels of the column factor were obtained by modifying these adjectives by *very, moderately,* and *not very,* shown on the horizontal axis.

The slope of each pair of curves reflects relevance weighting. When the adjective is irrelevant to the occupation, as *mechanical* is to *lawyer,* then the pair of curves should be horizontal. Non-zero slope is obtained only in the two cases in which the trait is important for proficiency at the given occupation.

The weighting interpretation rests specifically on the smaller vertical separation between the row curves when the column adjective is relevant. But for this difference, the data could be accounted for in terms of

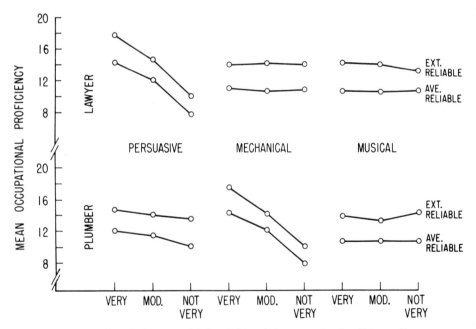

FIGURE 4.17. *Effect of relevance weighting. Subjects judge occupational proficiency of lawyers or plumbers, each described by two quantified personality traits in 2 × 3 designs as indicated. Slope is greater and vertical spread of curves is less for more relevant adjective, in accord with averaging model. (After Anderson & Lopes, 1974.)*

scale values alone. Since it is often useful to compare vertical separations in this manner, it is worthwhile making the model analysis explicit. From Eq. (22) of Section 1.6, the difference between the two row curves in each 2 × 3 design can be written

$$R_{1j} - R_{2j} = [w_A/(w_0 + w_A + w_B)](s_{A1} - s_{A2}). \qquad (7)$$

Here, s_{A1} and s_{A2} are the scale values of the row adjectives *extremely reliable* and *averagely reliable*, and w_A is their common weight. Because the weight of the column adjective, w_B, appears in the denominator, a larger w_B produces a smaller difference between the two row curves. Accordingly, the observed data imply that relevance affects the weight parameter.

The data also appear to show a slight negativity weighting. The pair of curves for each relevant adjective converge as adjective level decreases, a pattern similar to that discussed in Section 4.4.2. This effect was disappointing for it had been hoped that the adverbial quantifiers would yield equal weighting. However, the negativity interpretation does seem to have a straightforward rationale for judgments of occupational proficiency. If the plumber is not very mechanical, it may not make too much difference how reliable he is. As his mechanical ability begins to exceed a minimum level, his reliability becomes increasingly relevant. Such conjunctive integration rules can be represented in terms of differential weighting within the averaging model.

The main purposes of this experiment were actually to study adjective–noun integration and to compare information integration theory with the congruity formulations of Osgood and Tannenbaum (1955) and of Rokeach and Rothman (1965). These topics are discussed in the original report and in Section 3.5.1.

Dimension of Judgment. Even within a given person description, relevance weighting can appear across different dimensions of judgment. Oden and Anderson (1971) used descriptions of hypothetical naval officers, recent graduates of Annapolis Naval Academy, in the design shown in Figure 4.18. The row factor was a set of personality trait adjectives, High, Neutral, or Low in likableness value, listed by each curve. The column factor was class standing on an overall academy assessment, shown on the horizontal axis. Each officer was judged on three dimensions: command effectiveness, how much his men would respect him, and how much his men would like him. These data are shown in the three panels of Figure 4.18. The near-parallelism in each panel indicates that class standing is integrated with personality by a simple adding-type rule.[a]

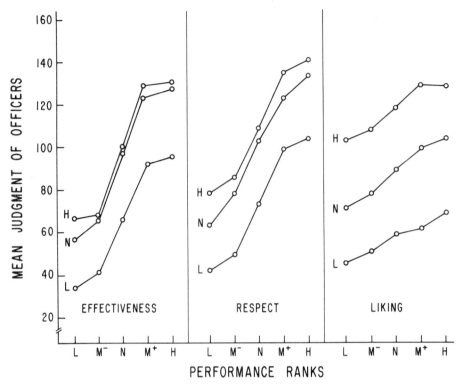

FIGURE 4.18. *Effect of relevance weighting. Subjects judge command effectiveness, respect, and liking for naval officers described by personality traits (curve parameter) and class standing (horizontal axis). Near-parallelism indicates that personality and class standing are integrated by averaging rule. Lesser slope and greater vertical spread of curves in right panel indicate that personality traits receive more weight in judgments of liking than in the other two judgments. (From Oden & Anderson, 1971. [Copyright © 1971 by the American Psychological Association. Reprinted by permission.])*

The main point of Figure 4.18, for present purposes, is that the curves have flatter slope and greater vertical spread for liking than for either command effectiveness or respect. This result agrees with common sense and with Eq. (7). The flatter slope means that the importance of the overall academy assessment is less for judgments of liking than for the other two judgments. In the same way, the wider vertical spacing means that the importance of the trait adjectives is greater for judgments of liking than for the other two judgments. Thus, the effective weights of the two stimulus factors are inversely related, in agreement with the averaging model.

As an incidental point, Figure 4.18 shows a nonlinear relation between scale values of the personality traits for the liking judgment and the other two judgments. The vertical spacing of the curves corresponds to the functional scale of the adjectives (Section 1.2.3), so the Neutral trait is midway between the Low and High traits in likableness but almost as high as the High trait for command effectiveness. That makes sense for Neutral traits such as *quiet* and *conventional* that were used in this experiment. Nonlinear relation between the scale values of the same stimuli across two judgment dimensions is not surprising. However, it illustrates that the traditional scaling orientation toward obtaining scale values as a methodological preliminary has limited usefulness for substantive inquiry because the judgment dimension, and hence the scale values, will depend sensitively on situational context.

Configural Weighting. A striking configural effect in moral judgment was discovered by Leon, Oden, and Anderson (1973), who obtained judgments of "badness" of groups of three criminals. A substantial minority of the subjects judged badness on the basis of the worst member in the criminal group, completely ignoring the other members. Thus, a person guilty of arson would have no effect on the badness of a group that contained a rapist, but a marked effect on the badness of a group that contained only fences or forgers.

This configural effect can be seen in the top curve in Figure 4.19, which represents judged badness of three-person groups with one person guilty of homicide. The other two persons were guilty of an offense listed on the horizontal. This homicide curve is essentially flat; increasing the severity of the offenses of the other two group members has virtually no effect on the badness of the group.

Similarly, the curve for arson represents three-person groups that contained one arsonist and two members guilty of offenses ranging from vagrancy to arson. The curve stays roughly horizontal until rape is added, at which point it rises vertically to meet the rape curve, as shown by the dashed line. There is a slight slope to the curve, but it is negligible compared to the configural effect.

In model terms, this judgment pattern can be described as configural weighting. It appears as though the subjects had assigned a weight of 1 to the worst member of the group, a weight of 0 to the other members. Although the majority of subjects did not exhibit such all-or-none effects, they were similar in that they placed exceptionally high weight on the more serious crimes. This result suggests that stereotypes of social groups may be overwhelmingly determined by a few extremists.[b]

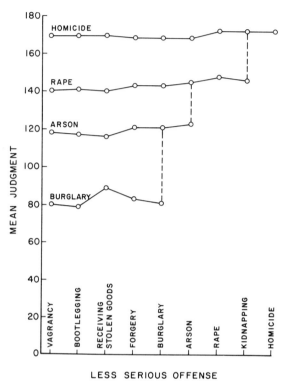

FIGURE 4.19. *Configural weighting in moral judgment. Subjects judge badness of group of three criminals, one guilty of offense listed as curve parameter, two guilty of offense listed on horizontal axis. Near-horizontal shape of curves implies that judgment is almost totally determined by badness of worst member of group, the two others having virtually no effect. (After Leon, Oden, & Anderson, 1973.)*

Self-Estimation of Weight. One of the more important problems facing integration theory, especially in naturalistic design, is the development of methods for obtaining self-estimates of the weight parameter. Two relevant sets of data will be discussed here.

The first set of data has already been presented in the lower half of Figure 4.14. Panel C showed the person impressions and Panel C' showed the corresponding importance ratings of the separate adjectives in each description. All descriptions had four adjectives, and each letter symbol in Panel C and on the horizontal axis of Panel C' stood for two adjectives of the specified value. After subjects had judged personal likableness, they rated importance of one or all of the adjectives in their likableness judgment.

Present concern is with the importance ratings in Panel C'. Comparison of the L and M⁻ curves shows an extremity effect, with the more negative L adjectives being rated as more important. This extremity effect is consistent with the deviation from parallelism in the lower part of Panel C, as discussed in Section 4.4.2, and thereby provides some support for the use of importance ratings as weight estimates.

A similar but smaller extremity effect appears in the comparison of the H and M^+ curves. Accordingly, a similar deviation from parallelism might be expected in the impression data in the upper part of Panel C. However, a power calculation in the original report indicated that the observed difference in importance ratings was too small to produce a detectable deviation from parallelism in the impression data.[c]

In another part of this experiment, the same subjects judged likableness of 32 person descriptions in a 4 × 8 design, with two, three, four, or six "equally accurate" adjectives all at one of eight graded levels of likableness. Following this likableness judgment, subjects rated "how important one (each) trait is in your impression of the particular person described by all the traits on that slip." The 40 subjects in Condition One rated one random adjective in each description; the 20 subjects in Condition All rated all adjectives in each description.[d]

The main results are in Figure 4.20. The left panel plots mean importance rating as a function of adjective value. The bow shapes reflect an extremity effect, with more extreme adjectives receiving higher ratings, both for positive and for negative adjectives. These data show the same pattern as Panel C' in Figure 4.14, including the positive–negative

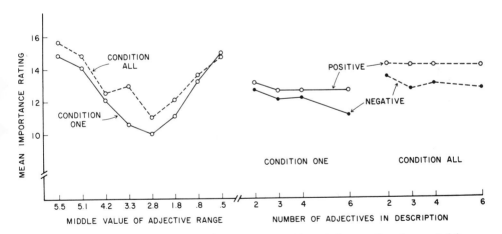

FIGURE 4.20. *Rated importance of single adjectives in likableness judgments. Bowed curves in left panel show that more extreme adjectives, positive or negative, are rated more important. Values on horizontal axis are middle value of short subrange of adjectives from master list with normative values ranging from 0 (bad) to 6 (good). Right panels show that rated importance is nearly independent of the number of adjectives in the description. Subjects in Conditions One and All rated importance of one or all adjectives in each description, respectively.*

asymmetry. Also of interest is the similarity in pattern between Conditions One and All, which suggests that obtaining the extra judgments does not cause serious bias.

The two right panels of Figure 4.20 plot importance rating as a function of number of adjectives in the description. These data show near-constancy, independent of set size. Since relative importance would decrease with set size, these data suggest that subjects judge absolute importance, an optimistic note for the study of self-estimation procedures.

4.5 Further Work

The basic experiments within which information integration theory was initially developed have formed the main content of Chapters 2-4. A few simple theoretical ideas have provided a good account of a complex array of data. However, this initial work was limited by its concentration on one task of person perception. This task has many advantages (Section 1.8.4), and concentration on it may be justified on the grounds that a closely interlocking data base is important in theory construction. Nevertheless, the value of the theory depends largely on its generality beyond this particular task. The next section looks at beginnings that have been made in various areas. Three additional sections discuss new directions that appear attractive.[a]

4.5.1 GENERALITY

Integration theory has been applied in a variety of areas, as illustrated in Chapter 1. No attempt will be made to review these studies here (see Anderson, 1968b, 1974a,b,d,e, 1976d, 1978b, 1980, 1981b), but a glance at the spectrum of applications is appropriate to the question of theoretical generality.[a]

Social Judgment. Substantial work has been done in most areas of social psychology, especially in attitudes, attribution, and moral judgment. Formation of attitudes has been conceptualized as informational learning, although this has been complicated by the operation of separate basal and surface components (Section 2.5.4) and by more than one memory system (Section 4.2). In attribution, the focus has been on multiple causation, which has led to an approach based on integration rules as causal schemes. Work on moral judgment and equity has found substantial evidence for a moral algebra, together with some results on its development in children. In addition, a few studies have been done in group dynamics.[a] Oddly enough, the assessment of generality in the base area of person perception by extensions to more complex and more realistic stimuli and settings has been neglected, in part because theoretical controversy became centered on the personality adjective task. Altogether, however, what has been accomplished may reasonably be offered as a unified foundation for a general theory of social judgment.

Psychological Decision Theory. Integration theory has provided a new basis for psychological decision theory, in large part the work of James Shanteau. A primary characteristic is the shift away from traditional normative and axiomatic approaches to a descriptive approach oriented toward cognitive theory (Section 1.8.1). An essential element in this shift has been the development of functional measurement methodology and its systematic experimental application. Algebraic models have been popular in decision theory, but they could not have become cognitive algebra without a capability for psychological measurement nor without systematic experimental analysis. The integration-theoretical approach may be epitomized by its solution to the classical problem of joint measurement of subjective probability and value (see Section 1.5.1). Substantial work has also been done on a variety of related issues, especially on causal inference, and the outcome compares favorably with other theoretical formulations.

Psychophysics. Considerable work has been done in psychophysics, largely oriented around the traditional question of the psychophysical law. An integration approach leads to a conceptual shift in focus of investigation—from the psychophysical law to the psychological law. In this shift of perspective, the psychophysical law loses much of its presumptive pertinence, and interest moves toward substantive problems of perception (Section 5.4).

Psycholinguistics. Various problems in psycholinguistics fall naturally into an integration-theoretical framework. Indeed, the personality adjective task can be viewed as essentially psycholinguistic, as can the problem of meaning constancy considered in Chapter 3. Also interesting are problems such as class belongingness and truth value, which have traditionally been represented in discrete propositional terms but have increasingly been recognized to require continuous semantic representations. Especially in the hands of Gregg Oden, integration theory has provided rigorous analysis of continuous semantic variables that has received good empirical support. Among other accomplishments,

Oden's fuzzy-propositional approach has been able to convert normative fuzzy logic into true cognitive algebra (Sections 1.8.2 and 1.8.3).

Developmental. The integration-theoretical approach has been very
successful with children, as shown in the finding that children have
markedly more cognitive ability in typical integration tasks than had
previously been recognized (see references in Sections 1.3.5 and 1.3.6).
Besides high integrational capacity, both a general purpose metric sense
and a general purpose adding-type rule appear to be present at least as
early as 4 years of age. Moreover, integration tasks reveal that young
children are sensitive to a multiplicity of stimulus cues in both their
physical and social environments. Besides their intrinsic interest, such
developmental studies provide evidence on the nature of cognition that
makes this work among the most exciting now being pursued.

4.5.2 MARRIAGE AS AN INVESTIGATIONAL SETTING

Marriage and the family, the most important of human institutions,
have been pretty well ignored in the mainstream of psychology, even in
the social personality area. Yet marriage and family offer notable advantages for psychological investigation. This section sketches some of
these advantages and suggests how information integration theory might
be adapted to obtain useful experimental analysis.

Investigational Advantages. The marriage setting constitutes a sociopsychological microcosm in which person perception is most prominent.
Some couples remain superficial acquaintances, a fact as striking as the
intimate mutual understandings that develop in other couples. Included
within spouse perception is causal attribution—of mood and motivation,
for example—about states of the spouse that are considered to account
for his or her behavior. Numerous moral issues also arise, including
fairness in division of household chores, treatment of relatives, and
varied problems in raising children. Attitude functioning and attitude
change are central in marital adjustment as both partners come under
pressure to adapt and modify their social activities and habits of food
and cleanliness. Above all, the family is a small group and provides a
prime opportunity for studying group dynamics. Everything of interest
to a social psychologist is present in the marriage setting.

Many nonsocial areas are also important in marriage. Most prominent
is decision making, especially in the often sensitive area of money. Instances range from daily consumer purchases for food, clothing, and
recreation to general issues of earning and spending. Learning is ecologically central, as reflected in various adjustment processes with spouse

and children. And, of course, many basic aspects of human motivation, from loving to quarreling to making up, are primarily operative within marriage.

The practical relevance of marriage studies hardly needs emphasis; they get out of the laboratory cubicle into a natural setting that has basic importance in the life of every person. Not less important are the theoretical advantages of a sociopsychological microcosm for both social psychology and general cognitive analysis. The person perception task studied in Chapters 2–4 has the advantage of tapping into cognitive skills and abilities already developed in daily life. Similar opportunities to study ecologically important and well-developed aspects of cognition are prominent in the marriage setting. Quite aside from issues of outcome generality, therefore, the marriage setting has advantages for process generality.

As these considerations show, marriage offers a remarkably diverse array of interrelated opportunities for experimental study. This variety and flexibility, coupled with social and theoretical relevance, make marriage an "endless frontier" for social personality psychology.

Embedding Method. The paucity of experimental studies of marriage stems in part from difficulties of obtaining experimental control within natural settings. The traditional method of control group–experimental group is limited in both practicability and appropriateness. Although this method can be useful for studies of social action, it lacks the precision that is desirable for analysis of social perception and cognition.

An embedding method may provide a useful alternative. The idea is to embed a general experimental design within the particular context of each individual marriage. That will often require stimulus preselection for husband and wife in such a way that the general design is realized in terms of personalized stimuli that are meaningful to the individuals. In studies of fairness in division of household chores, for example, the specific chores used as factor levels would be chosen to be meaningful and relevant within each household. The judgments are not abstract or nominal, therefore, but personally meaningful within the experiential background of each individual. Such stimulus preselection can increase both statistical efficiency and substantive validity of the investigation.

Among the advantages of the embedding method are that it facilitates within-subject design and the study of steady-state behavior. Between-subject experiments can suffer because subjects may get only half-adapted to the task, owing to lack of varied experience or time. Constraints from politeness or "company manners" might be present initially, for example, that would adapt out with more experience. Although such transient effects have their own interest, they would often constitute extraneous confounds (e.g., Anderson & Butzin, 1978, p. 604). Much behavior in natural situations, including marriage, is more or less habitual and represents well-learned skills operating in a steady-state environment. For the study of such behavior, the embedding method seems useful if not necessary.

A seeming limitation on the embedding method is that the judgments are to some degree hypothetical and may involve a certain amount of role playing. But many marital activities, such as planning a dinner party or purchasing a gift, involve major hypothetical components at the time the decisions are made. Moreover, present concern is less with outcome, which will certainly depend on the total context of everyday life, and more with process, which may be expected to have some generality (Section 1.8.2). If the embedding taps into some existing values and cognitive skills of each individual, it will be useful.

The integration approach is well suited to the embedding method because it allows for personal values within its focus on integration processes. Both the general theory and the various methods and procedures become applicable to natural settings. Where this approach can be employed, it can provide results that may not be otherwise obtainable.

Illustrative Design. The following design illustrates a general marital interaction task as it might be applied to moral judgments in child rearing. Husband and wife each get certain information, not necessarily the same, about a child who committed a harmful act. Based on only their own information, each makes initial judgments about how undesirable, naughty, and deserving of punishment the child's action was. After a round of such initial judgments, each case would be reconsidered, with an exchange of information and opinion and a revision of the initial judgment.

Many questions can be studied within this design. The set of initial responses allows assessment of moral algebra and moral values individually for husband and wife. The revised responses would reveal changes that might be produced by the information exchange and interaction. At the same time, comparison of the initial and revised responses would reflect mutual influence power.

An attractive starting place would be the intent–damage question, in which information is given about the intention behind the child's act and the damage caused by the act. Existing evidence points to a moral algebra (Leon, 1980). Moreover, moral issues may be expected to evoke considerable differences of opinion and hence also considerable husband–wife interaction.

In any actual study, one main design decision would concern the degree of reality of the stimulus children. That could range from the generalized information that has characterized previous work on the intent–damage question to realistic case histories to an embedding of one of the couple's own children.

This general design could be used in many ways. The study of moral judgment could be extended in several directions to questions of inter-personal and social attitudes and actions. Alternatively, the focus could be changed to consider decision making in household expenditures. Various forms of information exchange could be used to analyze interaction processes on a within-couple basis. Also important is the potential for priming husband and wife with different initial information, which could often be done in a natural way to provide a measure of experimental control. An initial experiment is described in Note 4.5.2a.

4.5.3 INTUITIVE PHYSICS

The process of learning and understanding physical relations offers many opportunities for psychological research. One line of inquiry concerns the correspondence, or lack thereof, between intuitive physics and nature. People get extensive experience with motions and collisions of objects in their environment, and some degree of understanding them is prerequisite to their continued existence. Moreover, humans' phylogenetic legacy includes great ability in many perceptual–motor tasks, as illustrated in various games with balls. Some internalization of the dynamics of the physical world clearly occurs, but not much is known about whether the laws of intuitive physics mirror the laws of actual physics or follow some logic of their own.

For purposes of illustration, consider a pendulum mass that is released from a point on its arc and strikes a ball, driving it up an inclined plane. The mass and release angle of the pendulum are varied in a two-way factorial design, and the subject predicts how far up the incline the ball will roll. Many other stimulus variables and many similar tasks could be used, but this example will illustrate some main points.

Predictive Principles. Subjects can make sensible predictions in such tasks, based on their pre-experimental knowledge. This background knowledge is extensive, but rarely will it be directly related to the given task. Rather, it will consist of bits and pieces of information about components of the physical event. In this example, there are three main components, namely, the falling pendulum, the collision, and the uprolling ball. One theoretical problem is to delineate the elements of pre-experimental knowledge and determine how they are put together to obtain a prediction.

Three qualitative knowledge principles may be hypothesized for the given task. The first is an obvious monotonicity principle, which states that the outcome is a monotone function of each stimulus variable. The second is a shape principle, which states that the response curve has one of a few basic shapes, especially straight, bowed up, or bowed down. In addition to these one-variable principles, a two-variable principle is also needed. This is a pattern principle, which states that the factorial graph has one of a few basic patterns, especially parallel, divergent, or barrel-shaped.

These qualitative principles may capture a major part of the prediction process but they need quantitative supplements, including slope and curvature parameters. Fortunately, the graphs of subjects' responses provide visible representations of their pre-experimental knowledge. The distance response seems likely to be a linear scale, so adding and multiplying rules, in particular, will manifest themselves as parallelism or as linear fan patterns in the data. Since many physical laws have algebraic forms, it would be of interest to study to what extent they may have been internalized during past experience.

These three principles are linked in a more primitive collision scheme that connects the three components of the physical event in a causal sequence. One attraction of intuitive physics is the opportunity to study such causal schemes. In this task, the causal scheme directs the assemblage of background knowledge that is utilized in the valuation operations associated with each component of the physical event. This task may provide a useful, general paradigm for schema analysis.

Learning. Physical prediction tasks also present interesting problems for learning studies. Nature herself imposes a standard of correctness, whereas traditional learning tasks, from the nonsense syllable to the operant bar, impose a correct response that is typically arbitrary and unrelated to subjects' prior experience. Moreover, stimulus and response variables are typically continuous in physical tasks, typically discrete in traditional learning tasks. There is a corresponding difference in what is learned: functional relations and rules in the one case, discrete associations in the other. Even the rules of concept studies typically involve arbitrary, discrete classifications. To a considerable degree, moreover, these characteristics of traditional learning have been carried over into modern information-processing theory (Section 1.8). Physical prediction tasks thus provide an interesting complement to traditional approaches to learning and transfer.

Four loci of learning may be distinguished. One concerns the underlying causal scheme already discussed. Two concern function shapes of the one-variable and two-variable functions that correspond to the valuation and integration operations, respectively. In addition, there are various calibrations associated with stimulus weighting parameters and with the unit of the response scale. In particular, the subject's pre-experimental rule could exhibit the correct pattern but be off by a multiplicative constant corresponding to the unit of the response scale. These three aspects of learning need not be independent, but they may be affected differently by informational feedback. For example, a single physical observation could markedly affect calibration without affecting function shape.

Because of the central role of functional relations in intuitive physics, traditional learning procedures are not always appropriate. Accuracy is the traditional standard, but it tells little about response pattern. Response pattern can exhibit correct functional relations even though each individual response is seriously in error. For tasks involving functional relations, therefore, the normative accuracy criterion may be uninformative and even misleading (Anderson, 1968c).

Similarly standard, trial-wise knowledge of results or even event observation may not be too helpful in learning functional relations. Intuitive physics dating back to Aristotle says that the heavier ball will fall faster, an inference that relates naturally to the monotonicity principle noted above. Nineteen centuries of observation did not correct this error, for otherwise there would have been no need for Galileo's experiment on the leaning tower of Pisa. Galileo's method of pair comparison illustrates one useful kind of informer, but others that facilitate acquisition and transfer of functional relations need to be developed.

Response Functions. Intuitive physics has the further attraction that the response need not be restricted to point predictions. Instead, the response may be explicitly a function of one or more stimulus variables. The subject could be given a sheet of graph paper and instructed to draw a predicted curve as a function of one variable. For example, subjects could be asked to draw a curve to show how far the ball would roll up the incline as a function of the angle at which the pendulum is released. This response procedure could be extended to have subjects draw a complete factorial graph with separate curves for various pendulum masses.

In this visible form, the response process itself would be expected to assist subjects in utilizing their bits and pieces of pre-experimental background knowledge. This mode of learning could be expanded to include a Socratic element—for example, by asking the subject whether the curve should be straight, bowed up, or bowed down. Such directed questioning could help to develop or firm up subjects' background knowledge. An initial experiment is described in Note 4.5.3a.

4.5.4 COGNITIVE STRUCTURE AND PROCESS

Whereas the two preceding sections were concerned with new experimental areas, this final section returns to standing theoretical issues that require more extensive study. The first three subsections consider some processing questions for the three basic operations. The last three subsections point up some relations of integration studies with motivation, learning, and memory.

Operations. Molar analysis offers the possibility of exact theory independent of finer process details (Sections 1.1.5 and 1.6.5). This is a great advantage, for it bypasses many intricate problems. Once a molar model is established, however, it leads on to more-detailed process questions that have as yet received little study. This section offers a few suggestions about the three basic operations, namely, valuation, integration, and responding.

Valuation operations involve a variety of processes, foremost among which are inferences. The following examples with the personality adjective task illustrate some bases for inference. First, of course, is common meaning, as in the *generous* person being *unselfish* or, less directly, in the *dependable* person being *punctual.* However, common meaning does not seem to account for the *sensible* person being *friendly.* This inference may involve an instrumentality relation, in that being friendly is sensible behavior. Related to this is presupposition, as in the *confident* person being *intelligent*—because unintelligent persons are unlikely to possess confidence. Other inference rationales include halos, as in the *handsome* person being *intelligent,* and stereotypes, as in the status value of speech accent. Finally, the *tall* person being *heavy* illustrates the important case of empirical association, the two terms being correlated across persons. In practice, most inferences would presumably result from joint action of several such processes so their experimental analysis does not seem overly inviting.

An alternative approach is to suppose that the stimulus and response adjectives both set up an image or schema of a person and that the inference results from a similarity comparison between the two. This allows the various inference subprocesses to be treated as a molar unit, thereby avoiding sticky problems of attempting to isolate and verify what may, in part, be merely rationalizations for a more basic judgment of similarity. The similarity averaging formulation of Section 5.6.5 would have two advantages. First, it would allow for the effects of a super-posed schema, as when age, sex, or occupation of the person is specified. Second, it can account for stimulus–response asymmetry of inference, as in the *crafty* person being *intelligent* but the *intelligent* person being *not crafty*.

The network of inferential relations among the adjectives and other stimuli that can describe a person is known as *implicit personality theory.* Similar inference networks appear in other domains, but not a great deal is known about their structure. Multidimensional scaling, which might seem well suited for structural analysis, appears to have difficulties even with the fact of asymmetry. By recasting the task in an integrational format, it becomes possible, in principle, to get functional scales of the inferences that would be helpful in mapping out cognitive structure (Sections 1.8.3 and 3.4.2).

Another class of processes is also required in valuation, namely, comparative judgments with reference standards. The scale value of a stimulus may depend on its similarity to or distance from certain com-parison stimuli. End anchors are a prime example, but other comparison standards are no less important. A small car is larger than a large cat because such relative adjectives are evaluated with respect to compari-son standards specific to the dimension of judgment defined by each noun. A similar phenomenon appears also with nonrelative adjectives, as with *happy-go-lucky*, which may assume positive or negative values depending on the situation. In social judgment, in particular, the same physical stimulus may have different values, depending on age, sex, occupation, and many other factors.

Similarity judgments play a basic role in valuation, as in the compari-sons with standards or prototypes already discussed. Two standard conceptualizations of similarity are in terms of distance in a dimensional space and in terms of common and noncommon sets of features. How-ever, intensive study of similarity judgments has mainly shown their complexity and general theory seems elusive (see Gregson, 1975). Judgment-decision theory requires task operations, moreover, that do not seem amenable to standard dimensional and set-theoretical formula-tions. This issue is pursued in Section 5.6.5, which also presents a new approach to similarity judgments.

For integration operations, it is intuitively clear that typical tasks do not involve the mental arithmetic represented in the formal model. Analog processes are plausible, but the same formal model can be em-bodied in more than one process. Averaging (arithmetic mean) is in-teresting because it can be accomplished by diverse processes, each of which may have psychological reality in certain tasks (Anderson, 1974d, p. 253).

Mixing substances illustrates one averaging process. Each stimulus would correspond to a beaker of substance with its weight parameter represented by volume or number of elements. The elements themselves may be positive, negative, and neutral, and their net concentration represents the scale value of the beaker. Mixing the contents of several beakers together yields a mixture with the average concentration.

Mixing can be considered as a general integration process. For the purposes of psychology, each stimulus would be represented as a set of psychological elements or features, with mixing represented by joint activation of the several sets of features. In the personality adjective task, the features of each adjective would be its elemental meanings and implications. Representations of judgments about a known person would include an internal stimulus field representing prior integrations of prior information.

A different averaging process is obtained by representing each stimulus as a point mass on a massless plank. Weight and scale value correspond to mass and location along the plank, respectively, and the average corresponds to the balance point. This process has been studied experimentally by asking subjects to estimate center of gravity for a row of dots (Anderson, 1968c). Something similar may be involved in weighing evidence or striking a balance among the pros and cons.

A third averaging process appears in serial integration, in which subjects adjust their response after each additional piece of information. Each successive response may then be represented as an average of the prior response and the present information (Section 2.5). The purest form of this process appears in number averaging, as in the experiment shown in Figure 2.10. Of course, serial integration may be employed as a strategy for handling larger masses of information presented simultaneously, with processing order determined by attention and salience.

A closely related issue concerns two processing modes that may be denoted V-I and and I-V. In the V-I mode, the valuation operation precedes integration. This mode may be expected with serial integration, in which the dimension of judgment is specified beforehand and the information is given to be integrated one piece at a time. An obvious procedure would be to evaluate each stimulus as it is received and update the cumulative resultant, as in number averaging. The evidence suggests that this processing mode may also operate in the personality adjective task.

In the I-V mode, the integration operation precedes or mediates the valuation operation. This mode seems likely when the nature of the response is not specified beforehand. In the personality adjective task, for example, subjects could be asked to write paragraphs describing the hypothetical person in their own words before being informed about the dimension of judgment. The I-V mode also seems likely when dealing with familiar concepts and issues. Judgments about a friend might be expected to follow the I-V mode because an internal representation of the friend already exists as an integrated resultant of previous information.

The I-V mode allows the resultant of the integration to be a general knowledge structure, like an attitude in the distinction between attitude and attitudinal judgment of Section 1.8.1. The initial integrated resultant would be an attitude in the sense of a knowledge structure (Section 1.8.2), including features and structural relations among features. Attitudinal judgments would then be obtained by evaluation of the structure along specified task dimensions (Anderson, 1967b, p. 164, 1971a, p. 173).

The I-V mode poses the question whether the valuation operation occurs at some feature level or holistically. Valuation at a feature level would entail a secondary, response-specific integration of the values of the separate features. Holistic evaluation might be obtained on the basis of a general impression or in terms of structural units, as perhaps in face perception. In practice, hybrid valuation might be expected with response based on a general impression augmented by various particulars.

Feature-type models have been popular in theories of semantic memory and of similarity judgment, but judgment-decision theory may require different conceptions of features than has been typical of work in those areas. The evidence for two memory systems (Section 4.2) indicates that features cannot generally be identified with natural semantic representations of the stimuli. Failure to obtain expected effects of ostensible inconsistency and redundancy in person perception demonstrates similarly that the feature and network representations commonly employed in semantic memory may not be too useful in judgment-decision theory. A different problem is posed by source reliability, which is not an autonomous feature, but a weighting operator on some other feature. The need for this and other cognitive operators, which process the contents of more passive memory systems, is characteristic of judgment-decision theory (Sections 1.8.3 and 3.6.1).

Adding and subtracting mechanisms have not received much attention, in part because these two rules have been empirically infrequent. They have intrinsic interest, however, and may have considerable importance if the general purpose adding-type rule discussed in Section

1.3.6 involves adding rather than averaging. Serial integration, already discussed for the averaging rule, suggests a natural analog mechanism for adding and subtracting. For these two rules, in contrast to averaging, the response adjustment produced by each successive piece of information would be independent of amount of prior information.

Next to averaging, the multiplying operation has been most frequently observed in integration studies. At least some of these instances, however, illustrate "as-if" rules. For example, judging area of a rectangle seemingly does not involve any multiplication of base and height, but rather a perceptual addition of unit areas. Other cases in which an adding rule can produce a linear fan form appear in Shanteau's reaction time model for memory search (Section 1.5.7) and in Wilkening's (1981) eye movement strategy for judgments of travel distance based on information about travel time and velocity. The Adverb × Adjective model of Section 1.5.4 carries this point further, for the underlying processes may be unrelated to any algebraic rule.

However, the following analog process for multiplying seems likely to have psychological reality. Each level of one factor is represented as a distance along a graphic scale. Each level of the other factor is represented as a proportion or fraction operating on those distances. Multiplication may thus be accomplished by location of an initial distance and fractionation of that distance. Besides its psychological plausibility, this process has statistical interest, for it implies that response variability will be approximately constant along the scale, in agreement with reported results (Anderson & Shanteau, 1970).[a]

Finally, processes underlying the response operation also need consideration. Numerical responding, in particular, is a remarkable and extremely useful ability that is not well understood. One clue appears in the concept of general purpose metric sense (Section 1.1.6) that arose in the study of children's judgment. However, this internal metric must be adapted to the operative dimension of judgment and calibrated to the operative range of stimuli and to the range of response. For the rating response the end anchors play a key role. Indeed, a simple response model may be obtained by treating the response to any given stimulus as an average of the end-anchor values with weights that reflect similarity of the given stimulus to the corresponding anchors. The output process could then be realized as a graphical analog device much like that used in various experiments.

Response systems other that ratings also need to be considered. Examples of other metric responses are given in Sections 1.3.8, 1.5.2, 1.5.5, and 1.5.7 but there are many others, especially various behavioral measures common in animal research and in developmental psychology. Also of interest are motor skills, including perceptual and muscular factors exemplified in catching and throwing a ball. In addition, there is the important class of nonmetric responses, for example, writing a paragraph to describe the person in the personality adjective task. However, the study of integration processes in such tasks will require substantially new theoretical developments.[b]

Causal Inference. Valuation and integration constitute two major classes of inference processes. Valuational inference was considered in the previous subsection, but integration deserves more specific comment, especially with respect to the important problem of causal inference. Since the relations between integration theory and Bayesian theory in decision making have been discussed elsewhere, the present section will consider social judgment in which causal inference is also important.

"Why did the person act that way?" is a frequent question in daily life and it epitomizes much of social attribution theory. Answers in daily life, and in many attribution theories, are typically either–or, attributing the action either to a disposition of the person or to a force in the environment. Person and environment are joint causes, of course, and it seems unlikely that either–or responses adequately reflect the answerer's processing. Instead, each cause will have a strength that is an integrated resultant of prevailing information; the overt response itself results from a further integration of these causes. Attribution is thus information integration. However, more detailed analysis must face the two problems of categorical response and communicational constraints.

Categorical response can be handled in terms of a decision criterion. Any given cause must be above some minimum strength in order to be seen as sufficient to produce the action and so be acceptable as an answer to the question. When one cause is weak, the other strong, this amounts to picking the likely cause. When both causes are weak, one may need to be imputed. When both causes are strong, of course, the answer may include both. By use of decision criteria, an analysis in terms of response strength may be able to account for categorical response (Anderson, 1974a, pp. 246–249, in press, Section 3.10).

Communicational constraints are important both in questions and answers. An answerer is under constraint to convey information that is desired or needed by the asker and to avoid conveying information that is already known. Thus, the response strengths of the various potential answers depend on assumptions, explicit and implicit, about the asker's purpose and goal and about what the asker already knows (Anderson, 1974b, 1978b).[c]

Questions may, of course, have social functions other than conveying information. In general, however, answers depend not only on the specific statement of the question, but also on various context factors and on a complex of background knowledge. Although context and background can never be known exactly, they can be exactly taken into account by virtue of the unitization principle. Development of a cognitive algebra of causal inference should thus be useful for analysis of context and background.

Cognitive Organization. Stimulus interactions are interesting as cognitive processes and as reflections of cognitive organization. Assumptions about interactions have been popular, but solid evidence can be hard to get. Often the assumed effects do not actually exist; when they do, they often have alternative, noninteractive interpretations (e.g., Sections 1.6.6, 3.2, 3.3, and 4.1.2). The difficulty of pinning down interactions and organizing processes does not lessen their interest, however, and two related directions for further study will be noted.

One direction involves tasks that require organizing processes, such as field construction tasks. For example, subjects could be presented with a field of information, some of which is presumed to be unnecessary, unreliable, or even incorrect, and instructed to select a subfield that best fulfils some function. In the personality adjective task, for example, the subject could be given a set of adjectives and told to select a subset that best describe a person, as illustrated in the total discounting condition of Section 3.4.1. Similarly, the subject could be told to use some or all of the given adjectives to construct descriptions of two persons. This construction task could be extended in various ways; for example, by including photographs or other nonverbal information.

This task requires the subject to organize the information, using the interrelations among the field elements to construct subfields that satisfy certain criteria. The flexibility in choice of field elements facilitates the isolation of various organizational processes. Numerous task variables can be manipulated to bring out the operation of various organizing processes; for example, organization based on redundancy or inconsistency. Specification of a base field to serve as a nucleus for the organizational processes can provide a helpful measure of experimental control (Section 3.4.2).

This task requires the formation of organized wholes, the central concern of the gestalt movement. For visual perception, gestalt psychologists have discussed organizing principles such as closure, persistence, and constancy (Boring, 1942, pp. 252ff). Analogous organizing principles are readily hypothesized for the personality adjective task. Most obvious is the well-known principle of *consistency*, which avoids inclusion of contradictory elements. More general is the principle of *homogeneity*, which avoids inclusion of outliers and acts to decrease variance in the subfield. Also relevant is a principle of *informativeness* that requires that an added element should add new information and avoid redundancy. For person descriptions, at least, there is also a principle of *completeness* that requires that a person have some breadth, for which purpose an unrelated element might be chosen.

Field construction tasks seem generally suitable for applying judgment-decision theory to the study of cognitive organization. Principles of organization will appear in the relations of the various parts to the whole, and numerical response scales can yield direct measures of these relations. Integrated responses to the whole field can provide additional evidence on the function of the various parts (e.g., Section 3.4.1). The same methods that have been employed in the present work thus become applicable to the study of organizing processes.

The other direction involves complex stimulus fields, which present problems of organizing the sequential processing of the information. One problem of this kind has already appeared in the primacy–recency studies on effect of presentation order (Section 3.2). Although results to date point to noninteractive integration, more interactive processes could no doubt be obtained by manipulating information that affects the dimension of judgment. For example, information about age, sex, or occupation in the personality adjective task would control the valuation operation. If such information is not given initially, some working assumption may be required to allow the valuation of the adjectives to proceed. When the information is finally received, it may be too late to allow reevaluation, since the verbal materials may be lost (Section 4.2).

More generally, some organizing hypothesis may be needed to guide the search for and selection of information from complex stimulus fields. Not all possible information will be processed, of course, for some cost is always involved even when the information is readily available. The information selected will, accordingly, depend on the person's initial state. Different persons in the same situation will face effectively different stimulus fields, therefore, and analysis can be difficult without knowing the effective field. To some degree, this problem can be bypassed with the unitization principle, which allows for individual differences in valuation. This approach can provide boundary constraints on theories about the organizing process (Section 1.8.3), but it does not address the question of detailed process structure.

This class of tasks has been largely avoided in previous work on integration theory for two reasons, one associated with attentional factors, the other with the use of numerical response measures. The strong influence of attentional factors, as illustrated in the primacy–recency studies (Section 3.3), points to the usefulness of experimental control over the effective informational field. That was a major consideration in the initial work, which was concerned with developing linear response scales. Possession of methods for obtaining linear response scales can greatly facilitate the study of interactive processes (Section 1.2.6). Although presently available response methods need continuing study, they seem adequate for pursuing these substantive problems.

Problems of information search have been considered in many areas of psychology, most extensively in decision making. Also notable is the acquaintanceship process, whether in interviews, casual social interaction, or marriage. As Argyle (1967) emphasizes, social interaction is a feedback process in which incoming information continually modifies the momentary direction of behavior. The injunction to "put your best foot forward" reflects the fact that, in getting to know another person, initial information will guide the remainder of the information search. These problems will not be considered here, but they touch on important substantive areas within which to pursue the study of cognitive organization.

Learning. Information integration involves modification of behavior through experience and so falls within customary definitions of learning. However, traditional learning tasks, which typically involve repeated trials with the same stimulus or class of stimuli, are rather different from the tasks that have characterized integration studies. Learning about issues or persons, for example, even in the simple tasks of serial integration in Section 2.5, requires a more cognitive approach than has been typical of learning theory.

This difference appears in the distinction between what is learned and the actual response. In the stimulus–response view, the learned response is just a latent form of the actual response. In the integration-theoretical view, this distinction parallels that made between attitude as cognitive structure and attitudinal judgment as a specific response in Section 1.8.2. What is learned is a knowledge structure. The observed behavior results from interaction between this knowledge structure and the prevailing situation, which includes internal motivational components and external stimulus cues and constraints. These situational factors set up valuation and integration operations that process the knowledge structure to produce the behavior.

This view is similar in spirit to that held by Tolman and other, more modern cognitive theorists, but it has the advantage of standing on a

firm base in judgment theory. Because of this involvement with judgment tasks, however, learning as information integration has received
little systematic attention. Some studies have been made in probability
learning, in which the outcome has been a shift from an associationistic
conditioning view to a cognitive view (Anderson, 1974d, pp. 266–267).
Some studies have also been done on various tasks of serial integration.
As indicated by the discussions of intuitive physics, motivation, and
memory, however, there are many opportunities for new studies of
learning.

Motivation. A functional view of motivation is implicit in integration
theory, which is concerned with motivation as it becomes manifest in
behavior. Two functions may be discerned. One is general, embodied
within the direction of activity in relation to a goal. The other is specific,
appearing in the valuation processes associated with various aspects of
the goal and with various courses of action to reach that goal.

This approach takes cognizance of the multiplicity of motivation.
Consider going out to dinner. The general goal arises from motivational
habits connected with time of day and day of week, from various social
pressures and, perhaps, from hunger. Among the relevant motivations
in selection of a specific restaurant are taste preferences, diet, and
novelty, together with additional habits and social influences as well as
ancillary motivations from cost and time constraints. In selecting from
the menu, a similar aggregate of motivations is operative.

Such multiplicity of motivation seems characteristic of much human
behavior. Accordingly, the study of motivation is bound up with problems of multiple causation or, in present terms, information integration.
With this integration view goes a corresponding conceptual outlook.
Typologies of motivation lose much of their presumptive importance
because motivation is embodied within a complex of response patterns
that are unlikely to exhibit any clear elemental structure. It remains
useful to speak of motivations to blame and to avoid blame, for example,
but these need not be considered as more than provisional labels for
interesting areas of investigation; less explanations than what needs to
be explained. Similarly, drive theories recede in importance because
there is only indirect, limited linkage between need states and
satisfiers–annoyers, which are central in learning, and motivation,
which is embodied within aggregates of response patterns.

The present view has similarities with cognitive theories of motivation, especially in its emphasis on informational processes. It also has
kinship with judgment-decision theory, especially in the close relationship between motives and values. Indeed, the good–bad evaluative ele-

ment that is so pervasive in human judgment may derive directly from the goal-oriented nature of behavior.[d] However, previous approaches have lacked a capability for analysis of multiple causation, which, by the above reasoning, is basic to the study of motivation. A capability for handling multiple causation thus offers a new perspective on the psychology of motivation.

Memory. The constructive view discussed in Section 1.8.2 reflects the basic role of memory in judgment theory. Judgments on a somewhat unfamiliar issue, for example, will depend on memory access to prior values and attitudes and to particular incidents and events. Even the valuation of stimuli in the immediate external stimulus field generally requires reference to long-term memory.

Principles of memory that have emerged from the research on information integration include the finding of Section 4.2 that words and their meanings may be stored in different memories. This outcome points to a certain complexity in the structure of memory. It must include verbal symbols, their meanings and implications, values, and partially formed judgments and attitudes on a variety of issues. Both storage and retrieval, moreover, will depend on the operative dimension of judgment.

Also of interest is the finding that attitudes may have two qualitatively distinct components, namely, a labile surface component and a semi-permanent basal component (Section 2.5.4). These may perhaps correspond to short-term and long-term memory, although the merit of this analogy has not been explored. In any case, the result itself appears to represent an important distinction in the structure of memory as it relates to judgment.

An additional principle involves the concept of what may be called an operating memory for performance of judgmental operations. The immediate origin of this concept lies in the finding that valuation and integration are independent in some important cases (Section 3.6.1). It follows that the integration operations occur elsewhere than in long-term memory.

Perhaps the same conclusion holds for the valuation operation. Valuation must operate on long-term memory to obtain the psychological values of the physical stimuli, but only in a few well-practiced tasks is this likely to require simple retrieval of stored values. In general, a constructive process is required that involves effects of context on the dimension of judgment as well as comparison processes within that dimension. This constructive process may also require an operating memory that is distinct from long-term memory.

Some problems in relating judgment to memory can be illustrated with the concept of redundancy. Repetition has been the premier independent variable in traditional memory studies, in which it has strong effects. In standard judgmental tasks, however, repetition typically has small effects; most of the relevant information is extracted the first time, and subsequent presentations are discounted. Accordingly, the repetitions that are so important for strengthening verbal associations are not so interesting in judgment research.

In a different way, however, memory is basic to the operation of redundancy. To be able to discount repeated information, it is necessary to recognize the repetition. This requires reference to memory, whether for verbal material or for meanings.

To complicate the analysis, redundancy cannot be conceptualized solely in terms of semantic or inferential similarity. Source factors are also involved. The identical message will generally have more effect from two different sources than repeated twice by a single source. This and other source effects require judgmental operations that are distinct from memory itself.

Judgment research has taken memory for granted despite its importance for valuation processes, which are immersed in background knowledge. Memory is also central in forming concepts, as in the coherence of diverse pieces of information about various individual persons. At the same time, memory research has tended to pass by judgment theory, as illustrated in the present discussion of operations and in earlier discussions of continuous representations. Not only does judgment theory require a coordination of memory to a superstructure of operations, it may also require more complex representations than have been characteristic of memory theories. These interrelationships between judgment and memory reflect the many opportunities for cooperative inquiry.

Notes

4.1.1a. This hypothesis of adding with contrast was suggested by a number of persons during colloquium presentations of the initial adding–averaging studies. I am indebted to Sherman L. Guth for his helpful comments regarding use of the task of component judgments.

4.1.2a. A rather different hypothesis would be that contrast results from cognitive, rather than perceptual, interaction. No contrast would then be expected in the Word condition, only in the Person condition. This hypothesis was eliminated by the positive context effect in the Person condition before it could gain currency. It is mentioned here to re-emphasize the perceptual origin and character of the standard contrast hypothesis.

4.1.2b. Griffitt and Jackson (1970) used an interesting task variation in which two independent person descriptions were listed side by side, one with three adjectives, the other with a single adjective. The three-adjective description showed the positive context effect discussed in the text. The likableness judgment for the one-adjective description was unaffected by the value (high or low) of the adjacent three-adjective description, further evidence for lack of contrast. (The interaction reported in this article reflects the presence of the positive context effect in the three-adjective description and the lack of any effect on the one-adjective description.)

Wyer (1974a, p. 242) has suggested that the parallelism in the graphs of the positive context effects presented in the text could result from averaging out real nonparallelism in the different stimulus replications. If so, then the Component × Context × Stimulus Replication interaction should be statistically significant. These interactions have been reported and they were not significant (Anderson, 1971c, p. 80; Anderson & Lampel, 1965, p. 433).

It is still possible that these global interaction tests would miss relatively small interactions among specific adjectives. This possibility may repay further investigation with adjectives chosen to embody expected interactions. However, any such effects would be far too small to explain the central phenomenon, namely, the large positive context effect.

4.1.5a. A few additional details of procedure should be recorded for these three unpublished experiments. Experiments A and B both used the same stimuli, four independent replications of the basic 4 × 4 design, each with a different selection of adjectives. Experiment C used two independent selections of 16 component adjectives and 16 context pairs. Each selection was used to construct four stimulus replications, balanced in a greco-latin square type of design. The experiment shown in Figure 4.2 used the same stimuli as Experiments A and B but was mildly complicated by an additional balance of each adjective group across component and context. All choices were made at random from the H, M^+, M^-, and L sublists.

In Experiments A and B, the test adjective was always lowest on the card; it was rated last in Experiment B. In the other two experiments, the adjectives were typed on the cards in random order, and the components were rated in random order when more than one was rated. The deck of cards was shuffled separately for each subject to randomize order of presentation. Each subject judged two different stimulus replications of 4 × 4 sets in Experiment C, one replication of 4 × 4 sets in the other experiments. There were 40 subjects in Experiment A, 20 in Experiment B, and 24 in each condition of Experiment C. This work owes much to the patience of Anita Lampel and Ann Jacobson.

4.1.6a. Wyer and Watson (1969, p. 31) interpreted the positive context effect in judgments of an individual member of a group in terms of change of meaning, as though the adjective describing that group member changed its meaning in the context of the adjectives describing the other group members. However, the group existed only by arbitrary, minimal instruction, with no relation among the members. Hence there is no consistency constraint among the adjectives describing the different group members such as may be expected when the adjectives describe a single person. The alternative interpretation in terms of a group halo is consistent with the moderate-size effect obtained by Anderson, Lindner, and Lopes (1973).

4.1.7a. Further evidence on the halo interpretation has been presented by Nisbett and Wilson (1977b), who showed videotapes of a college instructor with a foreign accent who projected either a warm or cold orientation toward students. Subjects who saw the "warm" videotape subsequently made more favorable ratings of the instructor's physical appearance, mannerisms, and accent, even though these were ostensibly the same in both videotapes. This of course agrees with the present theoretical interpretation. The information from the videotape is integrated into a general impression of the person in which the components no longer have a distinct existence. When asked for a judgment on some particular dimension, the general impression is integrated as one determinant of the judgment. The halo integration model thus accounts directly for the observed result.

4.1.8a. Snyder and Uranowitz attempted to support their interpretation in terms of memory reorganization by showing that control subjects who did not read the case history but were only told about Betty K.'s sexual life style "in no way whatsoever produced the same differential pattern of answers [p. 948]" as in the experimental conditions. But inspection of their Table 1 shows that the control conditions did produce the same pattern of answers. The effect was not significant in the control condition, but the null hypothesis probably should not be accepted in this case. If the information about Betty K.'s sexual life style produces stereotypic implications in the experimental conditions, it should do the same in various control conditions.

4.2.2a. As Anderson and Hubert (1963) pointed out, work on verbal learning provides some support to the conclusion that separate memory systems exist for words and for ideas. English, Welborn, and Killian (1934) found that subjects forgot specific details more quickly than general ideas. Although this result does not require separate memories, it is suggestive and possibly testable with an integration task, as illustrated in the text. Recent work on person memory is discussed by Hastie, Ostrom, Ebbesen, Wyer, Hamilton, and Carlston (1980).

4.3.1a. Kaplan's assumption that subjects with positive and negative dispositions place the same weight and value on external information should not be expected to hold in general. It evidently worked well for the adjective descriptions about hypothetical persons, but, in general, personality dispositions could affect s as well as I_0 (see also Section 4.3.3).

4.3.1b. The set-size effect in the data of Figure 4.11 shows an asymmetry that has theoretical interest. When only sociable adjectives are in the set, the response is substantially greater with eight than with four adjectives. Furthermore, this set-size effect is greater for the unsociable than for the sociable group. This is as expected, on the grounds that the scale value of a single sociable trait should be near the upper end of the response scale for both disposition groups and that the initial impression is lower for unsociable disposition. By Section 2.4, the set-size effect will be greater the greater the difference between the initial impression and the stimulus value.

When the sets contain all changeable and no sociable adjectives, then the set-size effect appears only for the sociable disposition. This is consistent with theory on the assumption that the scale value on the sociableness dimension of a single changeable adjective is roughly equal to the initial impression for the unsociable disposition. The sociable disposition, having a more positive initial impression, would show a set-size effect. The lack of a set-size effect for unsociable disposition suggests that the effective zero in this 7-point, unsociable–sociable response scale was around 2, below the scale midpoint. Such asymmetrical usage of the response scale could arise because of the corresponding asymmetry in the adjective sets, none of which contained unsociable adjectives. This argument is post hoc and speculative, but it illustrates the level of precision that may be obtained from the theory.

4.3.1c. The linear divergence in Figure 1 from Bossart and DiVesta (1966) also illustrates this prediction of linearity from the averaging model.

4.3.1d. The strong discounting in importance judgments is not consistent with the marginal discounting in person impressions (center and left panels of Figure 4.12, respectively). This difference may reflect the design structure. Each curve of importance judgments is on a within-subject base. In contrast, the person impressions come from a between-subject design with four independent groups of subjects. A more sensitive assessment of dispositional discounting might be obtained by making instructions a within-subject variable, for example, by attributing each trait adjective to a different source and saying that all or only half of the sources for each description are reliable.

4.3.1e. The conclusion that personality disposition produces active discounting depends on the comparison between the two instruction conditions. It is not enough to show that the two disposition groups have different weight parameters; that only suffices to demonstrate individual differences in values. That the difference observed under naturalistic instructions is proper discounting follows from the lack of corresponding difference under equal accuracy instructions (see also Section 3.4).

4.3.2a. The conclusion of Ostrom *et al.* (1978) that their subjects presumed innocence ($I_0 = 0$) differs from that of Anderson (1959b), who found I_0 near .5 in an experiment that used testimony from a bigamy trial. In this latter experiment, however, subjects were instructed not to act as jurors, but to give their personal opinions about guilt or innocence. In contrast, subjects in Ostrom *et al.* were instructed to act as jurors.

4.4.1a. Different processes will of course be involved in the various determinants of weight. Thus, relevance requires a comparison process, whereas the others do not. Quantity can be viewed as an integration of molecular units, whereas salience rests on a valuation operation. Moreover, configural weighting such as discounting will require other kinds of processes.

4.4.2a. Other studies relevant to the negativity effect include judgments of social class (Himmelfarb & Senn, 1969, p. 50), person perception (Figure 4.14; Schmidt & Levin, 1972, see Section 2.3.5), clinical judgment (Anderson, 1972b), occupational proficiency (Figure 4.17), moral judgment (Leon *et al.*, 1973), and studies cited in Note 4.4.4b. Hodges (1974) reports results like those in Figure 4.14 but with a more extensive coverage of stimulus values, though he also suggests that his obtained set-size effect might disagree with the averaging model.

Negativity effects have also been reported by Ostrom and Davis (1979), who attempted to extend the study of differential weighting to the individual level. Each of 40 subjects judged likableness for 40 two-adjective person descriptions constructed from ten 2×2 designs on each of 5 successive days. The majority of these 400 individual interaction tests were statistically significant, and it was concluded that these interactions reflected differential weighting at the individual level. But as Ostrom and Davis recognized, this conclusion depends on the assumption that the response scale is linear. Otherwise, nearly all the nonparallelism in the 2×2 designs could be eliminated by monotone transformation. Unfortunately, neither of the two lines of evidence that they present provides adequate evidence for response scale linearity.

The first line of evidence rested on the statement that response scale nonlinearity could be dismissed if the data satisfied a test of disordinality (p. 2031). This test is actually just a crossover interaction in the 2×2 design. But the primary value of the crossover test is that it rules out additive models despite response scale nonlinearity. Even if the crossovers were genuine, they would not resolve the question of response scale linearity.

The second line of evidence rested on selecting two of the ten 2×2 designs for each subject, one that showed no significant interaction and one that did, so related that the two upper row curves from the two designs showed a crossover and similarly for the two lower row curves. The essential idea was that the response scale could not be transformed so as to eliminate the nonparallelism in one design without simultaneously destroying the parallelism in the other design. This idea is correct in principle, but the 2×2 design is too small to make it effective. As may be seen from inspection of their Figure 2, it is straightforward to find a monotone transformation that simultaneously renders both 2×2 data tables arbitrarily close to parallel. Accordingly, this intersection test does not warrant any conclusion about the reality of the interactions. Thus, neither line of evidence presented by Ostrom and Davis speaks to their question about response scale nonlinearity. Their conclusions about differential weighting at the individual level, which rest on the assumption of response linearity, are accordingly not as secure as would be desired.

The effort of Ostrom and Davis to apply an integration approach to study individual differences in values certainly deserves to be followed up. The ideographic aspect of integration theory provides a useful alternative to traditional approaches to individual and cultural differences (Lopes, 1976a; Singh, Gupta, & Dalal, 1979) because it allows analysis of the individual at the individual level. The problem of establishing response scale linearity could be approached with a two-operation model or with the method of interlaced designs (Section 1.7.2; Anderson, in press, Section 3.2). Helpful collateral information might be obtained from self-estimated weights (Section 4.4.1).

4.4.2b. This uncertainty over the reality of the negativity effect illustrates the general problem of interpreting small deviations from parallelism that can arise from residual nonlinearity in the rating response. This concern emphasizes the need for care in the use and further development of response methodology.

4.4.3a. Since each source is separately characterized for reliability, an alternative integration rule could be considered in which a separate evaluation is made for each adjective by averaging each separately with the initial impression and then integrating these two evaluations. This latter integration could not be equal-weight averaging as that would imply parallelism, contrary to the crossover of Figure 4.15. However, if each evaluation was weighted by its absolute informational weight, namely, $(w_0 + w_{source})$, then Eq. (5) would become

$$R = C_0 + (w_A s_A + w_B s_B + 2w_0 s_0)/(w_A + w_B + 2w_0).$$

This formulation cannot be distinguished from Eq. (5) on the basis of data in the text. It would, however, predict a different set-size curve, and on that basis it can tentatively be ruled out because of the results of Section 2.4.3. This question has theoretical interest for its bearing on processing flow, especially in serial integration, in which some form of the alternative rule might be necessary.

4.4.3b. The vertical spread of the three curves in the right panel is about half that in the left panel. This implies that the initial impression has approximately the same weight as a single adjective, a result that agrees with other evidence (Section 2.4.3). This implication follows fairly readily by comparing the effects of equal increments in w_A and in w_B on the slope function, $w_A/(w_A + w_B + w_0)$.

4.4.3c. An algebraic expression for the change in response produced by a change in the weight of one adjective can be obtained as follows. Let w_B and $w_B + x$ be the initial and changed weights. Then the initial and changed responses are

$$R = (w_A s_A + w_B s_B + w_0 s_0)/(w_A + w_B + w_0),$$
$$R' = [w_A s_A + (w_B + x)s_B + w_0 s_0]/(w_A + w_B + x + w_0).$$

Straightforward algebra yields

$$R - R' = x(R - s_B)/(w_A + w_B + x + w_0).$$

Thus, the change is positive or negative according to whether $R - s_B$ is positive or negative.

4.4.3d. In Wyer's Table 9.4, the value of the less extreme adjective varies systematically down the column. When reliability of this adjective is changed from low to high, the person impression predicted by the adding model should change by an amount proportional to the value of this adjective. The adding model thus implies that the change in person impression response should also change systematically down the column. Entries in the two columns labeled "*Difference*" should thus decrease in magnitude down the column, and the decrease predicted by the adding model would be substantial in view of the stimulus values in this experiment. Failure to observe a systematic trend is prima facie evidence against the adding model, although the power of this test is limited by the prevailing variability.

As an incidental comment, it may be added that the specific manipulation of source reliability may have introduced a confounding. The low reliable source was characterized as being 55% accurate, "slightly better than chance." Such instructions could produce an active discounting that would complicate the theoretical interpretation.

4.4.3e. In magnitude, the observed source effect agrees with theoretical expectation. To illustrate, suppose that $s_0 = s_M = 10$, $s_L = 2$, $w_0 = 1$, and that the weight parameters corresponding to the three source conditions are .5, 1.0, and 1.5. Then the low–medium combination yields predicted means of 6.93, 7.43, and 7.79; these agree fairly well with the observed values of 6.98, 7.30, and 7.63.

4.4.3f. The assumption by Lichtenstein *et al.* (1975) that subjects average the "regressed" values of the cues could be interpreted to mean that each cue is integrated separately with the prior expectancy and then averaged together. This interpretation has theoretical interest, as discussed in Note 4.4.3a.

4.4.3g. Fiske (1980) reports a close relation between weight parameters of captioned photographs in person perception and a behavioral measure, namely, time spent looking at each photograph. Levin, Ims, and Vilmain (1980) consider the role of information variability and reliability in grading practice.

4.4.4a. An unexpected and still unexplained deviation from parallelism was obtained in this experiment (Oden & Anderson, 1971). The descriptions of Figure 4.18 each contained four trait adjectives that were themselves combined according to a 2 × 3 design. Pairs of adjectives were used in each factor level because single adjectives would have been overweighed by the information on class standing. The three curves in each panel of Figure 4.18 represent the three-level adjective factor averaged over the two-level factor.

Although the interactions between class standing and adjectives were either small (for command effectiveness) or nonsignificant (for liking and respect), the interaction between the two adjective factors was substantial for all three dimensions of judgment. This result was disappointing because it vitiated the planned estimation procedure, which allowed for differential weighting only on one dimension. Furthermore, it disagreed with the usual finding of parallelism or near-parallelism with adjective combinations.

Some obvious procedural explanations can be ruled out. Quite similar results were obtained from two different selections of adjectives, indicating that the interaction had some generality. Active discounting seems unlikely because in this experiment the adjectives had been chosen to avoid inconsistency and also to avoid redundancy with the class-standing information. Attentional factors would not be a problem with five short stimuli presented on a card. The shape of the interaction was consistent with a negativity weighting interpretation, as noted in the original report. However, negativity weighting would also imply interaction between class standing and the adjectives, and those interactions were small or nonsignificant. Although the critical tests in this experiment provided qualitative support to the averaging hypothesis (Sections 1.6.1, 2.3.4), the adjective-adjective interaction remains a problem.

4.4.4b. A similar configural effect for moral actions of a single person has been reported by Richey (e.g., Richey, Richey, & Thieman, 1972) and by Birnbaum (1973, Figure 1c; Riskey & Birnbaum, 1974). These results indicate a partly irreversible effect of very negative information, as though no number of good deeds will ever wash out the stain of one evil deed. The parametric data of Riskey and Birnbaum indicate that the first few good deeds do have substantial ameliorative effects. Consequently, the irreversibility cannot easily be explained within the averaging model by assuming that bad deeds have extremely high weight. A two-factor interpretation may deserve consideration; the response is a composite of two underlying moral dimensions, a bipolar dimension that can incorporate both good deeds and bad deeds and a unipolar dimension that incorporates only bad deeds.

4.4.4c. The upward slopes of the M^+ and M^- curves in Panel C' of Figure 4.14 mean that ratings of the M^+ and M^- adjectives are significantly higher in homogeneous than in heterogeneous descriptions. This could point to possible discounting of moderate traits in the context of extreme traits. However, the effect is small and not overly attractive as a line of research.

The importance ratings in Panel C' of Figure 4.14 are averaged over one adjective in each condition for each of 60 subjects. This was the one rated adjective for the 40 subjects in Condition One, and the first rated adjective for the 20 subjects in Condition All. There was virtually no order effect for the four successive ratings in Condition All for homogeneous descriptions. In the heterogeneous descriptions, the extremity effect was somewhat less for Condition All than for Condition One.

4.4.4d. The data of Figure 4.20 were not included in the original report (Anderson & Alexander, 1971) and so some details of procedure will be given here. Most of these details were the same as those used to obtain the published data reported in Panels C and C' of Figure 4.14. Eight subranges of the master list of 555 trait adjectives were chosen, and adjectives low on meaningfulness were eliminated to leave 12 adjectives for each range. Each subject selected 6 words of equal likableness from each group of 12, and the experimental sets were constructed separately for each subject, with the 6 adjectives balanced across set size.

Each set of adjectives was written in random order on a slip of paper that also contained a numerical, 0–10 scale for judging person likableness, and a 2-inch line by each word for rating importance (similar to the left panel of Figure 1 of Anderson & Alexander, 1971). The experimenter handed each slip to the subject, who read the adjectives slowly aloud. After a 4-second wait, the subject made the impression response. The experimenter then pointed to one adjective, and the subject rated its importance. This adjective was determined by a written random schedule, subject to the restriction that a different adjective be rated in each of the four sets at each value range. Subjects in Condition One rated importance of just this one adjective. In Condition All, the remaining adjectives were treated similarly. No order effects were obtained, and the first adjective in Condition All yielded essentially the same results as the later adjectives.

4.5a. The importance of further work on theory and methodology of the response scale may also be emphasized. For study of interaction and organizing processes, in particular, possession of a linear response scale can be almost indispensable. One advantage of studying cognitive algebra in simple tasks is the opportunity for developing methods to obtain linear response scales.

4.5.1a. Many interesting studies of information integration have had to be omitted from this volume, but a few of the most important may be briefly noted. Louviere (1977) has edited a compilation that includes several papers on applied psychology and decision theory. Kaplan, Bush, and Berry (1979) have demonstrated the usefulness of integration models as a validational criterion in the development of a medical health index. Ebbesen and Konečni (1975) have applied cognitive algebra to bail setting in the courts. The usefulness of model analysis also appeared in an article by Brehmer and Slovic (1980) that provides cogent support for the validity of self-estimated scale values in a prediction task.

An ingenious thesis by Clavadetscher (1977) gave a clear demonstration of joint operation of asssimilation and contrast effects in geometrical illusions and showed how these effects could be incorporated in an integration model. Interesting work on the functional role of psycholinguistic quantifiers is given by Borges and Sawyers (1974). Work on group dynamics includes an article by Anderson and Graesser (1976) as well as a notable thesis by Graesser (1977) that compared a social averaging theorem with a formulation based on social decision schemes. Studies of time judgment are given by Svenson (1970) and by Blankenship and Anderson (1976). Bettman, Capon, and Lutz (1975) discuss the role of cognitive algebra in applied work with attitudes. The report by Norman (1976b) explores the effect of informational feedback on parameters in integration models, an important area that deserves intensive study. Wilkening (in press) presents a conceptual analysis of the development of interrelations among travel time, travel distance, and velocity of moving objects that provides a significant advance over Piagetian views.

A striking instance of social learning was discovered by Leon (in press), who found in individual analyses that mothers and sons had similar integration rules in moral judgment. Other work on moral algebra is given by Lane and Anderson (1976), Leon (in press), Hommers (1981), Verdi (1979), Anderson (1976d), Farkas (1977), and Butzin (1978).

4.5.2a. An illustration of the embedding method is provided by the following initial study of marriage perception, which was designed to test the averaging model and to compare the use of actual and hypothetical incidents. The subjects were 16 divorced women; the first part of the session was devoted to recall of specific incidents in which the husband had exhibited various degrees of appreciation, affection, or understanding. Typical incidents for C. C. were: Husband was very appreciative of her concern for his mother (H value); husband was understanding about her reaction to the pill (M^+); husband was offended when she went to get food stamps (M^-); husband wouldn't take just any job when they were very poor and needed the money (L). Subjects generally found this incident listing to be interesting, although a few had trouble remembering specific negative incidents along the given dimensions. One woman who could recall almost nothing germane was dropped from the study.

Two groups of four incidents were selected to form the factors of a 4 × 4 design for each woman. Some sets with one or three incidents were also included as tests of the averaging model. The subject was given a set of incidents, told to imagine that they occurred very close together in her marriage and to rate on a 1–20 scale how satisfied she would have been with her marriage at that time.

An exactly parallel design was also used in which the factors were abstract dimensions of appreciation, affection, and understanding, each quantified at four levels by a bar graph display. The design for each woman was constructed from the two factors that she rated as most important.

Results are shown in Figure 4.21, with judgments about actual incidents in the left panel. The four solid curves in the lower left panel represent the main design. Some deviation from parallelism is visible, presumably a negativity effect in which negative incidents are averaged in with greater weight. The dashed curve represents judgments to the column incidents alone and provides the standard test between averaging and adding (Section 2.3). Although the dashed curve shows greater slope than the solid curves, the crossover is not significant. However, the upper left panel shows a clear crossover that embodies the same logic. The dashed curve is the same in the upper and lower left panels, but the solid curve in the upper left represents judgments based on the column incident plus a near-neutral M^+M^- pair of incidents. This crossover eliminates adding interpretations and supports the averaging rule for judgments of satisfaction in marriage.

Nearly identical patterns of judgments were obtained for actual and hypothetical inci-

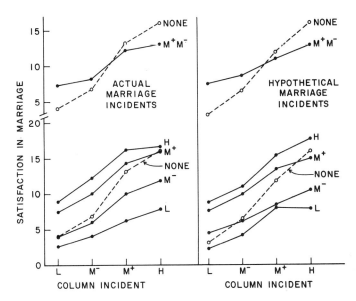

FIGURE 4.21. *Divorced women follow averaging model in judgments of marriage satisfaction. Left panel show judgments based on actual incidents from each marriage. Shapes of these factorial plots support averaging model, with greater weighting of negative incidents. Right panels show similar data based on hypothetical levels of husband's understanding and appreciation. Similarity of shape in left and right panels supports generality of experiments that employ hypothetical incidents. (From N. H. Anderson, unpublished experiment, 1980.)*

dents, shown in the left and right panels, respectively. The only notable difference is that the crossovers are clearer with hypothetical incidents, a consequence of the greater flexibility in choice of levels.

Moreover, the patterns in both sets of judgments about marriage satisfaction are much the same as have been obtained in the personality adjective task. To some modest extent, therefore, this experiment illustrates the feasibility of the embedding method and the likelihood of generality of previous results of this research program.

4.5.3a. In the following study of intuitive physics, subjects judged the time it would take a ball to roll down an inclined plane. In the condition reported here, the stimulus display was schematic, so no actual ball rolling occurred. A 3 × 5 design was used, with three angles of inclination and five distances along the plane. Subjects made time judgments on a graphic scale defined with end anchors in the usual way (Appendix A).

The left panel of Figure 4.22 shows the initial set of point predictions in which subjects made separate time judgments for each of the 15 inclination–distance combinations. These data show a marked law of diminishing returns, especially for the lower curves, which represent steeper inclinations.

Following the initial round of point predictions, subjects sketched a 3 × 5 factorial graph on a sheet of graph paper. The mean of these factorial graphs is shown in the center panel of Figure 4.22. These curves are nearly linear as a function of distance, showing only a trace of the diminishing returns in the left panel.

Finally, the right panel shows a replication of the point prediction task. These data

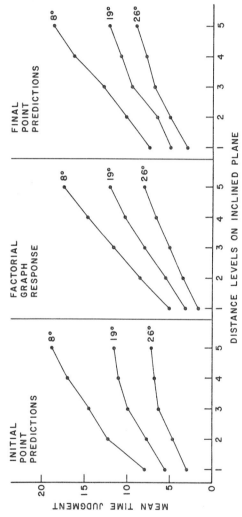

FIGURE 4.22. *Intuitive physics on an inclined plane. Subjects judge descent time for ball located at various distances from base of a schematic inclined plane. Left panel shows mean point predictions for the 15 distance–inclination settings in the initial phase of the experiment; right panel shows similar data from the final phase. Center panel from intermediate phase shows mean of factorial graph response in which subjects sketched three curves on a sheet of graph paper. Sixteen college students served as subjects in all three phases. Curve parameter is angle of inclination of the plane. (From N. H. Anderson, unpublished experiment, 1980.)*

appear less regular but seem to retain some of the increased linearity from the factorial response phase.

All three panels are fairly close to the linear fan form although the valuation function for distance appears to change from phase to phase. Björkman (1965), who gives a thoughtful discussion of problems of intuitive physics, suggested that estimated times from his experiments were proportional to the distance along the plane, which would agree with the factorial response in the center panel but not with the point predictions in the left panel. However, Björkman's interpretation rested on indirect analysis that is hard to compare with the present results.

Mention may also be made of another condition in the present experiment in which subjects estimated actual rolling times from direct observation. A reasonable range of times was obtained by using a plane of length 2 meters with inclinations of only a few degrees. Initial point predictions prior to any ball rolling showed a pattern similar to that in the left panel of Figure 4.22 but with considerably less separation between the three curves, presumably because the three inclinations were not greatly different. Direct judgments of the actual rolling times showed a similar pattern, with substantially greater separation among the three curves and with greater diminishing returns in the lower curves, less in the upper curve. Point predictions following this event observation showed a nearly identical pattern. Thus, subjects learned and transferred two ideas from observing the actual events: First, that inclination angle had greater effect than had been recognized in their pre-experimental knowledge; second, that motion time followed a diminishing returns function of distance, an accentuation of their pre-experimental knowledge. This latter result has special interest because it implies that direct observation of physical events may lead to incorrect physical knowledge even in simple situations.

4.5.4a. A fractionation process for analog multiplication has been studied experimentally by Lopes (e.g., Lopes & Ekberg, 1980).

4.5.4b. Study of paragraph responses in the personality adjective task has been neglected owing to the dominant concern with the meaning-constancy hypothesis of Chapter 3. That such paragraphs involve a cognitive illusion about the integration process does not decrease their interest as aspects of consciousness or as verbal behavior.

4.5.4c. Also relevant to the problem of communicational constraints are the discussion by Norman and Rumelhart (1975) and various references cited by them.

4.5.4d. The good–bad sense could derive from two aspects of goal-oriented behavior. First is the good–bad quality of the goal itself, as in Thorndike's satisfier–annoyer distinction. Second is success or failure in goal attainment. Both aspects could become encapsulated in the general metric sense discussed in Section 1.1.6.

Measurement Theory

This chapter presents brief, critical discussions of a number of approaches to psychological measurement. Its main purpose is to compare functional measurement with other approaches. In addition, some misconceptions about functional measurement are discussed.

Issues in measurement theory have evoked sharply different points of view. The point of view in this chapter is personal, not disinterested, and should not be expected to portray other approaches adequately. In general, moreover, critical discussions tend to focus on weak points and pass over positive contributions. It should be recognized, therefore, that the present view, even where critical, is much indebted to the work of others (Section 1.8.1). An apology is in order because the critical purpose and brief space do not allow adequate recognition of positive contributions by investigators working within other approaches.

5.1 Preliminary Comments

5.1.1 THE PROBLEM OF THE LINEAR SCALE

Many different problems of measurement arise in psychology, but one has caused unique difficulty. This is the problem of the linear, or "equal-interval," scale, which is the central, if narrow, concern of what has come to be called measurement theory in psychology. Common examples include the centimeter scale for length, the Celsius scale for temperature, and the erg scale for work or energy. Such linear scales are usually taken for granted in physical science.

In psychology, however, linear scales have been elusive. The mental sensation of loudness has become a classic example. Intuitively, loudness ought to be measurable. Different sounds can be rank ordered in loudness to yield a monotone scale. Furthermore, loudness is a smooth, continuous function of physical intensity. It seems almost necessary, therefore, that loudness is a metric quality that is in principle measurable on a linear scale. The problem of the linear scale, then, is to find a method that will assign numbers to sounds in such a way that these numbers are a linear function of loudness (Anderson, 1979a).

But loudness is a subjective sensation, not an objective, observable quantity. Given some assignment of numbers to sounds, what criterion can be found to decide whether or not these numbers constitute a linear scale of loudness? If two investigators obtain different assignments of numbers to the same sounds, what criterion is available to decide which one, if either, really measures loudness? Attempts to answer this question have led mainly to controversy. No general agreement yet exists about linear scales even for simple sensory qualities such as loudness.

Loudness is a good example because it seems intuitively clear that a linear scale must exist. Loudness and other sensory qualities have consequently received much attention. However, more cognitive qualities are not less important. These include motivation, expectancy, value, and risk as well as social and moral qualities such as likableness, obligation, fairness, and blame.

Psychology is perhaps unique among the sciences in its need to deal with such subjective qualities (see also Krantz, 1972b, p. 1434). Regardless of whether such concepts can ultimately be reduced to physicalistic terms, they appear to exist in their own right and they demand analysis at their own level.

5.1.2 MONOTONE AND LINEAR SCALES[a]

In the functional view, measurement is woven into the evolving network of substantive knowledge. It can be convenient to focus on certain strands of this network as "measurement," but it is dangerously artificial to do so. The network of knowledge is an evolving entity, and the separate strands develop and attain meaning as part of that whole. A local view is convenient, often necessary, to reduce immediate problems to manageable size. However, measurement theory cannot be understood divorced from its working role in substantive inquiry.

This point is well illustrated in the distinction between monotone and linear scales. Monotone scales are adequate for many quantitative purposes (Anderson, in press, Section 1.1.9). However, the qualitative nature of what is being measured is not less important than its quantification. Monotone measurement represents the definition and isolation of some one-dimensional concept or quality, and that may reflect an accumulation of knowledge far greater than might be required to transform the monotone scale into a linear scale. The major stage in measurement is thus a monotone scale, at least for such qualities as can be given a one-dimensional interpretation.[b]

Monotone measurement is often taken for granted. If people are instructed to judge loudness, for example, it is generally believed that they will do so. Similarly, people can be instructed to judge "likableness," say, or "subjective probability," and can be constrained to do so on a one-dimensional scale. Such instructions and procedures can incorporate and utilize the vague, uncertain knowledge of introspection and common experience. They thus provide an invaluable first guess about the definition and measurement of some psychological quality or concept. However, they do not guarantee the existence of that concept. A major reason for seeking linear scales lies in their contribution to construct validity.

By the time that investigators begin to feel reasonably confident that they have something approaching a monotone scale, they will typically also feel reasonably confident that they have something approaching a linear scale. Ratings of loudness and likableness seem to be more than mere rank orders. They may not be exact linear measures, but they should not be greatly off. To assess this is not easy, however, for there is a qualitative difference between a monotone and a linear scale. To establish a linear scale requires new leverage beyond what may be needed to develop a monotone scale.

In the functional measurement view, this new leverage comes from quantitative models. The mere statement of such a model presupposes the existence of metric scales, and to establish such a model entails metric scaling. Model and scales are facets of the same theoretical entity and it may be artificial to separate them. In some sense, however, the model is more basic, for it provides the scaling frame on which to metricize the response and stimulus variables.

Such metric scaling is also part of an evolutionary network. Experimental reports, it is true, often take the convenient local view that they have proved or disproved a model. It is generally understood, however, that establishing any model requires an interconnecting network of evidence, with confidence developing gradually as the network expands and stabilizes. A major advantage of metric scaling lies in its contribution to construct validity; that is, to establishing the psychological reality of the concepts in the model (Section 1.8.2).

5.1.3 INDUCTIVE ORIENTATION IN MEASUREMENT THEORY

As the preceding discussion has shown, the functional view sees measurement as bound up with substantive knowledge. This reverses the traditional approach, exemplified in Thurstonian scaling theory, in which measurement is preliminary to substantive investigation. It also reverses the mathematical approach, exemplified in conjoint measurement, in which measurement theory is a mathematical discipline logically prior to, or independent of, substantive inquiry. In both these approaches, measurement theory has had limited interaction with experimental analysis.

A major symptom of the problem with such approaches was given by Torgerson (1958), who quoted the philosopher Nagel to the effect that the end of measurement, at least in physics, is the establishment of certain equations and theories. Torgerson (1958) continued: "Yet it seems that, when one gets into a discussion of the logic of measurement, there is a tendency to lose sight of these considerations. There seems to be a tendency to become so involved with one or another preferred method of developing a scale of measurement that the means itself becomes the end [p. 12]."

This isolation from experimental analysis has in some ways increased since Torgerson wrote. No theory of measurement attempts to live in an empirical vacuum, of course, but the predominant direction in the two decades following his statement has been increasingly toward abstract, mathematical analysis with less and less relation to experimental analysis.

In the functional view, scales of measurement derive from substantive theory of which, indeed, they are an organic part. They may be considered as summarized knowledge, but their proper referent is the knowledge summarized. That includes the experimental conditions that presently surround the observations themselves. It also includes ill-defined knowledge of general experience from which many scales originate. This bulk of background knowledge is often usefully ignored, but it all enters into the inductive definition of the concept and of its measuring scale.

Functional measurement provides a logically sufficient basis for psychological measurement. Technical problems of data analysis have been approached by using and extending existing statistical theory. Although various technical questions remain, current methods are able to handle the main problems, especially those of monotone analysis and goodness of fit (Anderson, in press). Substantive applications have been made in many areas, hand in hand with the development of practical experimental procedures. Epistemologically, such substantive applications form the essential foundation for measurement theory, for there is no measurement apart from empirical laws. The work on cognitive algebra illustrated in Chapter 1 demonstrates the validity of this inductive approach. These empirical applications show that functional measurement has, in fact, provided a logically sufficient basis for psychological measurement theory.

5.2 Fechner's Method of Additive Units

The problem of the psychophysical law has been prominent in psychological thought since the time of Fechner more than a century ago. This "law" is the function that relates subjective sensation s and physical intensity S. In terms of the functional measurement diagram (Figure 1.1), the psychophysical law is a special case of the valuation function, V, and may be written

$$s = V(S).$$

If both S and s were known for each stimulus, V would be determinate. Determination of the psychophysical law thus rests on two problems of measurement, one physical, one psychological. The science of physics provides the theory and instruments for measuring physical magnitude S. If the science of psychology could provide analogous theory and instruments for measuring psychological magnitude s, then Fechner's problem would be solved. The long controversy over the psychophysical law has been basically a controversy over psychological measurement.

Fechner himself hit upon the ingenious idea of using the just noticeable difference, or *jnd*, as a measuring unit. Let S_0 be some given physical intensity and let S_1 be the intensity that is just noticeably greater than S_0. Similarly, let S_2 be just noticeably greater than S_1, S_3 just noticeably greater than S_2, and so on. Then the sequence $S_0, S_1, S_2, S_3, \ldots$ shows an increase of 1 *jnd* at each step. Assign any arbitrary value a_0 to S_0. Then, according to Fechner, the numbers

$$a_0, a_0 + 1, a_0 + 2, a_0 + 3, \ldots,$$

would constitute a linear scale of the sensation values of the corresponding physical intensities. Fechner assumed, in other words, that each *jnd* represented an equal increment in sensation. If Fechner's assumption was correct, then it provided an additive unit method for measuring sensation.

Fechner went farther. Research by Weber had shown that the physical size of the *jnd* was approximately proportional to the reference stimulus intensity. Weber's empirical law, coupled with Fechner's theoretical assumption, led to a logarithmic formula for the psychophysical law,

$$s = c_0 + c_1 \log S,$$

where c_0 and c_1 are constants. This logarithmic law implies that subjective loudness is a linear function of decibel value of physical intensity. That seemed to be at least approximately true, so Fechner's law made practical sense, at least for energetic dimensions such as loudness and brightness.

For the best part of a century, Fechner's logarithmic law held the center of attention. It promised a simple solution to the problem of the linear scale, replacing the uncertainty of measuring subjective value, s, by straightforward measurement of physical value, S. No less impressive was the appearance of an exact mathematical law in a realm that seemed overwhelmingly vague. This law, moreover, asserted a direct relation between the mental and physical worlds, truly a psycho–physical law.

Adding to the fascination was the continued lack of agreed criteria for putting Fechner's assumption to the test. The idea that *jnd's* are all effectively equal was very attractive, but it continued to hover just beyond proof. The perplexity, not to say confusion that prevailed over the century following Fechner is well summarized in Boring (1950). Well might Fechner conclude his final answer to his critics:

> Der babylonische Thurm wurde nicht vollendet, weil die Werkleute sich nicht verständigen könnten, wie sie ihn bauen sollten; mein psychophysisches Bauwerk dürfte bestehen bleiben, weil die Werkleute sich nicht werden verständigen können, wie sie es einreissen sollen [Fechner, 1877, p. 215].

> The tower of Babel was never completed because the workers could not agree on how they should build it; my psychophysical edifice will remain standing because the workers will never agree on how they should tear it down.[a]

5.3 Thurstonian Theory

The method of paired comparisons developed by L. L. Thurstone (1927a,b, 1959) is a remarkable attempt to develop a true theory of psychological measurement. His method requires only simple choice responses and avoids the biases that can afflict ratings and other numerical response measures. His farsighted conceptualization of the stimulus as a distribution anticipated signal detection theory. But most important, Thurstone's formulation was testable. It required the data to satisfy certain consistency criteria, and these criteria provided the validational base that is necessary for a theory of measurement. For the first time, as Thurstone emphasized, it seemed possible to resolve the long muddle over Fechner's assumption that all *jnd*'s are equal.

5.3.1 THURSTONIAN SCALING

Paired-comparison scaling is applicable only under limited conditions, including the important restriction that the stimuli be imperfectly discriminable. The basic datum is a proportion of choices between two stimuli, but this is subjected to a normal deviate or similar transformation that is undefined for choice proportions of 0 and 1. That is as it should be, since a choice proportion of 1, although it says that one stimulus is definitely preferred, is silent about degree of preference. However, this requirement that the stimuli be close together and imperfectly discriminable imposes serious limitations on paired-comparison scaling.

Psychophysical Stimuli. Typical psychophysical stimuli have two properties that make them ideal for paired-comparison scaling. They are continuously variable, so pairs can readily be chosen close together and imperfectly discriminable. And they are not separately identifiable, so the same pair can be presented many times to an individual subject without problems from response fixation or memory of previous responses. It is entirely feasible, therefore, to use paired comparisons to scale psychophysical dimensions at the level of the individual.

Thurstonian methods were thus directly applicable to Fechner's problem of determining the psychophysical law. These methods should either have failed or have proved their worth by finally determining the measure of sensation. Surprisingly, nothing much happened. For reasons that are only partly clear, Thurstonian scaling had little impact on psychophysics.

Some factors that contributed to this historical peculiarity are discernible. Thurstone himself was interested in social, rather than psychophysical, judgment. Not long afterwards, moreover, he turned to factor analysis, which largely occupied the remaining two decades of his life. Furthermore, scaling over any reasonable stimulus range can be very tedious; Guilford (1954, p. 157) used 5600 paired comparisons to scale a 30-gram (1-ounce) range of lifted weights. Finally, there is a problem of monotonic indeterminacy (Bock & Jones, 1968, p. 66; Eisler, 1965). For example, the usual Case V assumptions, in which the discriminal dispersion is constant, imply that subjective length is a logarithmic function of physical length. The Case VI assumptions, in which the discriminal dispersion is proportional to subjective value, lead to a more plausible linear psychophysical law for length. However, the logarithmic and linear laws cannot be distinguished within the paired-comparison approach.

For these and other reasons related to historical currents of interest, the Thurstonian development had little impact in just that area to which it was most suitable. In more recent times, of course, the Thurstonian approach has been overwhelmed in the popular imagination by S. S. Stevens's (1957) conception of psychophysical measurement.

Verbal and Symbolic Stimuli. Paired-comparisons scaling is not generally possible at the individual level for verbal and symbolic stimuli. To provide choice proportions different from 0 and 1, the individual must be uncertain in his choices and must maintain this same uncertainty over repeated presentations of the same choice pair. But definite preferences are the general rule for verbal and symbolic stimuli. Even for close choices, any initial uncertainty is likely to disappear across repeated presentations.

This limitation of paired comparisons is well known (Bock & Jones, 1968, p. 4; Torgerson, 1958, p. 167). In their definitive treatment of the Thurstonian approach, Bock and Jones refer to individual scaling by saying, "No suitable data representative of [this] case could be found in either the literature or other sources available to us [p. 2]."

Innumerable applications of paired comparisons have been made to construct scales of social attitudes, for example, and scales of preferences among foods, political candidates, handwriting, and so forth. In these cases, usable choice proportions are obtained by the practice of pooling choices over groups of individuals, each of whom ordinarily makes a single choice for each stimulus pair. This pooling would not help, of course, if all individuals had definite preferences and the same rank order; all choice proportions would then be 0 or 1 and no scaling would be possible. Pooling works because different individuals have different rank orders and so their pooled data provide usable choice proportions.

But these group scales have a peculiar status. The basis for their construction is the real individual differences in rank order that are lost in the pooling. Such group scales cannot adequately represent the individual scales. Such group scales may have practical information about the proportion of individuals that favor various choices, but they are not scales in any psychological sense.

Method of Categorical Judgment. Thurstone's conceptual analysis for paired comparisons can be extended to category ratings (see Bock & Jones, 1968; Torgerson, 1958). Individual scaling becomes possible if stimuli that are close in rank order overlap in the distributions of their ratings on repeated trials by the same individual. This requires a fairly dense array of stimuli, however, and would not generally be practicable. Bock and Jones (1968) indicate that no such applications had been made at that date. As a practical matter, the raw ratings might well do the same job more simply.

5.3.2 RELATIONS BETWEEN THURSTONIAN THEORY AND FUNCTIONAL MEASUREMENT

Similarities. Thurstone's system has similarities with functional measurement that appear most simply in Case V of paired comparisons. This paired-comparison analysis rests on an algebraic model, a monotone transformation, and factorial-type design. The model has a subtracting form although it is an "as–if" model, derived from other basic assumptions and not intended to mirror degree of preference on any choice. The monotone transformation is the normal deviate, which theoretically transforms the observable choice proportions to a linear scale. Finally, paired comparisons ordinarily lead to a design that can be considered as a triangular half of a complete factorial.

Although the subtracting model has only an "as–if" status, its two-variable constraint, implemented in the triangular design, allows a test of goodness of fit. Thurstone's approach, therefore, in contrast to that of Fechner, satisfied the first criterion of measurement theory; namely, that of providing a test of goodness of fit. Indeed, the Case V assumptions lead to a graphical test of parallelism. Because of these similarities, Thurstone's formulation may be viewed as a precursor of the functional measurement approach.

Differences. Thurstonian scaling theory and functional measurement take opposite approaches to measurement. In the Thurstonian approach, scaling is a methodological preliminary to substantive inquiry. There is just one basic scaling model; its role is to provide stimulus values for substantive analysis in every area. In the functional measurement approach, by contrast, scaling is usually derivative from some substantive model, and that model is employed as the base and frame for measurement within its own situational context.

The idea that stimulus scaling is a necessary methodological preliminary is natural enough, for that seems to be the practice in physics. The grip of this idea can be seen in Hull's (1952) attempts to adapt Thurstonian methods to his needs for quantification. Ironically, Hull's theory was expressed in algebraic form, and these theoretical equations could have led to a functional measurement approach (Anderson, 1962b, p. 408, 1978a, p. 360).

Response scaling, which has obvious importance for behavior theory, has the primary role in functional measurement. By contrast, response scaling is essentially ignored in the Thurstonian approach, which is concerned with stimulus scaling. Response scaling facilitates individual analysis, whereas Thurstonian methods face the difficulties already discussed.[a]

The Thurstonian approach is striking in its conceptual parsimony. It rests on a bare minimum of theoretical assumptions, all quite reasonable, and it requires only the simplest kind of response, namely, a discriminative choice. Functional measurement is more demanding, for it typically rests on some substantive algebraic model and usually employs a numerical response. It is these properties, however, that make it possible to obtain functional scales within the situational context. The analysis of the weight–size illusion of Section 1.3.7, for example, depends on ability to measure the heaviness value of visual appearance. That is straightforward with functional measurement, but would be difficult or impossible with Thurstonian methods. And although previous applications of functional measurement have rested on the validity of some algebraic model, they offer the prospect of developing experimental procedures that will yield linear scales of numerical response for general use where the models no longer hold.

Comment. Despite its extended popularity, the overall outcome of Thurstonian measurement theory has been disappointing. It does not compare in achievement to signal detection theory, for example, even though Thurstone's descriptive framework seems superior in its psychology to the normative framework used in signal detection theory (Anderson, 1974a, p. 246). Thurstone himself (1959, p. 15; see also Mosteller, 1958) commented on the neglect of his simple methods for subjective measurement as compared to the many applications of the far more demanding techniques of factor analysis.

A few attempts were made to extend the Thurstonian approach beyond scaling to investigate substantive problems, as in the study of gift combinations by Thurstone and Jones (1959; see also Bock & Jones, 1968), and in Cliff's (1959) treatment of the Adverb × Adjective model. From the present perspective, however, these attempts mainly exhibit the inappropriateness of Thurstonian methods (Anderson, 1974d, p.

279; Leon, Oden, & Anderson, 1973; Shanteau, 1975a). But regardless of the outcome, Thurstone's formulation deserves tribute as a major intellectual achievement in measurement theory.

5.4 Stevens's Method of Magnitude Estimation

In his paper, "On the Psychophysical Law," Stevens (1957) remarked that Fechner's formulation had been almost universally criticized for a century, but still held the center of attention and had no serious competitor. Stevens continued:

> The lesson of history is that a bold and plausible theory that fills a scientific need is seldom broken by the impact of contrary facts and arguments. Only with an alternative theory can we hope to displace a defective one.
>
> The purpose here is to try to do just that—to show that there is a general psychophysical law relating subjective magnitude to stimulus magnitude [p. 153].

With these words, Stevens introduced his method of magnitude estimation, in which subjects are instructed to assign numbers to stimuli in proportion to their subjective magnitudes. This method led to a striking result called the "power law." On almost every conceivable stimulus dimension, subjects' magnitude estimations formed a power function of physical stimulus intensity. The simplicity and apparent regularity of this empirical result made it seem to be a major discovery.

But the meaning of the power law rested entirely on one uncertain assumption. Stevens claimed that magnitude estimation provided a linear (or ratio) scale of subjective magnitude.[a] If Stevens's claim is justified, then the power function is the psychophysical law, the long-sought solution to Fechner's problem. Indeed, if Stevens's claim is justified, then magnitude estimation provides a solution to the problem of the linear scale.

On the other hand, Stevens's claim may not be justified; magnitude estimation may be a biased, nonlinear scale of subjective magnitude. Then Stevens's power law is not the psychophysical law, but an expression of response bias. A power law of response bias would still be interesting, but it would not have the significance that Stevens attributed to it.

The essential issue is conceptually simple. Some validational criterion must be obtained to test Stevens's central assumption that magnitude estimation is unbiased and provides a true linear scale. This issue has been emphasized by many writers, who have also pointed out that Stevens failed to provide the necessary validational criterion. The controversy that surrounded magnitude estimation was largely concerned with this issue of validity.

5.4.1 FUNCTIONAL MEASUREMENT AND MAGNITUDE ESTIMATION

Epistemological Relation. Functional measurement and magnitude estimation lie at different epistemological levels. Magnitude estimation is not a theory but a method, a specific experimental procedure for obtaining numerical responses. It contains within itself no answer to the central issue of validity. Whether magnitude estimation is biased or unbiased must be assessed at a different level.

Functional measurement is impartial in the controversy about bias in magnitude estimation (Anderson, 1970b, p. 166, 1975a, p. 475). However, functional measurement procedures can test whether magnitude estimation does yield a true linear scale. These procedures provide a validational criterion that can adjudicate the essential claim of response linearity.

Magnitude Estimation versus Ratings. The central necessity for a validational criterion is made clear by the sharp disagreement between the method of magnitude estimation and the method of ratings. The two methods are rather similar in procedure. Both ask for direct numerical estimates of subjective magnitude, and subjects find both methods simple and natural. Introspection gives little indication that one method might be preferable or that the two typically give different results.

How different these results can be is illustrated in Figure 5.1. Each curve represents judgments of grayness as a function of physical reflectance of Munsell gray chips. Magnitude estimation yields a curve that is bowed up; the rating method yields a curve that is bowed down. These two curves are quite different, and they suggest quite different conclusions about sensory function. Both cannot be true linear scales of sensation. Both cannot be the psychophysical law. Which, if either, is correct?

Parallelism as a Validity Criterion. The parallelism theorem of Section 1.2 provides a validational criterion that can be used to test Stevens's claim that magnitude estimation yields a linear scale of subjective magnitude. The logic is straightforward. Observed parallelism provides joint support for the model and also for the response scale. This logic applies not only to magnitude estimation but to any other response measure. If

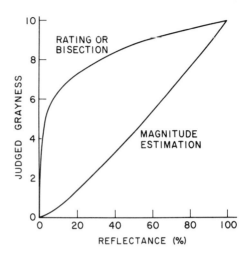

FIGURE 5.1. *Idealized judgments of grayness from rating (or bisection) response and from magnitude estimation as a function of reflectance (see Anderson, 1976c, 1977b; Stevens, 1975, p. 15). These two numerical response measures yield very different results.*

parallelism is obtained experimentally, that is prima facie evidence for the linearity of whatever response measure was employed.

However, initial applications of parallelism analysis supported the rating method, not the method of magnitude estimation. One example from social judgment (Anderson, 1962a) was presented in Figure 1.2; the observed parallelism supports the linearity of the rating response. According to Stevens (1966, 1975), ratings and magnitude estimation are nonlinearly related for this kind of social judgment. If the ratings form a linear scale, then magnitude estimation cannot.

One example from psychophysics (Anderson, 1970a) was shown in the judgments of lifted weights in Figure 1.10. Here again, the observed parallelism supports the linearity of the rating response. However, Stevens has repeatedly shown that ratings and magnitude estimation are nonlinearly related for such psychophysical judgments. If the ratings form a linear scale, then magnitude estimation cannot.

Parallelism analysis was used by Weiss (1972) in a direct comparison between ratings and magnitude estimation. Subjects were instructed to make intuitive judgments of average grayness of two Munsell gray chips. The instructions imply an averaging model, of course, and so parallelism is expected—if a linear response scale is used. Weiss used both ratings and magnitude estimation. Both have equal opportunity to satisfy the parallelism criterion. Both might fail. Both cannot succeed since they are nonlinearly related.[a]

Figure 5.2 shows that the ratings succeeded (left panel) and that magnitude estimation failed (right panel). Indeed, the vertical spread of the curves for magnitude estimation varies by more than 2:1 from left to

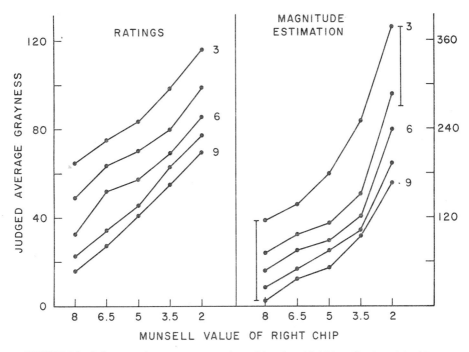

FIGURE 5.2. *Judgments of average grayness of two chips of specified Munsell value, right chip on horizontal, left chip as curve parameter. Parallelism in left panel provides joint support for averaging model and linearity of rating response. Nonparallelism in right panel, emphasized by the two equal-length vertical bars, implies that magnitude estimation is biased and invalid. (After Weiss, 1972.)*

right. Judged by the parallelism criterion, it must be concluded that magnitude estimation does not provide a true linear scale.[b]

These parallelism tests between ratings and magnitude estimation are impartial. The success of the rating method is important, but the essential point is that the issue of response validity need not remain a matter of opinion. The parallelism theorem places the issue of response validity on an empirical basis.

Of course, no one experiment or group of experiments will be conclusive. An adequate evaluation must look at the network of evidence. Accordingly, it is necessary to take up the evidence more directly associated with magnitude estimation itself.

5.4.2 EVIDENCE ON MAGNITUDE ESTIMATION

Bias Hypothesis. The following overview of evidence on magnitude estimation will be easier to understand if the bias question is first made explicit. Stevens's central claim is that magnitude estimation is unbiased and provides a linear scale of subjective magnitude. However, many writers have been concerned that magnitude estimation is biased and provides a nonlinear function of subjective magnitude (e.g., Attneave, 1962; Garner, 1958; Poulton, 1968; Zinnes, 1969).

One form of bias has particular relevance. Under this bias hypothesis, subjects use numbers in accord with a law of diminishing returns. It is as though the subjective difference between 110 and 120, say, is less than the subjective difference between 10 and 20. To express equal differences in stimulus value, therefore, the subject must use larger numerical differences at higher levels. Thus, the subjective difference between 10 and 20 would equal the subjective difference between, say, 110 and 130.

But magnitude estimation takes these numbers at face value. Hence it implies that subjective magnitude increases faster than it really does. Nearly all the differences between ratings and magnitude estimation can be readily understood in terms of such diminishing returns bias.

Diminishing returns bias is plausible from daily experience. A price differential of a few dollars would typically be more important in purchasing new tires, for example, than in purchasing a new car. Of course, more than a plausibility argument is needed. Indeed, the bias might go the other way, with the rating method being afflicted by a bias of increasing returns. Thus, the question returns again to the necessity of validational criteria.

The Power Function. Magnitude estimation attracted so much interest because it led to the so-called power law. If the response, R, is expressed as a function of physical intensity, S, it typically obeys a power relation of the form

$$R = c_1 S^n,$$

where c_1 is an inessential unit parameter and n is an exponent. As early as 1965, Ekman and Sjöberg could conclude: "The power law was verified again and again in literally hundreds of experiments [p. 467]." Even more hundreds of experiments since then have largely reached the same conclusion.

The attractiveness of the power function is clear. It has a simple mathematical form across diverse sensory dimensions that can hardly be fortuitous. It carries the tantalizing, persuasive implication that it arises from some deeper regularity of sensory functioning.

To be sure, any such substantive interpretation is totally dependent on the assumption that the response measure is valid. Otherwise, the power function is no more than a regularity in response bias. However, the very simplicity and seeming generality of the power function argue persuasively for the validity of the response measure. This is an appealing argument and, certainly, a legitimate hope. Nevertheless, it still requires validational support.

Stevens himself repeatedly asserted that magnitude estimation does provide a valid response scale and that the power function is therefore a true substantive law. Stevens's power law is thus intimately bound up with magnitude estimation. If magnitude estimation is valid, then the power function will provide useful information about sensory functioning. But if magnitude estimation is invalid, then so is Stevens's power law.

Psychophysical Integration. The most relevant evidence on magnitude estimation comes from psychophysical integration in which an algebraic model provides a validational base. Although functional measurement is neutral in the controversy, results obtained from many applications in over a decade of work have pointed to major biases in magnitude estimation.

One extensive line of evidence comes from the work on information integration theory (Anderson, 1974a, 1981a). Illustrative experiments are given in Figure 5.2 and in Figures 1.10–1.12. This work has employed the parallelism test as a validity criterion, and the rating method has been quite successful. By the reasoning of Section 5.4.1, this implies that magnitude estimation is biased and invalid.

Of special interest is the cross-task validation that results when the same stimulus scale is obtained from different tasks (Anderson, 1972a, 1974c, 1976c). For example, the psychophysical law for grayness of Figure 5.1 was verified with three different integration tasks—averaging, differencing, and bisection (see Figure 5.3). This grayness scale thus satisfies both within-task and cross-task consistency criteria, and so it seems reasonable to consider it valid. Accordingly, the comparison of the two curves of Figure 5.1 points to a strong diminishing returns bias in magnitude estimation.

Many other investigators have adopted algebraic models for scaling purposes (see Anderson, 1970b, p. 168, 1974a, p. 291, 1975a, p. 480). Notable among these is Garner's (1954a) joint use of two nonverbal tasks, fractionation and bisection, to obtain a scale of loudness. Also notable are studies by Marks, Curtis, Rule, Sjöberg, Birnbaum, and Weiss. But this work with algebraic models, which is discussed in Section 5.6, has generally been unfavorable to magnitude estimation.

In addition, there are a number of studies specifically intended to

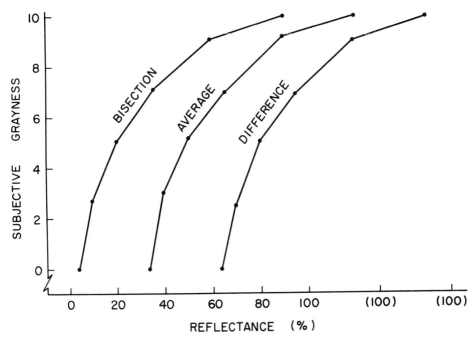

FIGURE 5.3. *Cross-task validation of functional scales of grayness. Each curve plots the psychophysical law as determined from one of three tasks: (a) bisection, (b) rating of average grayness, and (c) rating of difference in grayness. All three curves have virtually identical shape, thereby showing the same subjective values. (Curves displaced horizontally.) (From Anderson, 1977b.)*

assess magnitude estimation in simple integration tasks. For example, Beck and Shaw (1967) obtained magnitude estimations for differences in loudness of two tones, but found that these judged differences were not linearly related to differences calculated from Stevens's *sone* scale of loudness. Stevens (1971) tried to resolve this disagreement by claiming that the difference task itself, despite the use of the magnitude estimation response, produced a bias. An alternative interpretation is that the bias is in the method of magnitude estimation. Indeed, Stevens's own Figure 4, which replots the data of Beck and Shaw, shows rough parallelism—with a log response scale. By virtue of the parallelism theorem, Stevens's graph suggests that the difference model was correct, that magnitude estimation is biased, and that an approximate log transformation is needed to remove the bias (Anderson, 1974d, p. 272).

Other pertinent studies of difference judgments include Dawson (1971) on loudness and area and Marks and Cain (1972) on area, heaviness, and roughness. If magnitude estimation is indeed a valid scale, then the difference between the judgments of the two separate stimuli should equal the judgment of their difference. This prediction did not hold, which suggests that magnitude estimation is not a valid scale. Marks (1974b) summarizes this and related work by noting that all these experiments agree in their finding of nonlinear relations between the obtained and predicted difference judgments (p. 257). Judged by such validity criteria, magnitude estimation fails.

Social Judgment. The method of magnitude estimation has also been applied in the social realm to assess esthetic preference, seriousness of offenses, attitude toward religion, and similar judgments. These stimuli have no physical metric, and so a power function cannot be fit to these data. However, Stevens (1966, 1968, 1975) stressed that magnitude estimations of social stimuli were nonlinearly related to ratings.

This disagreement implied that at least one of the methods did not yield a linear scale of subjective magnitude. The pattern of disagreement was the same as had been obtained with psychophysical stimuli. Both social and psychophysical stimuli show that ratings are roughly logarithmic functions of the magnitude estimations for most stimulus dimensions. If the ratings are valid, magnitude estimation suffers diminishing returns bias. Stevens claimed that magnitude estimation yielded the true social metric, but he presented no validational criterion. [a]

The opposite conclusion comes from experimental studies on information integration theory. The integration model provides the necessary validational base, and extensive work on social judgment over two decades has provided exceptionally strong support for the rating method. Judged against this validational criterion, magnitude estimation fails.

Inconstant Exponents. The power function was attractive because it implied that the operation of each sensory system could be characterized by a single parameter, namely, the exponent. That was Stevens's (1957, 1960, 1961) original claim; namely, that the exponent was an invariant characteristic of the sensory transducer. This requires the exponent to be constant, independent of various stimulus conditions. Observed constancy would provide validational support for the power law. In fact, however, exponents are not constant but inconstant.

The first sign of inconstant exponents was obtained by Stevens himself in his initial article on magnitude estimation of loudness. This article took up various aspects of method, including the choice of a standard stimulus, relative to which the other stimuli are to be judged. "Use various standards," advised Stevens, "for it is risky to decide the form of a magnitude function on the basis of data obtained with only one standard [1956, p. 6]." But Stevens's own data (e.g., Figure 4) showed that two different standards yielded markedly different exponents. Both could not represent the true magnitude function; at least one must be biased. Neither then nor later, however, did Stevens provide a basis for deciding which exponent was correct or how the correct value was to be ascertained.

Exponent values also show strong dependence on the range of stimuli presented for judgment. The greater the range, the smaller the exponent, much as though subjects tend to use a fixed range of response numbers. This is not a sensory effect but a response bias, as Stevens (1971) acknowledged. As Poulton (1968) remarks in his review, this inconstancy of exponent values is not compatible with Stevens's claim that the exponent is an invariant characteristic of the sensory system.

These and other biases in magnitude estimation have been discussed by many writers (see Jones, 1974; McGill, 1960; Marks, 1974b, 1978b; Poulton, 1968, 1979; Stevens, 1974). Stevens's position is that they represent procedural problems that can be resolved experimentally. To resolve them requires a validational criterion for determining the correct value of the exponent. Stevens did not provide such a validational criterion.

Ratings Yield Power Functions. Category ratings also yield power functions for many psychophysical dimensions, as was shown by Marks (1968). This result created a problem for Stevens, who had argued that ratings are biased and invalid and that the power function obtained with magnitude estimation testified to the validity of the technique. If ratings also yield power functions, then the same reasoning testifies to the validity of the ratings.

Different power functions were obtained from ratings and magnitude estimation, of course, because the two are nonlinearly related, as illustrated in Figure 5.1. This result re-emphasizes the dilemma. Since both response methods yield power functions, some validational criterion is necessary to determine which, if either, is correct. Stevens (1974) resolved the issue by calling the rating exponents *virtual* in contrast to the *actual* exponents obtained from magnitude estimations. This terminology is arbitrary and evades the issue.

Power Functions as Curve Fitting. There is a simple statistical explanation for the power function. It is well known that it is a flexible tool for fitting a general class of curves (e.g., Mandel, 1964). It is not surprising, therefore, that a power function has been found to fit almost any judgment of psychophysical magnitude. As noted above, different standards yield power functions, different stimulus ranges yield power functions, and even ratings yield power functions. But these power functions all have different exponents. Both curves in Figure 5.1 are power functions, but the exponents are .2 and 1.2 for the ratings and magnitude estimations, respectively. *The* power function can hardly be a substantive law, therefore, when the true exponent is not determinate. The empirical ubiquity of the power function is merely a reflection of its capability as a curve-fitting device.[b]

Ecologically, the ubiquity of the power function has an obvious origin. Sensory systems, vision, for example, need to be sensitive at extremely low intensities, as at night, yet still function at extremely high intensities, as in the midday sun. The sensory systems accomplish this through a law of diminishing returns. As a purely statistical matter, most reasonable laws of diminishing returns will be well fit by a power function. To say that sensory systems follow a "law" of diminishing returns seems trivial. Observed power functions say little more.

Stevens's power law claims to say much more, of course, for it claims to represent subjective magnitude. If this claim were true, then Stevens's power law would provide a useful descriptive tool. But this claim rests on the assumption that magnitude estimation is a true linear scale of subjective magnitude. If magnitude estimation is invalid, then so is Stevens's power law.

Cross-Modality Matching.[c] Stevens attempted to validate magnitude estimation by coupling it with cross-modality matching. Since cross-modality matching at present seems to be the main argument for magnitude estimation, it deserves specific consideration. The subject gives magnitude estimates for two sensory dimensions, loudness and brightness, for example, and power functions are obtained in the usual way

$$R_L = S_L^{n_L}, \text{ and } R_B = S_B^{n_B},$$

where L and B index loudness and brightness and the inessential unit constants are ignored. In addition, the subject adjusts loudness to match each brightness, and the two corresponding sets of physical intensities provide a third power function

$$S_L = S_B^x,$$

where x is the cross-modality exponent.

The loudness–brightness matching has the advantage of dealing directly with the sensations themselves. Hence it is unaffected by any bias in the magnitude estimations. Moreover, it seems to provide a validational criterion: If magnitude estimation is a ratio scale, then the ratio of the two magnitude estimation exponents, n_B/n_L, equals the cross-modality exponent, x. To understand this, consider a matched loudness–brightness pair. If cross-modality matching is valid, then both stimuli have equal subjective magnitude and so elicit equal magnitude estimation responses. For each matched pair, therefore, $R_L = R_B$ and, from the two initial power functions

$$S_L^{n_L} = S_B^{n_B}.$$

Hence

$$S_L = S_B^{n_B/n_L}.$$

The two equations for S_L yield

$$x = n_B/n_L.$$

This relation is testable because the three exponents can be calculated from the observed data. Verification of this relation would naturally seem to provide strong support for the validity of magnitude estimation. Unfortunately, that support turns out to be hollow.

One problem, which various writers have pointed out, is that the same prediction holds even if magnitude estimation is not a valid ratio scale of sensation (see e.g., Anderson, 1972a). Under the hypothesis of diminishing returns bias, for example, the true sensation, r, may be considered as a power function of the observed magnitude estimation response, R. If a is the bias exponent, then

$$r_L = R_L^a = S_L^{n_L} \text{ and } r_B = R_B^a = S_B^{n_B}.$$

For each matched loudness–brightness pair, $r_L = r_B$, and so

$$R_L^a = R_B^a,$$

or

$$S_B^{n_L/a} = S_B^{n_B/a}.$$

Upon taking the ratio of these two exponents, the bias factor a cancels out

$$x = (n_L/a)/(n_B/a) = n_L/n_B.$$

Thus, the cross-modality exponent, x, is predicted even if magnitude estimation is biased. Empirical verification of the cross-modality prediction therefore does not validate magnitude estimation (see also Luce & Galanter, 1963).

A suggested way out of this dilemma appeals to the fact that magnitude estimates of length, duration, and angle are roughly proportional to their physical values (e.g., Marks, 1974b, p. 30). On the plausible assumption that subjective sensation is proportional to physical value on these three dimensions, it would follow that magnitude estimation is unbiased for these three dimensions. From this it is inferred that magnitude estimation is unbiased for all dimensions.

This inference does not follow. It rests on an implicit assumption that magnitude estimation is a uniform procedure across all dimensions. The three nonenergetic dimensions of length, duration, and angle are exceptional, however, in that ratings and magnitude estimation generally agree on them. Since ratings and magnitude estimation do not generally agree on other dimensions, at least one must be nonuniform; that one may be magnitude estimation. Even if magnitude estimation is unbiased for these three dimensions, therefore, it may still be biased for other dimensions, which indeed appears to be the case.

In short, the studies of cross-modality matching have not provided the sought-for validational base for magnitude estimation. At the same time, the cited studies of stimulus integration indicate that magnitude estimation is generally subject to a diminishing returns bias.

5.4.3 VIEWS ON MAGNITUDE ESTIMATION

The Issue of Validity. The several lines of evidence surveyed here all tend to a uniform conclusion. Magnitude estimation has repeatedly failed to satisfy the validational criteria that have been proposed, whereas other numerical response methods have frequently succeeded. It seems fair, perhaps only prudent, to conclude: "Magnitude estimation . . . must be biased and invalid [Anderson, 1972a, p. 389]." This conclusion agrees with earlier discussions by Garner (1954a, 1954b, 1958) and by Poulton (1968).

What is curious is that this conclusion seems to remain very much a minority view. Current journals are replete with experimental studies that unquestioningly employ magnitude estimation as though it yielded a ratio scale and meaningful exponents. In mathematical psychology, Luce and Galanter (1963, pp. 280ff) express concern over Stevens's claim that magnitude estimation yields a true ratio scale, and Krantz (1972b, p. 1432) suggests that Stevens's claim is a pun on the use of the word *ratio* in the instructions to the subject. Nevertheless, Luce (1972) comments on "the striking regularities of the data obtained by magnitude methods [p. 98]," and Krantz comments similarly. Other prominent workers in measurement theory have also expressed favorable views (e.g., Tukey, 1969, p. 88), while Krantz, Luce, Suppes, and Tversky (1971), Krantz

(1972a), and Shepard (cited in Krantz) have tried to find an axiomatic basis for magnitude estimation. Even the National Science Foundation (1974) has hailed extensions of magnitude estimation from psychophysics to social science, saying, "This research may well improve the precision of measurement in the social sciences by several orders of magnitude [p. 33]."

The preceding survey of evidence, although brief and simplified, does not seem greatly different from Stevens's own evaluation, although his interpretation is different. Stevens attributes the inconstant exponents caused by changes in the standard to bias induced by the standard, not to bias in magnitude estimation itself (see Section 5.4.2 and Anderson, 1974a, 1974d). Similarly, he attributes failure of magnitude estimation to satisfy the models for bisection and for difference judgments to partition bias induced by these tasks, not to bias in magnitude estimation itself (Stevens, 1971, pp. 430–433). Indeed, Stevens has either ignored or disavowed nearly all the validity criteria that have been proposed for magnitude estimation. Only cross-modality matching remains admissible, but Stevens apparently never discusses the well-known fact that this task does not actually provide a validational criterion.

Stevens's position is not, perhaps, logically refutable. For example, the choice of standard could cause a bias, and it is reasonable to try doing without an explicit standard. Similarly, the failures of the bisection and difference models to support magnitude estimation could mean that those models, not magnitude estimation, are invalid. But this stand leaves magnitude estimation without a validational base. As a previous review (Anderson, 1974a) concluded:

> Without a validational base, however, the controversy over magnitude estimation becomes meaningless. From that standpoint, the claims that have been made for magnitude estimation are not so much invalid, as devoid of meaning or content. Stevens (1971) appears to recognize this problem in his discussion of validity since he concludes that "the answer becomes a matter of opinion. . . ."
>
> To the many workers who have employed one or another of the algebraic models, response validity is not "a matter of opinion." The model provides a criterion against which to validate the response measure. Magnitude estimation and rating scales have equal opportunity to meet this criterion. Judged on this ground, magnitude estimation fails [p. 290].

The New Psychophysics. Magnitude estimation was heralded as the basis for a "new psychophysics" characterized by this method of direct scaling. Instead of the threshold discriminability data of Fechner and Thurstone, subjective sensation was to be assessed directly with verbal estimates of magnitude. In itself, the search for a direct scaling method was desirable and important. Quite naturally, but yet unfortunately, this search became centered on single-variable analysis. Some many-variable studies were conducted, but, as already noted, these were virtually all disavowed by Stevens.

The reason for concentrating on single-variable tasks is clear. Application of magnitude estimation was imbued with the traditional view of measurement as a methodological preliminary to substantive inquiry. Since contextual stimuli can affect the subjective magnitude, the overt response, or both, it seems desirable to eliminate such stimuli, with the hope of reducing the subjective magnitude and its measure to their purest form.

This point may be illustrated in terms of the functional measurement diagram of Figure 1.1. Magnitude estimation claims to provide a direct measure of subjective magnitude, s. This requires that the overt response, R, be a linear function of the implicit response, r. It further requires that r and s be equivalent, or, in other words, that no integration of contextual stimuli distort perceived magnitude of the focal stimulus. The obvious way to do that is to eliminate all but the focal stimulus, thereby adopting a single-variable approach.

This traditional approach to measurement theory, and the associated concentration on single-variable tasks, characterize the bulk of work on magnitude estimation. Primary concern has been with scaling and the psychophysical law, or valuation function, in Figure 1.1. But because of the concentration on single-variable tasks,

> The experimental base of traditional psychophysical scaling with direct, numerical response methods is inherently too narrow to support a solution to its problems. Tasks based on psychophysical information integration provide a broader and potentially simpler approach to scaling. . . . Scaling does not decline in importance, but instead assumes its proper place in the development of substantive theory [Anderson, 1970b, pp. 153, 168].

Stevens's dream of a new psychophysics characterized by a method of numerical response is shared by workers in information integration theory. Their approach is different from Stevens's conception, however, for it employs the rating method that Stevens condemned. More important, this new psychophysics lays its foundation at a deeper level of substantive inquiry, replacing the single-variable approach with the many-variable analysis of stimulus integration. Many-variable analysis allows the study of algebraic models that in principle provide the necessary validational criterion for the numerical response measure and in practice have had substantial success with the rating response.

As a consequence of this focus on many-variable analysis, the psychophysical law loses much of its interest and former significance (Section 1.8.1). The focus of inquiry moves to the integration function, or psychological law. Scaling and measurement are no longer primary, but derivative from substantive laws of stimulus integration. Psycho–physics itself merges into the greater field of perception.

5.5 Conjoint Measurement

Functional measurement (Anderson, 1962a,b) and conjoint measurement (Luce & Tukey, 1964) have several similarities. Functional measurement emphasizes the use of factorial design by its joint variation of two or more stimulus variables; so does conjoint measurement. Functional measurement makes considerable use of adding and multiplying models; so does conjoint measurement. Functional measurement incorporates monotone response analysis, thereby allowing joint scaling of stimuli and response; conjoint measurement aims to do the same by using ordinal analysis. These characteristics are also shared with optimal or best-fit scaling (Section 5.6.1).

However, functional measurement and conjoint measurement both differ from optimal scaling in one basic way: Both recognize the necessity for tests of goodness of fit aimed at establishing some model. Luce (1967, p. 37), Krantz and Tversky (1971a, p. 167), and Krantz, Luce, Suppes, and Tversky (1971, pp. 33, 435) emphasize that ANOVA, MONANOVA, and other optimal scaling methods are not measurement theories because they do not focus on establishing model validity. Such methods, as they point out, simply assume a scaling model and impose it on the data. They always yield a set of scale values, but they leave open the question of scale validity, which is essential to measurement theory.

The term *conjoint measurement* is used here to refer to the axiomatic formulations of Luce and Tukey (1964) and Krantz *et al.* (1971). This point may need emphasis because some writers have used the term as though it referred to optimal scaling, especially with additive models in factorial-type design. That seems inappropriate because optimal scaling has a quite different, nonaxiomatic basis that originates from the work of R. A. Fisher and other statisticians. Moreover, as the cited comments indicate, optimal scaling is not measurement theory because it merely assumes the validity of the scaling model.[a]

Functional measurement, in contrast to optimal scaling, is centrally concerned with model validity. The model serves not only as the scaling frame but also as the base for scale validity. Conjoint measurement similarly emphasizes the essential need to test and establish a model.

Because of these similarities, functional measurement and conjoint measurement have frequently been confused. At the same time, persistent misconceptions about functional measurement have arisen. This section will try to clarify the distinction between the two approaches and present a defense of functional measurement.[b]

The four main points may be briefly summarized. First, conjoint measurement is not necessary. Functional measurement provides an alternative logical foundation based on integration functions rather than on the axioms of conjoint measurement. Second, conjoint measurement is logically insufficient; the basic representation theorem is not valid for finite designs that are required in practice. Third, conjoint measurement has minor practical usefulness. Finally, conjoint measurement rests on an inappropriate epistemological base, for the axioms are not substantive in nature.

5.5.1 LINEAR RESPONSE SCALES

Krantz and Tversky (1971a) have criticized the use of rating scales in functional measurement with the statement that such use rests on the assumption (i.e., presupposition) "that the numerical values of the dependent variable can be regarded as interval-scale measures of the relevant psychological attribute [p. 166]." This statement reflects a misconception, but it has been widely accepted, even by authorities on measurement (e.g., Cliff, 1973, p. 480; Falmagne, 1976, pp. 65–66). This issue requires discussion because it relates to a basic difference between functional measurement and conjoint measurement.[a]

Logic of Parallelism. Suppose that numerical responses are obtained for an experiment in two-factor design and that the data exhibit parallelism. This obviously demonstrates that an additive representation exists, for it subsists in the parallelism. To appeal to conjoint measurement is neither necessary nor useful.

To apply conjoint measurement, the data would have to be treated as rank orders for the axiom tests. Reduction to rank order can only lose information, of course, and so can only yield weaker conclusions. When the observed data are additive, conjoint measurement cannot properly reach any conclusion different from parallelism analysis. With only ordinal methods, however, and without a test of goodness of fit, conjoint measurement may not be able to reach any conclusion (Note 5.5.2a).

Functional measurement allows analysis of variance methods to assess whether observed deviations from parallelism are statistically significant. This requires no assumption about linearity of the response scale

(Anderson, 1961a, in press, Sections 1.1.9, 2.1), only certain distributional assumptions about observable responses. If these distributional assumptions are not met, distribution-free tests of interaction components can be employed. In general, therefore, valid and sufficient tests of fit are available.

These tests do not assume linearity, they assess it. If the parallelism test fails, that raises a question about response linearity; if the test succeeds, that supports response linearity (Section 1.2). In the beginning, there were certainly good reasons to doubt that this numerical response methodology would succeed. But it has succeeded, as illustrated by the many experimental applications in Chapter 1, and has thereby provided a foundation for measurement.

It deserves re-emphasis that the metric quality of the numerical response measure does not rest on any assumption that the response numbers are a linear or interval scale. The metric quality derives solely from the algebraic structure of the integration model. Hence there is no need for prior measurement. In accord with the terminology of Luce and Tukey (1964), therefore, this application of functional measurement may be considered to be a "new type of fundamental measurement [p. 1]."

Representation Theorem. The role of the parallelism theorem as a representation theorem may deserve mention since such theorems constitute one of the main concerns of mathematical measurement theory (Krantz *et al.*, 1971; Suppes & Zinnes, 1963). The parallelism theorem may be considered a representation theorem for adding-type models because it specifies testable and sufficient conditions for an additive representation of error-free data. These conditions are also necessary if the response scale is linear. If the response scale is only monotone but the underlying integration process is additive, then monotone transformation procedures will linearize the scale and produce parallelism.

The parallelism theorem is not ordinarily considered as a representation theorem, in part because it is mathematically so elementary. The fact is, of course, that every model analysis is perforce concerned with representation; that is, whether the data can be represented in the form prescribed by the model. An advantage of the parallelism theorem is that it readily allows for incorporation of monotone transformation and response variability. It thereby provides effective analysis even for ordinal data.

5.5.2 THE AXIOMATIC APPROACH

Role of the Conjoint Axioms. The major accomplishments of the work on conjoint measurement are the representation theorems. For simplicity, present discussion considers the additive model, which has been of main concern in conjoint measurement. The representation theorem asserts that if certain axioms are satisfied, then the data can be represented in additive form. This theorem is important, however, only because of self-imposed limitations of conjoint measurement. This theorem is not necessary to measurement theory.

To understand this issue, it is first desirable to see why the representation theorem is important for conjoint measurement. Suppose that data from a two-factor design are available and that the investigator wishes to determine whether or not the data can be represented in additive form. In attempting to answer this question, conjoint measurement limits itself to ordinal comparisons of response pairs. The independence axiom involves rank order comparison of two response pairs from a 2 × 2 design; the two rows (and columns) should not cross over for independence to hold. With error-free data, independence can be assessed by inspecting all possible 2 × 2 subdesigns. No crossovers is clearly a necessary condition for an additive representation, but it is not sufficient (W. H. Kruskal, cited in Scheffé, 1959, p. 98).

In a search for sufficient conditions, conjoint measurement proceeds to add the assumption or axiom of double cancellation, which can be viewed as an ordinal implication of parallelism in two linked 2 × 2 designs (Krantz et al., 1971, pp. 250–253; Scheffé, 1959, p. 98). Since the adding model implies double cancellation, this condition is also necessary. The question is whether the two conditions of independence and double cancellation are sufficient.

The answer to this question is a conditional *yes*, as shown by the basic representation theorem of additive conjoint measurement. Independence and double cancellation, together with certain more technical assumptions, are sufficient to guarantee that the data can be expressed in additive form. And implicit in additivity is the existence of linear scales of the stimulus factors. Mathematically, this representation theorem is impressive, for it is not at all obvious that such simple ordinal conditions could suffice to reveal an underlying additive structure. Indeed, only the independence axiom is needed with more than two factors. Furthermore, these conditions provide a potential basis for data analysis. However, three limitations on the representation theorem need to be kept in mind.

Limitations on the Axiomatic Approach. The first limitation concerns the difficulties of analyzing real data. The representation theorem applies only to error-free data; it says nothing about handling the response variability in real data. Because conjoint measurement has not solved these problems of data analysis, few, if any, applications have actually measured something. What applications have been made are virtually all subject to the cited criticism of optimal scaling methods, namely, that they assume the validity of the model.[a]

The second limitation is that the representation theorem is nonconstructive. Although the theorem may imply that the scales exist, it does not provide any way to get them. Additional procedures are required for actual measurement.

The third limitation of the axiomatic approach is that the representation theorem is not sufficient for finite designs (excepting special cases of minor usefulness). Even though all the axioms are satisfied, even though the data are error-free, the conclusion of the theorem may be false in a finite design. Tests of independence and double cancellation are thus logically insufficient to establish the model. A numerical example is given by Krantz et al. (1971, p. 425), who discuss this matter in more detail.

Actual applications are limited to finite designs, of course, so it is not enough to show that the axioms are satisfied. It is also necessary to demonstrate that a certain system of inequalities has a solution. For error-free data, as Krantz et al. (1971) point out, the system of inequalities can be analyzed using standard mathematical theory with numerical methods of linear programming. But this approach renders the axioms and theorems superfluous.

Nature of the Axioms of Conjoint Measurement. The axioms of conjoint measurement are different from the usual kind of scientific axioms. Scientific axioms ordinarily refer to theoretical processes, whereas the conjoint axioms do not. The conjoint axioms cannot, therefore, provide a basis for theoretical deduction. Instead, their essential function is to provide a basis for testing goodness of fit. This point requires discussion because it is sometimes thought that the axioms and theorems have substantive significance.

The usual kind of scientific axiom refers to some empirical concept. In physics, axioms refer to physical concepts such as force, mass, and acceleration in Newton's laws. Psychological axioms, similarly, should refer to psychological concepts. Examples include the concept of stimulus as a distribution in Thurstonian theory and in signal detection, the concepts of motivation and valence in behavior theory (e.g., Section 1.5.5), and the concepts of subjective probability and value in utility theory (e.g., Section 1.5.1). The measurement axioms, in contrast, refer neither to psychological concepts nor to psychological processes involved in any actual response.

To illustrate this contrast, suppose that the response on any trial can be represented as the sum of the stimulus values for that trial. Substantive axioms should refer to the psychological processes that operate during that particular trial. It is these processes that generate the stimulus values and integrate them to yield the overt response. Functional measurement focuses on these processes. Indeed, the very statement of an algebraic model involves hypotheses not merely about the integration operation but also about the reality of the constructs represented in the terms of the model (Section 1.8.2).

The measurement axioms do not refer to processes on a single trial. Instead, they refer to rank-order relations among overt responses on different trials. Thus, the independence "axiom" refers to comparisons among the four responses in a 2×2 design. This comparison is made by the investigator; in general, it has no referent in the psychological processes involved in any one response. The same holds true for the various cancellation axioms. These axioms merely refer to statistical properties of observable data, much like the parallelism property.

The measurement axioms are sometimes called empirical laws, but that seems inappropriate. The absence of crossovers in a set of data may be a useful fact, but it hardly qualifies as an empirical law. There is, however, another sense in which the measurement axioms may be viewed as empirical laws. If they hold, they imply an additive representation of the data together with the associated scales of measurement, at least for infinite designs. From this standpoint, the model and scales might appear to be derivative from the axioms and the axioms therefore more fundamental than the model and scales.

But the same argument could be made for the response transformation procedures, together with the associated error theory developed in functional measurement. These procedures can construct and validate an additive representation and the scales. From a substantive standpoint, however, it hardly seems appropriate to say that the model and scales are derived from the response transformation. From a substantive standpoint, the axioms, like the transformation procedures, are more fundamental than the model and scales only in a data-analytic or statistical sense. Substantively, the model, which represents the structure of the data-generating process, is more fundamental than the data generated. The model may not be the final level of analysis, but any deeper analysis should seek for substantive processes, not rank orders of observable data.

It thus appears that the content and potential value of the measurement axioms and theorems is mainly for testing goodness of fit. The representation theorems are not theoretical deductions in the customary sense. They lack substantive significance and cannot of themselves confer any confidence in the model or any likelihood on the existence of measurement scales. These are empirical issues and can only be decided on a substantive basis. Contrary to popular impression, the representation theorems do not prove that the scales exist; they can do no more than test whether the data are consistent with the operation of, say, an adding model.

That seems to be the essential contribution of the conjoint measurement approach. It provides a novel basis for ordinal data analysis that is roughly analogous to monotone regression in functional measurement. But techniques for testing goodness of fit are conceptually insufficient as a foundation for measurement theory. The axioms and theorems of conjoint measurement have no deeper epistemological role.

5.5.3 EMPIRICAL ANALYSIS

Data Analysis. Applications of conjoint measurement are rare, as has been remarked by numerous writers (e.g., Cliff, 1973, p. 478; Estes, 1975, p. 273; Falmagne, 1976, p. 65; Rapoport & Wallsten, 1972, p. 133; Tukey, 1969, p. 88; Zinnes, 1969, p. 453).[a] The rarity of empirical applications stems from the inability of conjoint measurement to handle response variability. Valid statistical analysis of the axioms themselves is possible with multiple comparisons procedures (Anderson, in press, Sections 5.4, 5.5). The axiom tests are not sufficient, however, as already noted.

Beyond the first step of obtaining a valid significance test lie other questions of data analysis that are no less important. These include problems of power and various problems of estimation that are essential to measurement. These problems are difficult to study under the limitations imposed by the axiomatic approach. Conjoint measurement has no place for the rating scale or for the simple parallelism property that have proved so useful in functional measurement.[b] The remarkably useful analysis of variance, being numerical, also has no place within the ordinal approach of conjoint measurement. Even monotone regression methods such as MONANOVA (Kruskal, 1965) and ADDALS (de Leeuw, Young, & Takane, 1976) are classed as numerical and not admissible within the axiomatic approach (Krantz *et al.*, 1971, p. 33; Krantz & Tversky, 1971a, p. 167). This also sacrifices a valuable tool to the axioms.

To an empirical worker, it seems that conjoint measurement shuns just those techniques that are most useful. Ramsay (1975) comments, similarly, that the conjoint measurement procedure of standard sequences for scale construction does not make practical sense. Ramsay points out the practical superiority of the optimal scaling procedures that have been used in functional measurement and calls for further psychometric work in this direction. Some contributions to this endeavor are given in the companion volume on methods (Anderson, in press, Chapter 5).

Qualitative Tests. One qualitative test has played a central role in the development of information integration theory. This is the crossover test for 2 × 2 designs, well-known for its capability of ruling out additive models. This crossover test, which appeared in the averaging–adding issue discussed in Section 2.3, is robust in the sense that it requires only a monotone response scale. No less important, it rules out a large class of models that are not strictly additive, such as additive models with diminishing returns.

Writers on conjoint measurement have suggested that direct tests of qualitative axioms, such as the crossover test (independence axiom), can be more informative about failures of a model than can overall factorial analysis, with or without monotone transformation. The crossover test has long been used for just this reason, of course, as illustrated by the averaging–adding studies referred to in the preceding paragraph.

Conjoint measurement has pursued a systematic study of other qualitative tests, collectively known as cancellation tests. These tests can be used for diagnosis of more complex models (Krantz & Tversky, 1971a). These cancellation tests have potential applications for rejecting models in the same manner as the crossover test.

But rejecting models, however useful it may be, allows no measurement. Moreover, the crossover axiom for model rejection is prior to and independent of any representation theorem. Use of cancellation tests for model rejection is also independent of the representation theorem, as illustrated by the fact that these tests remain necessary for finite designs for which the representation theorem does not hold.

Accepting a model, which is required as a base for measurement, is a very different matter from rejecting it. This is well illustrated by the 2 × 2 design: A crossover will rule out additivity but lack of crossover gives little basis for accepting additivity. In the averaging–adding issue, it was not rejection of the adding model that established the averaging model but other tests of quantitative kind.

Averaging Theory. The averaging model has fundamental importance for measurement theory in psychology. One reason is the empirical ubiquity of averaging processes (Sections 1.6, 2.3). Another reason has to do with two aspects of the algebraic structure of the model. The averaging concept necessitates a two-parameter, w–s representation of the stimulus (Section 4.4). Associated with this two-parameter representation is the inherent nonadditivity of the model.

Conjoint measurement does not appear capable of analyzing the averaging model. With differential weighting, the model does not have a simple polynomial form. It is even nonmonotone as a function of the stimulus. Hence the averaging model will in general exhibit crossovers, thereby violating the axiom of independence. Without an axiom of independence, it seems doubtful that the ordinal approach of conjoint measurement can handle the averaging model. If not, conjoint measurement is too narrow to meet the demonstrated needs of measurement theory in psychology.

5.5.4 EPISTEMOLOGY

The foregoing differences between functional measurement and conjoint measurement stem from one basic epistemological difference. The two approaches embody fundamentally different conceptions about the nature of measurement theory. Functional measurement treats measurement theory as an organic component of substantive theory. Conjoint measurement treats measurement theory as an abstract mathematical discipline that is a preliminary to substantive inquiry.

This difference is not absolute, of course, for workers on conjoint measurement have been sensitive to the problem of putting it to work. Nevertheless, the difference is deep-seated and genuine. As Estes (1975) remarks:

> One reason for the relative paucity of connections between [conjoint] measurement theory and substantive theory in psychology may arise from the fact that models for measurement have largely been developed independently as a body of abstract formal theory with empirical interpretations being left to a later stage. The difficulty with this approach is that the later stage often fails to materialize [p. 273].[a]

The difficulties facing conjoint measurement that were discussed in preceding sections flow from this epistemological assumption that measurement theory is an abstract, formal discipline, committed to a certain kind of axiom and to certain algebraic methods of proof. This formalization has been pursued with high ability and with faithful adherence to first principles. It has led to better understanding of certain points in measurement theory and to cancellation tests that can be useful in rejecting models. Even the formal development, however, remains seriously incomplete in its treatment of finite designs and response variability.

And because of the a priori formal commitment, divorced from substantive inquiry, conjoint measurement has neglected empirical findings that have significance for the conceptual nature of psychological measurement. Judged against these needs, the potential of conjoint measurement as a foundation for measurement theory seems in doubt.

5.6 Other Approaches to Measurement Theory

5.6.1 OPTIMAL SCALING

Optimal scaling refers to procedures that transform the data to improve their fit to some statistical criterion: to maximize additivity, for example, or to obtain normal distributions. The ordinary use of transformations in analysis of variance is one familiar kind of optimal scaling. Kruskal's (1965) MONANOVA and the ADDALS program of Young, de Leeuw, and Takane (1976, p. 526) are also in the optimal scaling tradition.

It is generally recognized that such scaling techniques are not theories of measurement. They are techniques imposed more or less arbitrarily for practical utility rather than for theoretical description. For example, Krantz and Tversky (1971a) note of MONANOVA that, "Like other scaling techniques, these methods always provide numerical estimates of the scale values, regardless of whether the rule in question is valid or not [p. 167]."

A pertinent example of optimal scaling is provided by Tukey's (1949) test for nonadditivity. This nonadditive component turns out to be the bilinear component and provides a test of the multiplying model for independent scores. Tukey was not concerned with measurement theory, however, but with optimal scaling, and he accordingly advocated transformation to additivity. This will generally increase power for tests of main effects, a primary consideration in certain kinds of investigations. For measurement purposes, however, transformation to additivity of data from a multiplying rule would produce nonlinear scales and hence would normally be avoided.

Many developments in optimal scaling have been important in functional measurement. Even the analysis of variance might be considered as an optimal scaling procedure for the stimulus variables. Functional measurement is much indebted to workers in optimal scaling who have provided remarkably useful and flexible statistical methods that could be adapted in relatively simple ways (see Anderson, in press).

A related debt should be recognized, for the idea of using transformations to rectify the response scale also arose in optimal scaling, especially in the work of Fisher (e.g., 1958). The feeling has been widespread that a response scale that yielded additivity was somehow more natural. Justification for this feeling was lacking, however, and of course it cannot be justified on mathematical grounds alone. Functional measurement provided a justification because it placed the issue on a substantive basis. In so doing, it showed that the additivity criterion has narrow applicability. The averaging model, in particular, is additive only under the equal weighting condition. This substantive basis is vital to measurement theory, in which it is as important to know when not to tranform to additivity as to know how.

5.6.2 MULTIPLE SCALES

Marks (1974a, 1978a) has advocated a resolution of the controversy between ratings and magnitude estimation in which both are valid but measure different perceptual qualities. Magnitude estimates are considered to be measures of sensory magnitude of single stimuli, whereas ratings are considered to reflect relationships among two or more stimuli. Each would be valid, but for different kinds of tasks. In this view, the controversy would merely reflect an inappropriate assumption of scale invariance across different tasks.

One rationale for Marks's view comes from the task of judging differences between two stimuli. Difference judgments are obviously relational, and may be considered as judgments of dissimilarity between two magnitudes. Ratings might be appropriate for difference judgments, but magnitude estimation would not. That would remove the objection of Section 5.4.2 that magnitude estimation has failed to satisfy differencing models. A similar distinction between magnitude scales and discrimination scales has been advocated by Eisler (1963) and Treisman (1964).

Marks tentatively suggested this relational view might be extended to general integration tasks such as averaging, and even to the rating scale itself. Thus, averaging could perhaps be conceptualized in terms of equi-similarity. Again, ratings may have a relational quality because they require the given stimulus to be located relative to the end anchors.

The idea of two different sensory scales is not unreasonable. Discrimination between two stimuli commonly occurs subsequent to the sensory magnitudes themselves, and could well involve a nonlinear transformation on the magnitudes. Indeed, a more general view recognizes multiple processing stages and hence the possibility of multiple scales (Anderson, 1975a, p. 479, 1976c; Marks, 1978a; Section 1.8.1). The assumption of scale invariance cannot be taken for granted, therefore, but requires empirical demonstration. It seems doubtful, however, that

Marks's attempt to justify magnitude estimation in this manner will hold (Anderson, 1975a, 1981a).

What is most important about Marks's position is his recognition that magnitude estimation requires a validational criterion. Marks's position presumably relates to his work on sensory summation, a problem that is conducive to the use of algebraic integration models as a base for sensory scaling. Indeed, Marks's (1974b) concept of psychosensory law is quite similar to the present concept of integration function, and his general position is harmonious with the functional measurement approach: "The approach just outlined parallels the statements made by Anderson (e.g., 1970) about functional measurement [p. 31]. . . . In a fundamental sense, Anderson's position is quite close to that taken in this book [p. 277]."

From this position, the difference between ratings and magnitude estimation becomes open to empirical investigation. Both response methods suffer various biases, and the main goal is to improve the methods. Proponents of the two methods thus become protagonists engaged with common problems (Anderson, 1981a, in press; Marks, 1978b).[a]

5.6.3 TWO-STAGE FORMULATION

Input and Output Stages. Attneave (1962) accepted Stevens's view that the psychophysical function is a power function, but extended that view to include the number continuum itself. Thus, subjective number would be a power function of objective number, and vice versa. In this view, magnitude estimation responses are the end product of two successive power functions. In the input stage, the physical intensity is transformed into subjective number; this represents the true psychophysical law. In the output stage, subjective number is transformed into the observed number response; this represents the inverse of the psychophysical law for number. Hence an observed power function would actually be a composite of power functions at two distinct stages, one for stimulus input, one for response output.

This two-stage formulation has been pursued in an extensive series of studies by Curtis, Rule, and their associates (e.g., Curtis, 1970; Curtis, Attneave, & Harrington, 1968; Curtis & Rule, 1972; Rule & Curtis, 1978). Most of this work rests on judgments of sums or differences of two stimuli. In present notation, their model is

$$R = c_1 (S_1^k \pm S_2^k)^m + c,$$

where S_1 and S_2 are the physical values of the stimuli, k and m are exponents of the power functions associated with the input and output stages, and the plus or minus sign allows for sum or difference judg-ments. If this model is correct, then it allows the two stages to be sepa-rated.

Experimental Applications. The main goal has been to estimate the ex-ponent parameters, k and m. The main results from studies of a number of psychophysical dimensions can be summarized as follows. When magnitude estimation is used, m is not constant, but varies across indi-viduals and across psychophysical dimensions. This inconstancy of m does not support Attneave's hypothesis of a psychophysical law for the number continuum. However, it does support his view that magnitude estimation is a biased measure of subjective magnitude (see also Ander-son, 1974c, p. 230). Indeed, the values of m are usually greater than 1, which would correspond to the hypothesis of diminishing returns in magnitude estimation (Section 5.4.2).

When a rating response is used, the values of m are usually around 1. Since $m = 1$ corresponds to a linear output function, this result is consis-tent with the present view that ratings can provide valid linear response scales. The generality of this result is uncertain, however, since most of the studies used only magnitude estimation.

The values of k are perhaps less variable across individuals than the values of m. This is consistent with the two-stage conception in which the sensory system controls the input exponent, k, and learned number language controls the output exponent, m. However, the k values still show considerable individual variability, possibly due to real individual differences in the sensory system, possibly due to some shortcoming of the model.

Research on the two-stage formulation has been largely independent of the present research program, and so it deserves emphasis that both programs seem to warrant the same general conclusions about psychophysical judgment. Foremost among these conclusions is the lack of support for magnitude estimation. In addition, both approaches seem to agree on the psychophysical functions for lifted weight and for gray-ness. Applications of functional measurement by Curtis and Mullin (1975) and by Curtis and Rule (1977) have extended this agreement. To these may be added the study of grayness averaging by Curtis and Rule (1978), which includes a replication of results of Weiss (1972) noted in Figure 5.2.[a]

Relation to Functional Measurement. The two-stage formulation has a direct interpretation in terms of the functional measurement diagram shown in Figure 1.1. The input stage corresponds to the psychophysical law; the output stage corresponds to the psychomotor law. Similarly, the sum and difference models of the previous equation correspond to the integration function or psychological law. Thus, there is a direct correspondence between the two formulations.

The two formulations have had rather different research emphasis. The two-stage approach has been mainly concerned with fitting power functions to the input and output stages, basically following Stevens's conceptualization. In contrast, functional measurement has placed primary emphasis on the central third stage, namely, the integration stage. This third stage provides the base and frame for determining both the input and output functions, which are the two stages of the two-stage formulation.

Because of this emphasis on the integration function, functional measurement provides a more general analysis than the two-stage formulation. The two-stage assumption that the input and output functions are power functions is unnecessarily restrictive; arbitrary input and output functions can be handled with functional measurement. Moreover, functional measurement allows for stimuli that have no physical metric and are outside the two-stage approach because they cannot be expressed as power functions.[b] Finally, of course, functional measurement is not limited to sum and difference models but applies to more general integration functions.

5.6.4 ADDITIVE UNIT MEASUREMENT

In his influential writings on physical measurement, Campbell (1928) remarked on the odd contrast between the importance of measurement in physics and the near-total neglect of general theory of measurement. Physical laws have quantitative form, and the entire structure of physical theory rests on measurement. Yet physicists take physical scales more or less for granted and, although deeply involved with many difficult problems of practical measurement, have had little concern about logical foundations of the physical scales.

Campbell sought to place physical measurement on a rational foundation and claimed to find it in the process of physical addition. The length of a rod can be determined by laying unit lengths end to end to cover the rod and adding up the number of these units. That is exactly what the ordinary centimeter scale accomplishes. In analogous manner, the mass of the rod could be determined by adding up the number of unit weights required to balance it in a beam balance.

Measurement based on such physical addition reduces to a counting operation. This kind of measurement was called *fundamental* because it required no prior measurement. These fundamental scales could be used to measure other, derived quantities. Thus, the density of a uniform rod could be derived from its length and mass.

Campbell's formulation seemed to provide a rational basis for physical measurement. Almost all physical scales seemed to be reducible, directly or indirectly, to physical addition. Arithmetical operations on measured numbers were justified because they were isomorphic to physical operations. These measurement scales could thus provide a solid foundation for a superstructure of physical laws. This formulation was logically simple and elegant, and seemed to capture the essential nature of physical measurement.

Psychologists found little comfort in Campbell's formulation, for psychology seemed to possess nothing analogous to physical addition. Physical energy could be added, of course, but psychological sensation was clearly not proportional to physical energy. Fechner's idea of adding *jnd*'s (Section 5.2) was not satisfactory because the assumption that *jnd*'s were equal units was not resolvable. Some writers therefore concluded that true measurement of psychological quantities was not possible. Most psychologists refused to concede this but found it hard to avoid Campbell's formulation.

But Campbell's position carries no logical necessity. To say that it provides a rational basis for physical measurement does not imply that no other rational basis is possible. Some alternative formulation might be found that would do as well or better.

Functional measurement provides an alternative formulation. Substantive laws constitute the base and frame for measurement, as shown in Chapter 1. No analog of Campbell's physical addition is needed, for the scales are not constructed by any counting of units. Instead, measurement is inherent in and derivative from the algebraic structure of the substantive law. This view provides a different conceptual outlook from that of Campbell. It has provided a foundation for psychological measurement, not merely in principle but in the empirical development of cognitive algebra.

It may also be suggested that the functional measurement approach reflects actual practice in physics. Many laws of physics have an algebraic form, and so they provide a base and frame for physical measurement in the same way that cognitive algebra does in psychology. From this viewpoint, Campbell's method of additive units is a coincidence that masked the true nature of physical measurement as immanent in the structure of physical law.

To illustrate this point, it will be shown how functional measurement methods could be used to determine Galileo's law of the inclined plane and to scale physical time and distance. Consider a homogeneous ball that rolls, without slipping, a distance D starting from an elevation H on a frictionless plane. Galileo's law for the travel time, T, is

$$T = cD/\sqrt{H},$$

where c is a constant. For simplicity, this and other constants will be ignored in what follows. If D and H are varied in factorial design, the factorial graph of observed travel time will be a linear fan. This linear fan pattern implies that T obeys a multiplying rule

$$T = f(D)g(H),$$

where f and g are unknown functions of the corresponding stimulus variables. The marginal means yield functional scales of $f(D)$ and $g(H)$, and plots of these against physical distance will reveal that $f(D) = D$ and $g(H) = 1/\sqrt{H}$. Thus, Galileo's law can readily be induced from observation—if a clock is available.

But suppose that a clock is not available. (Galileo was severely handicapped because there was then no way to measure time intervals of the order of tenths of a second, but he found an ingenious way to resolve this problem [Drake, 1978, pp. 85–90].) The following discussion shows how simple paired choices suffice to establish the law and how it may then be used as a base and as a frame for scaling physical time and distance.

A two-operation model provides the needed constraint, but a preliminary attempt with a one-operation model will be instructive. As before, ideal physical conditions are assumed although the method is not sensitive in this respect. Identical balls are released simultaneously on two inclined planes and a human observer judges which reaches the bottom first. These judgments are well within human capabilities because small time differences are not needed. D and H are varied independently for one plane, which may be called the determining plane, but these two factors are treated as nominal without reliance on physical distance measures.

To rank-order the travel times for all determining planes in the factorial $D \times H$ design, it suffices to rank-order each pair of determining planes, which can be done by manipulating the second plane in any handy way so that it is faster than one plane of the given pair and slower than the other. These rank-order data can be transformed to additivity because the multiplying model is monotonically additive. Letting primes denote logarithmic quantities, this yields a law of the form

$$T' = D' - H'.$$

This adding model is hardly what is wanted, of course, but the investigator faces a problem of monotonic indeterminacy (see e.g., Anderson,

1974a, pp. 227ff; in press, Section 5.9; Ellis, 1966). The adding model is as simple and plausible as the multiplying model, and knowledge of the physical distance scales would not allow a resolution.

A significant advance is possible by using a determining plane with two successive segments, each independently and identically varied in D and H. As the ball reaches the bottom of the first segment, it trips a stop to release an identical ball on the second segment. The total travel times for these two-segment determining planes may be rank-ordered by using a second plane in the manner already described.

The unexceptionable convention is imposed that the total travel time on the determining plane is the sum of the travel time for the ball on the first segment plus the travel time for the ball on the second segment. Since only additivity is required, this avoids the requirement of equal units that is employed in Campbell's approach to extensive measurement. This adding operation may be used as a frame for monotone transformation in what is now a two-operation model (Section 1.7.2). The data are treated as a two-factor, Segment 1 × Segment 2 design. The monotone transformation on the ranks yields a linear scale of physical time because physical time is indeed additive across the two segments. With this transformed response measure, therefore, the factorial graph for each separate segment will have a linear fan form. This implies that T obeys a multiplying rule for each separate segment, as already noted. This two-operation procedure thus yields a law of proper form

$$T = f_1(D_1)g_1(H_1) + f_2(D_2)g_2(H_2),$$

where subscripts 1 and 2 denote the two segments of the determining plane.

Functional scales of the stimulus factors D and H are obtainable from the marginal means. If the corresponding physical scales are available, it will be found that $f(D) = D$ and $g(H) = 1/\sqrt{H}$. In the absence of auxiliary knowledge, however, it would be premature to take $f(D)$ as physical distance. It would be equally proper to take $g(H) = 1/H$, for example, so that $f(D) = D^2$. To be acceptable, a physical scale must have some invariance across more than one task. This one task does, however, provide functional scales for response time and for the two stimulus distance variables.

In this approach, the measurement scales are derivative from the structure of physical law. In contrast, Campbell's formulation does not seem truly adequate as a foundation for physical measurement. His method of additive units can measure length, for example, only over a small range of terrestrial distance. At the atomic level, and again at the stellar level (e.g., Hoyle, 1962, pp. 256ff), the concept and measurement of distance both derive from an interlocking network of physical laws that has little relation to counting up unit lengths. Although physics has a set of basic dimensions for its diverse concepts that has no analog in psychology (Luce, 1972), the present line of argument points to an important similarity between measurement theory in physics and in psychology. In both, measurement is derivative from substantive theory of which it is an organic component.

5.6.5 MULTIDIMENSIONAL SCALING

Multidimensional scaling aims to determine a metric space in which the dimensions correspond to meaningful and relevant attributes of the stimulus (e.g., Shepard, 1966; Torgerson, 1958). This is a central problem in many areas of judgment theory, especially in applications of multiattribute models. Multidimensional scaling is a potentially valuable method, but the actual contributions have been disappointing.

A simple thought experiment that leads to a new theoretical view of similarity as a judgmental process will be given. Similarity judgments provide the basic data for multidimensional scaling, but they have been treated as a means to an end, and the intrinsic interest and theoretical importance of the judgmental task itself have been neglected. A few aspects of this matter are noted here in the hope that the developments in multidimensional scaling can be brought to bear more directly on problems of judgment theory.

Model Analysis. Similarity judgments are the basic data in multidimensional scaling. The usual assumption is that judged dissimilarity of two stimuli, S and T, is given by the equation

$$\text{Dissim}(S, T) = \sqrt[n]{\Sigma |s_i - t_i|^n},$$

where s_i and t_i are the stimulus values or coordinates on dimension i, n is a constant, and the sum is over all dimensions of the space.

This equation can be viewed as either a spatial distance model or a dissimilarity integration model; this is clear for $n = 1$, the "city-block" metric. Then $|s_i - t_i|$ may be taken as the dissimilarity of S and T on dimension i, and the total dissimilarity is obtained by addition. The case of $n = 2$ represents ordinary Euclidean space and may be viewed similarly.

One reservation about studies with multidimensional scaling models is that, almost without exception, they have lacked proper tests of goodness of fit. The standard stress indices are not easy to interpret (see Cliff, 1973, pp. 482ff) and in any case do not test significance of the discrepancies from the model. Stress may seem low by usual standards even though the model is seriously wrong. Simple general tests of goodness of fit in multidimensional scaling can be obtained with the replications method described elsewhere (Anderson, 1977b, in press), but as yet no applications seem to have been made.

A more serious concern with multidimensional scaling models is that there is a certain sense in which they are untestable. As is well known, the scaling algorithm always yields a solution, regardless of degree of fit. If the stress is deemed too large, it can be made as small as desired by adding more dimensions. This is the standard practice. The specific form of the model might be varied, but the same problem remains. Common sense might balk at a large number of difficult to interpret dimensions, but there is no definite way to decide when to stop adding dimensions.

In this important respect, multidimensional scaling has been rather different from standard cognitive algebra. When the adding model for person perception fails, for example, it cannot be saved by adding another dimension. Some different structural representation must be sought.

Interaction and Valuation. Interaction and valuation are two representative problems that should be kept in mind in attempts to apply multidimensional scaling in judgment theory. The crossovers of averaging theory illustrate the interaction problem. In Figure 2.5, for example, the two dimensions of photograph and personality trait are built into the design. They correspond to the main effects, and are not of much interest since they are under experimental control. Multidimensional scaling would presumably be able to recover these two dimensions.

However, the crossover interaction creates a difficulty, for it disallows a two-dimensional representation of these stimuli. Application of multidimensional scaling to such data would require adding a third dimension. That hardly seems appropriate. The crossover results from an averaging process over the two given stimulus dimensions. It represents a nonlinearity within the integration model, not a third dimension in the stimulus representation.

Similar problems arise with other interactions such as may result from inconsistency or redundancy. Such interactions may be sought as systematic deviations from an algebraic integration model (e.g., Section 3.4.1). Indeed, a noninteractive model becomes an almost essential baseline for demonstrating the presence of interaction. The practice of adding dimensions to handle model deviations seems inappropriate. If the locus of interaction is in the stimulus parameters, either the scale values or the weights, adding another dimension does not provide a sensible representation. And, more generally, if the interaction depends on the relationship between the two stimuli, representation in a metric space may not even be possible.

The valuation operation presents a different kind of problem. Valuation is very sensitive to the dimension of judgment (Section 1.1). Two trait adjectives might be similar with respect to one dimension of judgment, dissimilar with respect to another. Furthermore, there are innumerable, diverse dimensions of judgment. To capture this dynamic character of the valuation operation does not seem feasible within a static spatial representation.

It might be suggested that the spatial representation can handle valuation by including all the response dimensions in the space. Valuation could then be viewed as a similarity operation and defined by proximity in this total space. However, there are two difficulties with this suggestion. The first is that the valuation operation may depend on interaction between the stimuli and the given response dimension. The preceding comments on interaction would then be applicable.

The second reason that the multidimensional scaling approach has difficulty with valuation is that valuation is asymmetrical. This asymmetry is well known from studies of the personality trait adjectives (e.g., Hendrick, 1968b, 1969). Thus, a *prudent* man would seem unlikely to be *cruel*, but a *cruel* man may well be *prudent*. The similarity models of multidimensional scaling require symmetry, a consequence of the basic spatial representation in which the distance from S to T must equal the distance from T to S. Accordingly, the pervasive asymmetries of the valuation operation are not amenable to analysis by the standard spatial representations of multidimensional scaling.

Similarity as Averaging. In multidimensional scaling, similarity judgments are primarily a means to determine a spatial representation. The models are chosen largely on rational considerations relative to such spatial representation. The integration-theoretical approach is more concerned with the judgmental processes and this concern leads to a different conceptual orientation.

Almost without exception, similarity models have been formulated in terms of addition of dissimilarities. This approach seems questionable in two respects, and it will be suggested that the judgments are based on integration of similarities, probably by an averaging rule.

The difficulty with the dissimilarity approach may be seen in the following two pairs

$$S_1 = C, \quad S_3 = A\ B\ C\ E\ F,$$
$$S_2 = D, \quad S_4 = A\ B\ D\ E\ F. \tag{27}$$

The addition of common letters seems to increase the similarity so that S_3 and S_4 are more similar than S_1 and S_2. This suggests that the hypothesis of additive dissimilarities is not tenable. Addition of common elements cannot decrease dissimilarity, or increase similarity, unless negative dissimilarities are allowed, and that is difficultly compatible with the representation of dissimilarity as distance.

On this basis, similarity analysis should be in terms of similarities, not dissimilarities. The next question is whether similarities are added or averaged. Presumably the addition of noncommon letters to S_3 and S_4 in (27) would decrease the similarity. To handle this, an adding model would need to allow for positive and negative similarity values. A critical test between averaging and adding is then possible on the same basis as in previous work on the averaging hypothesis (Section II, A,5). Addition of a near-neutral element should decrease either similarity or dissimilarity if it is averaged in. Some evidence that supports the averaging hypothesis for similarity judgments is given by Simmonds (1971).

The standard approach based on adding of dissimilarities seems to stem from the attempts to obtain a spatial representation. This approach has the mathematical convenience of providing a rational zero, namely, the dissimilarity of two identical stimuli. And it is directly associated with the treatment of dissimilarity as a distance measure in spatial models. This approach is not theoretically neutral, however, and may be misleading if the averaging hypothesis is correct [Anderson, 1974a, p. 256].

A number of writers (e.g., Anderson, 1974a, p. 254; Hyman & Well, 1967, p. 248; Lopes & Oden, 1977, 1980; Tversky, 1977) have suggested that multidimensional scaling has been overinvolved with continued elaboration of spatial models to the neglect of judgmental processes. This view has been explored in greatest depth by Tversky, whose formulation has affinities with the present approach. For example, Tversky treats similarity, not dissimilarity, as basic, in agreement with the argument given in the foregoing quotation. Again, Tversky's monotonicity assumption, that "similarity increases with addition of common features and/or deletion of distinctive features [p. 330]," mirrors the quoted thought experiment with letter arrays.

Other aspects of Tversky's treatment also seem generally consistent with the integration-theoretical formulation. Indeed, most of Tversky's criticisms of multidimensional scaling seem to have straightforward resolutions in terms of the concept of dimension of judgment (Section 1.1). Asymmetry in the valuation operation is a well-known result from studies of person perception and semantic inference, as noted in the previous subsection. Effects of context are also readily represented in terms of the dimension of judgment. An advantage of the present approach is that functional measurement may be used to obtain the operative stimulus values.

Empirical comparisons of the distance model of multidimensional scaling with the averaging model have been presented in an important investigation of judgments of kinship similarity by Lopes and Oden (1980). The design and analyses of this ingenious study require more detail than can be given here, but the main conclusion is that

> The results of the various qualitative and quantitative tests from both experiments are unanimous in supporting the differentially weighted averaging model over the multidimensional distance model. This support rests largely on the ability of the averaging model to account for the presence of crossover interactions in the data that violate the additivity assumption of the distance model [Lopes & Oden, 1980, pp. 228–229].

The authors point out that Tversky's (1977) formulation would also have difficulty in handling these crossovers.

No less interesting than this model analysis is the conceptual emphasis that Lopes and Oden place on judgmental processes. They point out that previous approaches to similarity have been dominated by the idea that judgmental process is determined by the structure or form of the hypothesized representation of the stimulus information in semantic memory. The distance model of multidimensional scaling, in which the judgment is determined by locations in semantic space, clearly illustrates this primacy of semantic structure over judgmental process. In the integration-theoretical approach, however, semantic memory plays a lesser role, and the main cognitive operations are dynamic, context-dependent judgmental processes.

5.6.6 UNFOLDING THEORY

One of the most persistent and dedicated attempts to place measurement theory on a nonmetric foundation is represented by the work of Coombs (1964). Coombs's unique contribution is his unfolding theory of preferential choice, which is based on the fundamental premise that stimuli and individuals can be considered as points in a common space that represents the underlying attributes of the stimuli. Each individual is represented in this space by an *ideal* point. The magnitude of preference for any stimulus is taken as the distance between that stimulus and the ideal for each individual. Different individuals may have different ideal points and hence also different rank orders of preference. However, all individuals are required to have the same underlying metric space so that interstimulus distances, in particular, are the same for all individuals.

Unfolding theory thus rests on a curious conceptual structure in which scaling depends on aggregate data from a group of individuals with two characteristics. They must be homogeneous in having the same metric space for the underlying stimulus attributes; they must be heterogeneous in having their ideal points distributed across the stimulus range. This heterogeneity is mathematically necessary, for otherwise the preference orderings of the stimuli, which are the basic data for unfolding analysis, do not provide sufficient constraints to determine the interstimulus distances. The analytical leverage of unfolding theory rests on this conceptual structure in which individual analysis is not possible.

Unfolding theory faces the same difficulties as multidimensional scaling plus others that stem from its reliance on aggregate group data. In practice, individuals will have inconsistent rank orders of preference, so it is necessary to select out a maximum subgroup of consistent individuals (Coombs, 1964, p. 94). This selection, however, lies at the mercy of several chance factors. Moreover, added dimensions can always be postulated to explain any discrepancies.

This problem may be illustrated with the person perception data of Figure 1.2, which shows a nonlinear relation between the adjective values for Subjects F. F. and R. H. Since the ideal point for likableness is presumably near the scale endpoint for both subjects, this nonlinear relation is inconsistent with unfolding theory if the underlying space is one-dimensional. It could be explained, of course, by postulating additional dimensions. Like multidimensional scaling, therefore, unfolding theory is in a certain sense untestable.

The functional approach can yield the scale values for likableness or for other judgment dimensions in a simple way. It does this by reliance on a behavior law at the level of judgment without concern for the structure of the underlying space of attributes. For scaling at this judgmental level, the functional approach is effective where unfolding technique is not.

Unfolding theory could find a role in structural analysis of the underlying attribute space. Perhaps it could even be extended to allow individual analysis by employing a variety of judgment dimensions that have their ideal points scattered across the stimulus range; for example, by using a number of personality adjectives both as stimuli and response. In any case, it would be far more effective to abandon the rank orders and begin with the functional values of preference.

5.6.7 BEHAVIORISM

A radical solution to the problem of psychological measurement is to deny its existence. Strict behaviorism adopts this solution in its credo that the only meaningful concern of psychological investigation is the physical law that connects observable stimulus and observable response. These must be measured in physical terms, and subjective measures are not admissible ideas.[a]

Strict behaviorism is attractive. It deals with facts. It avoids the uncertainties and failures of theoretical constructs. It avoids explanations that have only a verbal existence. And it avoids the not infrequent abuse of surplus meaning in common language terms.

Strict behaviorism has appeared in two distinct areas in psychology: perception and animal behavior. The close relation between perception and the physical world makes it attractive to seek an isomorphic formulation, even though this may require the introduction of higher order, relational percepts. And in animal behavior, the lack of introspective reports is naturally conducive to a behavioristic view.

However, strict behaviorism has not been adequate for a science of behavior. This is perhaps best indicated by the writings of the behaviorists themselves. In sensory psychology, for example, Graham and Ratoosh (1962) argue against subjective scales with the statement that "the concept of intensity of sensation is formally unnecessary if observations are restricted to the variables that must be used to define intensity of sensation [p. 505]." However, they also recognize the difficulties with this view, as in the concept of hue discrimination, when they speak of the "difficulty posed by this fact [that] hue varies with other factors than wavelength [p. 499]." As a consequence, the concepts of hue and hue discrimination cannot be replaced by the objective concept of wavelength discrimination.

That Graham and Ratoosh speak of this "fact" indicates that they recognize hue as a psychological entity. The "difficulty" is that banishing the term *hue* leads at best to cumbersome terminology. It is a reasonable working hypothesis that such subjective terms have some psychological reality and that they mirror processes by which the organism perceives. Banishing such subjective terms will accordingly hobble and distort the course of inquiry.

That strict behaviorism is inadequate for a science of behavior also appears in the evolution of the Skinnerian school. Herrnstein's (1970) *matching law*, which states that an organism will distribute its responses between two alternatives in proportion to their respective rates of reinforcement, may illustrate this evolution. In Herrnstein's notation

$$R_1/(R_1 + R_2) = r_1/(r_1 + r_2), \quad \text{(physical matching law)}$$

where R_1 and R_2 are the two response rates, and r_1 and r_2 are the two obtained rates of reinforcement. This may be called the *physical matching law*. It has the unique property that all four terms are physical observables, as is appropriate to a behavioristic approach. That makes it simply and immediately testable with actual observations. However, this physical matching law has had considerable difficulties in accounting for the data.

Accordingly, many workers in operant psychology have considered replacing the observable rates of reinforcement by unobservable psychological "values" denoted by v_1 and v_2. The preceding equation would then become

$$R_1/(R_1 + R_2) = v_1/(v_1 + v_2).$$ (psychological matching law)

These two equations look alike, but they are fundamentally different. The r values in the first equation are physical observables; the v values in the second equation are subjective, not observable. The second equation may accordingly be called the *psychological matching law*.

The psychological matching law has been criticized as tautological; not an empirical hypothesis but an untestable convention for how reinforcement value is to be measured. Since the v values are unobservable, the investigator has the liberty, if not the obligation, of choosing values that make the model fit the data. It has been argued that that is always possible (see Anderson, 1978a, p. 373). In fact, however, the psychological matching law is not tautological. Functional measurement analysis shows that the law imposes real constraints on the data that make it testable even while it allows for individual psychological values.

Exact tests of the psychological matching law can be obtained as follows. Let $R = R_1/R_2$ be the observed response and rewrite the equation as a multiplying model

$$R = v_1 v_2^{-1}.$$

Then the linear fan theorem of Section 1.4 applies. Experimentally, the reinforcement values of the two response alternatives would be varied in a two-way factorial design. No prior scaling is needed; the levels of each alternative may be merely nominal (see also Anderson, 1974d, p. 294, 1978a, p. 372; Farley & Fantino, 1978).

As this example indicates, functional measurement provides a way to put subjective terms on an objective basis. The internal stages in the functional measurement diagram are not merely a scaffolding to arrive at the physical law but a recognition of the existence of truly psychological entities. Functional measurement methodology can help define and measure these psychological entities within a conceptual network that is rigorously anchored to observable stimuli and response. Where this approach applies, therefore, it does much to resolve the behavioristic reservations about the use of subjective measures.

All science seeks to base itself on observables, and no one would quarrel with an objective emphasis. But the essential question is whether any subjective constructs are necessary. As soon as one is admitted, then strict behaviorism is no longer adequate but becomes nonessentially different from normal science. The sole issue then concerns the judicious use of theoretical constructs, a matter of degree, not of kind.

To belabor strict behaviorism may seem inappropriate since current behaviorism is pervaded with subjective and theoretical constructs. However, the behavioristic approach to theory construction is ambivalent, handicapped by its conceptual heritage. The needs that called the original behavioristic reaction into being are not less compelling today, but the original reaction is no longer appropriate. The need of the present time is for a theoretical behaviorism.

5.6.8 RELATED WORK

This brief treatment of measurement theory has necessarily passed over contributions of many workers. However, there are some who have particular relevance to this discussion and deserve particular mention.

Validity. The work of Garner occupies a special place in the classical problem of psychophysical measurement, namely, the problem of measuring loudness. Garner focused directly on the central issue of obtaining a validational criterion, an issue that was made starkly clear in his study of the method of fractionation. The standard sone scale then rested on fractionation data obtained by instructing subjects to set one sound at, say, half the loudness of another sound. If subjects made true half-loudness settings, then the data would provide a true scale of loudness. However, the validity of the method rested entirely on the arbitrary assumption that subjects really used the assigned fraction of one-half. In a test of this assumption, Garner (1954b) found dramatic effects of context on fractionation responses as well as large individual differences. The fraction that a subject actually used was sensitive to irrelevant factors that biased the results. That implied that the sone scale was invalid.

An alternative method for constructing a loudness scale used the bisection task, in which subjects are instructed to set one sound midway in loudness between two given sounds. If subjects made true midpoint settings, then the data would provide a true linear scale of loudness. However, the validity of this method rested entirely on the arbitrary assumption that subjects really bisected at the true midpoint. Bisection, like fractionation, is attractive because the judgment is perceptual and no verbal response is needed. However, both tasks rest on an arbitrary assumption and lack validational support.

But Garner saw how bisection and fractionation could be combined to obtain a validational criterion. Both tasks correspond to algebraic models. Under certain assumptions, fractionation yields a log linear scale and bisection yields a linear scale. If all assumptions are satisfied, then these two scales can be brought into linear relation by estimating two constants. This linear relation provided the sought-for validity criterion. Garner's approach succeeded (1954a, Figure 9), and his results have been confirmed in recent work. As Carterette and Anderson (1979) observe:

> Modern studies of loudness measurement begin with the work of W. R. Garner and S. S. Stevens in the mid-1950s. Garner attempted to employ the physical response methods of fractionation and bisection; Stevens began with fractionation but shifted to the verbal response method that he called magnitude estimation. These two approaches at once produced conflicting results. ... In the intervening time, Stevens' approach has flourished while Garner's approach has fallen into neglect. The present work supports Garner, both in result and in principle, and suggests that subsequent research took the wrong path [p. 277].

Measurement Structures and Qualitative Laws. Starting from a different background and different considerations, Krantz (1972b, 1974) has arrived at a view that has much in common with functional measurement. The two cited papers are concerned with what Krantz calls qualitative laws and measurement structures as a basis for measurement theory. Qualitative laws are typified by rank-order properties, especially the properties of independence and double cancellation from conjoint measurement. The term *measurement structure* is not entirely clear, but it may not be amiss to view it analogously to the present algebraic models. If that is correct, then Krantz's view is similar to the present view that the algebraic structure of the model provides the base and frame for measurement. Indeed, in his listing of advantages of measurement-theoretical formulations, Krantz (1974) remarks,

> One advantage, emphasized also by Anderson (1970) and Krantz (1972a), is that measurement is no longer regarded as something that must be struggled with prior to theory construction; rather, it is a consequence of the discovery of quantitative and/or qualitative laws [p. 171].

Taken as a whole, Krantz's work is notable for its elegant interplay of abstract formalism and concrete empirical analysis. This includes fundamental studies of color vision and, with Tversky, a variety of studies on foundations of multidimensional scaling.

A major difference in emphasis concerns the place of qualitative properties. Functional measurement makes less use of qualitative properties such as independence; they seem more helpful for eliminating models than for establishing them (Section 2.3). Krantz places primary reliance on the qualitative properties. In principle, qualitative properties can imply a representation of the data as an algebraic model, but practical difficulties are severe (Section 5.5). Krantz sometimes appears to identify the qualitative properties with the algebraic model, which seems inappropriate from the functional measurement view. However, Krantz (1972b) also recognizes that numerical response measures can establish an algebraic model and lead to true measurement. Despite some differences of orientation, therefore, the two approaches have a measure of common spirit.

The Swedish School. Special mention should be made of the Swedish school of G. Ekman and his followers, which has covered a diverse array of problems in psychophysics and measurement. This work is eclectic in nature and imbued with the spirit of science. Contributions have been made by a number of investigators, the most prominent being H. Eisler and L. Sjöberg.

Eisler's work has been much concerned with the relation between sensory magnitude of a single stimulus and sensory discriminability between two stimuli. This concern appeared in his early work on judgments of similarity as a ratio of magnitudes (e.g., Eisler, 1960; see also Eisler, 1975). Later work on this problem is concerned with the relation between magnitude scales and category rating scales, the latter being considered as discrimination scales. This led to Eisler's concept of a general psychophysical differential equation that expresses differential magnitude, dx/dy, of two sensory variables, x and y, as a ratio of their Weber functions. An early paper in this series is Eisler (1963), and the most recent are an impressive study by one of Eisler's colleagues, Montgomery (1978), and a theoretical paper by Eisler, Holm, and Montgomery (1979).

Sjöberg's many contributions begin with a series of papers on Thurstonian methods, including a further treatment of Hull's application to the scaling of excitatory potential (Sjöberg, 1965) and a generalization of Thurstone's method of categorical judgment (Sjöberg, 1967). This early work is in the traditional orientation, which considers scaling as a methodological preliminary to the study of judgment and behavior. A significant break with tradition appears in Sjöberg's (1966) paper in which scaling is conceived as part of the process of establishing the judgment model itself. This is practicable because the model provides a structural frame for scaling as well as a validational criterion based on internal consistency. Sjöberg's paper thus constitutes an independent development of essential ideas of the functional measurement approach. Sjöberg discusses various advantages of this approach and has made experimental applications in several areas, including ratio models in psychophysics (Sjöberg, 1971), the weight–size illusion (Sjöberg, 1969), and several studies in utility theory (e.g., Örtendahl & Sjöberg, 1979; Sjöberg, 1968).

Functional Measurement. Basic contributions to the development of functional measurement have been made by Michael Birnbaum, James Shanteau, and David Weiss. Their work is considered elsewhere in particular contexts, but a brief overview is appropriate here.

Birnbaum has made numerous and cogent contributions both to social judgment and to psychophysical judgment. In social judgment, his initial work was critical of the averaging model, but his later studies have provided strong support (e.g., T. Anderson & Birnbaum, 1976; Birnbaum, 1976; Birnbaum & Stegner, 1979). Moreover, these and related studies constitute an outstanding contribution to the theory of source effects. Work on psychophysical judgment includes an important paper on range–frequency theory (Birnbaum, 1974b) and an ingenious study of expectancy–contrast effects (Birnbaum, Kobernick, & Veit, 1974). Other studies have attempted to resolve certain monotonic indeterminacies in the vexing problem of distinguishing ratio models from difference models in psychophysical judgment (e.g., Birnbaum, 1978; Birnbaum & Elmasian, 1977; Rose & Birnbaum, 1975; see also Veit, 1978).

Shanteau has made fundamental contributions to several areas of decision theory. Especially notable is the first general solution to the problem of simultaneous measurement of subjective probability and utility (Figure 1.13; Shanteau, 1974, 1975a). No less important is the associated work on additivity and subadditivity (Shanteau, 1975a; Shanteau & Anderson, 1969).

Another program of research has compared Bayesian theory and information integration theory as models of serial integration (Shanteau, 1970a, 1972, 1975b; Troutman & Shanteau, 1977). Recent work has turned to applied problems with a combination of experimental soundness and ecological validity that is opening up important opportunities (e.g., Phelps & Shanteau, 1978; Shanteau & Nagy, 1976, 1979). Overall, Shanteau's work has provided a foundation for a new approach to decision theory.

Weiss has concentrated on psychophysical judgment, and one of his articles provided the test between ratings and magnitude estimation given in Figure 5.2. Other work has been concerned with problems in integration tasks involving length (Weiss & Anderson, 1969), area (Anderson & Weiss, 1971), and angle (Weiss & Anderson, 1972). The latter article provided an initial application of functional measurement with rank-order data, a line of work pursued in Weiss's (1973b) polynomial transformation program, FUNPOT, which followed up work of Bogartz and Wackwitz (1971). Weiss's most notable contribution is a complete and successful solution to Plateau's classic, century-old problem of grayness bisection (Weiss, 1973a, 1975). This series of studies played an important role in the extension of functional measurement ideas to psychophysics.

Notes

5.1.2a. In the present formulation, a measurement scale is a function relating two variables, ordinarily one observable and one nonobservable, with the observable being taken as a measure of the nonobservable. The terms (strictly) *monotone* and *linear* refer to this functional relation and are suggested as more appropriate than the traditional terms *ordinal* and *equal-interval*. Use of the term *equal-interval*, or *interval*, in particular, naturally leads to questions of whether the scale intervals or units are truly equal and to attempts to compare units at different places on the scale, thereby missing the essential meaning of measurement. It may be worth noting that the "quantity objection" that sensations are unitary and not divisible into part quantities does not arise in the present formulation (Boring, 1950, pp. 290ff).

5.1.2b. A carefully reasoned book by Ellis (1966) also emphasizes the primary importance of monotone measurement, although for reasons that are related to the logical primacy of an ordering relation rather than to the evolutionary character of measurement considered in the text. Ellis is primarily concerned with the problem of monotone indeterminacy in physics and his main conclusion is that even the physical scale of length rests on an arbitrary convention of simplicity or convenience. This conclusion seems unexceptionable and applicable to psychology as well. However, the problem of monotone indeterminacy has tangential relevance to the present discussion (see, e.g., Anderson, 1974a, p. 231, in press).

Ellis's views, incidentally, lead him to recommend that we change our way of thinking about psychological quantities as different from physical quantities when both are monotonically related, as with length and time. The concept, or quantity, in Ellis's terms, is the same: "A psychological scale of length is only one among many kinds of scales for the measurement of the *one* quantity, namely length [p. 45]."

This conclusion may reflect a misconception about the nature of psychological concepts. Although the perceiving organism may be attempting to measure physical length, the psychologist is attempting to measure the organism's percept, which is conceptually different. The need for this distinction is nicely illustrated by the weight–size illusion of Section 1.3.7, in which the visual appearance affects the felt heaviness. More generally, an integration-theoretical view implies that perception is not direct but constructive.

5.2a. An earlier translation of Fechner's *Nachwort* was given by Stevens (1957, p. 153).

5.3.2a. Another, more technical class of differences, concerning experimental procedure and statistical analysis, arose in a study by Leon, Oden, and Anderson (1973), who sought to compare functional measurement of seriousness of criminal offenses with the well-known studies by Thurstone (1927b) and Coombs (1967). A plot of Coombs's paired-comparison values against the functional measurement values was roughly linear with some suggestion of curvilinearity at the two extreme offenses, vagrancy and homicide. This degree of agreement was pleasant, and the curvilinearity was not of major concern. Nevertheless, there seemed to be no satisfactory way to test whether the curvilinearity was real because the variability in the paired-comparison values depended on individual differences whose role in the estimates was unknowable. Moreover, there seemed to be deficiencies of procedure in both paired-comparison studies. Seduction was judged worse than homicide by .06 of the subjects in Coombs's study, and similar trends were present in Thurstone's reported data. These evident reversals, presumably arising from confusion in collecting data in classroom batches (Anderson, in press, Section 1.2.6), would cause appreciable bias in the extreme normal deviates and vitiate assessment of the curvilinearity. There seemed to be no satisfactory way to compare the two sets of scale values, therefore, owing to unsatisfactory psychometric properties of the paired-comparison procedure. I am indebted to Robyn Dawes for comments on this matter.

5.4a. Stevens actually claimed that magnitude estimation yields ratio scales; that is, linear scales with known zeros. Linearity is the primary property, and no more is needed for most analyses. If the scale is not linear, it cannot be a ratio scale. Accordingly, the text will consider only linear scales unless otherwise indicated.

5.4.1a. This opposition between ratings and magnitude estimation seems reasonable on the grounds that they are surface variations in response language and procedure that can hardly affect basic sensations. This view is generally taken for granted, although some writers have suggested that the two response methods measure different things (see Section 5.6.2). In that case, both could be valid, a logical possibility in need of empirical support.

5.4.1b. The usefulness of parallelism analysis may be emphasized by comparison with "weak inference" methods (Section 1.2.8) that have been common in psychophysics. For this purpose, an additive or linear model was fit to the magnitude estimation data in the right panel of Figure 5.2. The linear model cannot fit very well, of course, because it predicts five parallel curves, whereas the actual curves are markedly nonparallel. It is surprising, however, how well the linear model seems to fit when assessed by correlation-scatterplot statistics.

Figure 5.4 plots the values predicted by the linear model as a function of the observed magnitude estimations. The correlation between predicted and observed is extremely high, $r = .983$. The points cluster closely around the diagonal line of perfect fit. Although some hint of systematic deviations can be seen, there is little sign of the serious shortcomings of the linear model that appear in the factorial plot of Figure 5.2. This is one of many illustrations that correlation and scatterplot are weak inference statistics for testing models (Anderson, in press).

5.4.2a. Actually, magnitude estimation does not even make sense for many social judgments because both positive and negative stimuli are involved. The instructions to judge ratios using only positive numbers makes sense on most psychophysical dimensions, but not on bipolar social dimensions with both positive and negative values. That subjects follow these instructions shows that they are doing something essentially different than Stevens assumed.

5.4.2b. Power functions may have limited usefulness even for curve fitting (see "cautionary note" in Carterette & Anderson, 1979, p. 280).

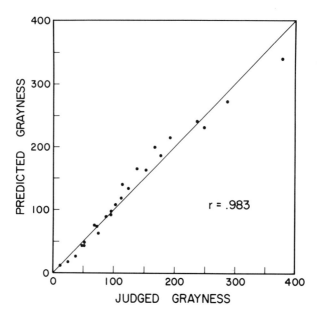

FIGURE 5.4. *Weak inference test of adding model with magnitude estimation response. Predicted values obtained from adding model are plotted as a function of the observed magnitude estimations from right panel of Figure 5.2. High correlation and clustering of points around diagonal line seem to support the model. Actually, the model is seriously incorrect, for it requires the five curves in the factorial plot in Figure 5.2 to be parallel. Correlation and scatterplot obscure what the factorial plot makes clear.*

5.4.2c. Cross-modality matching is an interesting experimental task for studying response processes independent of its use in magnitude estimation. However, it involves difficult statistical problems that have only recently begun to receive adequate attention as in Cross (1978).

5.5a. Writers on conjoint measurement may have contributed to this confusion in terminology by also using the term *conjoint measurement* in a second, nonaxiomatic sense to refer to procedures similar to those used in functional measurement. A pertinent example appears in experimental work by Tversky cited in Krantz *et al.* (1971, pp. 442ff). Tversky attempted to test a two-factor multiplying model for gambles using an ordinal analysis based on conjoint measurement theory, but this analysis was inconclusive owing to lack of a test of goodness of fit. To resolve this impasse, Tversky took logarithms of the raw data to reduce the multiplying model to additive form. These numerical data were then tested with regular analysis of variance—exactly as in functional measurement.

Appropriate terminology would seem to require that the term *conjoint* be reserved for applications that actually use the axioms and theorems that form the substance of conjoint measurement. The term *optimal scaling*, although it may have misleading connotations, seems reasonably fixed by past usage to refer to various statistical developments that largely bypass the matter that is essential for measurement, namely, establishing algebraic models. The term *functional measurement* is appropriately used in accord with its past applications, illustrated in Chapters 1–4, which allow numerical methods including monotone transformation procedures. The term *functional* refers to the use of the integration *function* as base and frame for measurement.

5.5b. Critical discussions of functional measurement besides those noted in the text include Gollob (1979), Krantz and Tversky (1971b), Matthäus, Ernst, Kleinbeck, and Stoffer (1976, 1977), and Schönemann, Cafferty, and Rotton (1973); see also Anderson (1971d, 1973a, 1979b).

5.5.1a. This misconception is sometimes stated in the alternative form: that functional measurement assumes (i.e., presupposes) the operative integration model. Of course, the parallelism test provides a joint test of model and response scale (Section 1.2).

5.5.2a. A pertinent illustration of the difficulties of applying conjoint measurement appears in the attempt by Krantz, Luce, Suppes, and Tversky (1971) to use axioms to reanalyze data from a study by Sidowski and Anderson (1967). In this study, subjects judged attractiveness of working at certain occupations in certain cities. Experiment 1 yielded an unexpected interaction that was replicated in Experiment 2, a nearly identical experiment that was run to verify the interaction.

Krantz *et al.* claimed that the interaction was not genuine, that it resulted from nonlinearity in the rating response, and that the integration process was truly additive. Their claim was based on application of the axiom of double cancellation to the mean data of Experiment 1. The means of the 3 × 4 design showed no violation of double cancellation and a monotone transformation was found that made these means parallel. They concluded that the interaction was not genuine and that the data were truly additive: "The interaction between city and occupation, therefore, is attributable to the nature of the rating scale, because it can be eliminated by appropriate rescaling [p. 446]."

This reasoning does not seem valid. The corresponding data of Experiment 2 exhibit a violation of double cancellation and so cannot be made parallel or additive. This violation is shown by the bold face entries in Table 5.1. Krantz *et al.* used the data of Experiment 2 to break ties in Experiment 1 but evidently neglected to test double cancellation in Experiment 2.

Of course this violation of double cancellation does not mean that the interaction is genuine; double cancellation could fail because of response variability. Without an error theory, double cancellation is inconclusive. However, the violation in Experiment 2 makes clear that the lack of violation in Experiment 1 is also inconclusive. The cited conclusion implicitly assumes the validity of the additive model.

5.5.3a. Perhaps the only bona fide empirical application of conjoint measurement is that of Falmagne (1976), who developed a novel test of goodness of fit. Few studies have

TABLE 5.1
Mean Rating of Attractiveness for City–Occupation Combinations

Occupation	Desirability of city			
	H	M⁺	M⁻	L
Lawyer or doctor	7.08	6.88	**6.20**	**5.14**
Teacher	6.95	**6.62**	5.80	**4.10**
Accountant	5.25	**5.05**	**4.42**	3.22

Source: After Sidowski and Anderson (1967, Experiment 2).
Note: Double cancellation (boldface entries) requires that the two inequalities 6.62 > 6.20 and 4.42 > 4.10 imply the inequality 5.05 > 5.14, which does not hold.

attempted to handle the necessary system of inequalities and, to the best of the writer's knowledge, none has presented a proper test of goodness of fit for the system. Even the axiom tests seem always to have relied on some arbitrary criterion that vitiates the significance level (Person & Barron, 1978). A number of studies in applied psychology have used optimal scaling and called it conjoint measurement, but these applications are subject to the very criticisms from workers in conjoint measurement cited in the text, namely, that they assume the validity of the proposed model.

5.5.3b. Krantz and Tversky (1971b) say that response linearity could, in principle, be established on the basis of ordinal tests. That has not been done, however, although attempts have been made to establish magnitude estimation as a true linear scale (Krantz *et al.*, 1971; Krantz, 1972a).

5.5.4a. Tukey (1969) commented similarly on the efflorescence of work on measurement axioms following the original paper of Luce and Tukey (1964): "So far as I have seen, this work has all been theoretical in the less satisfactory sense of that word—has explored the axiomatization or other mathematical aspects of such 'measurement.' For this emphasis I am sorry [p. 88]." Tukey adds that he was later told that "the real situation is better than these words would suggest [p. 88, n. 3]." Ten years later, however, the real situation remains essentially unchanged.

5.6.2a. A thoughtful attempt to resolve some of the perplexities of loudness measurement, especially the contradictory results obtained with rating and magnitude estimation, is presented in two recent papers by Marks (1979a, 1979b). These papers appeared too late to be considered in the text but two comments may be added here.

First, Marks seems to depart from his earlier view that the response language itself, namely, ratings and magnitude estimation, produces different loudness scales. Marks (1979b) rescales the ratings to agree with the magnitude estimations and concludes that they yield very nearly the same values of loudness (p. 437). Similarly, Marks (1979a) states, "The difference between the [loudness] scales resides in the perceptual relationship, not in the nature of a numerical response [p. 275]."

Marks employs the parallelism theorem as a validity criterion in both these papers. From Marks's present standpoint, therefore, the contrast between the two panels of Figure 5.2 constitutes evidence of bias in magnitude estimation of grayness.

Second, the question of whether there is more than one loudness scale may not be resolvable with the designs and analyses employed by Marks. Both ratings and magnitude estimation showed nonparallelism in some experiments, and Marks adopted the practice of transforming the data to parallelism, on the assumption that an additive model was operative. Unfortunately, monotone transformation is so flexible that it would be expected to make the data parallel even if the model was seriously nonadditive.

The severity of the transformation problem has only recently become clear (Anderson, 1977b; Carterette & Anderson, 1979). It appears that inappropriate transformation to additivity cannot generally be prevented by using two-factor design with a single score per cell in tasks for which no crossovers are theoretically expected. This set of conditions, which seems to hold in Marks's experiments, does not provide adequate power to prevent nonadditive data from being made to appear additive.

One way to obtain adequate power is with the two-operation method of Section 1.7.2 (see also Chapter 5 of Anderson, in press). This requires a change from two-factor to three-factor design. That would be feasible for Marks's task of loudness summation, in which subjects judged loudness of two tones separated in frequency by several critical bandwidths, by adding a third tone similarly separated in frequency. The hypothesis of loudness summation implies a three-factor adding model, $A + B + C$. The first adding operation would be used as the scaling frame; the data would be transformed monotonically to fit the adding model with the two factors, A and the compound variable (B, C). This transformation does not constrain the $B \times C$ integration—which therefore provides the needed test of additivity. If the adding model is correct, then the transformation will produce the true response scale—and the $B \times C$ factorial graph will exhibit parallelism. If the adding model is incorrect, then the transformation will produce a biased response scale—and the $B \times C$ factorial graph will generally be nonparallel. This two-operation method can provide adequate power together with a valid significance test of goodness of fit. On this basis, therefore, a rigorous test of the additivity hypothesis of loudness summation becomes possible.

5.6.3a. It should be noted that Curtis and Rule do not agree with all the conclusions in the text. In particular, they seem inconsistent in their view of magnitude estimation. Although they seem in some reports to consider magnitude estimation to be a biased response measure, it has remained their primary response measure. In at least two reports, moreover, they explicitly assume that magnitude estimation is a true ratio scale (Rule & Curtis, 1976, 1977).

Curtis and Rule have been attentive to statistical problems, and their work is among the best in the area. Nevertheless, there are still certain difficulties in their tests of goodness of fit (Anderson, 1972a, p. 393, 1974c, p. 230, 1977b, p. 208; see also Bogartz, 1980; Coleman, Graf, & Alf, 1981). As a consequence, their basic model, cited in the text, is not as well established in their data as would be desired.

5.6.3b. Later work by Curtis and Rule has employed nonmetric analyses in which the two-stage assumption of power functions is no longer required. However, that requires reliance on the integration stage, in the same way as in functional measurement.

5.6.7a. The *physical law* **P** is the function relating the observable response to the observable stimulus. In terms of the functional measurement diagram, **P** represents the composition of the psychophysical, psychological, and psychomotor laws: **P = MIV** (Anderson, 1974a, p. 284).

Appendixes

Appendix A: Experimental Procedure

A few aspects of procedure that may be of general interest are discussed briefly in the following comments. More extensive treatment is given in *Methods of Information Integration Theory* (Anderson, in press).

RATING SCALES

It cannot be taken for granted that numerical response measures are linear scales. This has been illustrated by the marked difference between ratings and magnitude estimation in Figure 5.1. The reliance on the rating method in this research program rests jointly on the validational properties of the parallelism and linear fan theorems, which are discussed in Chapter 1, and on certain experimental procedures, which are noted here. Two main aspects of rating procedure concern stimulus end anchors and number of rating categories.

End anchors are stimuli that are just noticeably more extreme than the regular experimental stimuli. Unless it would be inappropriate to the task under study, the end anchors are presented as part of the instructions, with subjects being told that these are the smallest and largest stimuli that they will receive. One purpose of the end anchors is to set up the frame of reference for relating the range of stimulus values to the response scale. A second purpose is to preempt the ends of the response scale, which are especially subject to nonlinear bias, so that the regular data come from the interior of the scale.

The 1–20 scale was initially adopted in the hope that using a moderately large number of response categories would avoid biases that can be produced by the distribution of stimulus values. This hope was fulfilled by the success of the models. Evidence on bias with small numbers of response categories is discussed in Anderson (in press, Section 1.1). A number of experiments have employed 1–10 scales, but fewer than 10 steps seems risky unless only a monotone scale is required.

Graphical rating scales are used in the most careful work. Although somewhat more time consuming, they appear to reduce residual biases that may arise from number preferences even with the most careful procedure. Moreover, they can readily be used with young children. A mental graphic scale may underlie scales with discrete categories and the linearity of both may derive from the general purpose metric sense discussed in Section 1.1.6.

Each experimental task has its own peculiarities that may interact with the procedure for response measurement. Although ratings have had wide success, other methods may be needed for certain tasks. The present discussion is not intended to set up standard procedures, but to offer suggestions for continuing development of response methodology.

INSTRUCTIONS AND PRACTICE

Judgment tasks are often fairly natural to subjects so elaborate instructions are not needed. The main problem in writing instructions is that subjects are easily led astray in what is to them an unknown and uncertain situation.

The following three principles have been found helpful in writing instructions. Begin with a general statement of the purpose of the task. That provides an element of certainty to the subject as well as a conceptual framework for organizing subsequent detail. Do not include unnecessary details because they compete for the subject's attention. For the same reason, no detail should be mentioned until it becomes necessary.

As an example, the opening sentence could be, "In this experiment your task is to make judgments about other people based on small amounts of information." Then typical stimuli would be shown and the nature of the response dimension defined. However, the rating scale itself would not ordinarily be mentioned until the paragraph beginning, "Here is how you make your judgments."

Practice allows the subject to develop a stable task set and helps reduce response variability. Practice is essential for establishing the usage of the response scale. This is ordinarily begun by presenting the end anchors as soon as the response scale is introduced. Additional practice trials that cover the stimulus range are then presented, including any chosen to test the subject's understanding. Although some tasks can cause trouble, subjects are typically facile in most that have been studied and around 10 practice trials may suffice, at least for group analyses. Often the first replication of the design is treated as additional practice (see also Notes 2.2.1a, 2.4.3a, 4.4.4d, 4.5.2a).

It may be added that these instructions place no demand on the integration rule. The instructions are intended to cause the subject to use the response scale in a true linear way, to be sure, but that deals with the response function, M, in Figure 1.1. That has no necessary or likely relation to the integration function, I, in Figure 1.1.

PILOT WORK

Pilot work is important in developing instructions and procedure. Even simple variations on established tasks require debugging. The recommended procedure is to run a few subjects individually with complete written instructions but with a reduced design. Specific stimuli are then selected, and subjects are asked to verbalize their reactions. This may not reveal much about underlying process, but it can be an effective way of discovering misunderstandings. Several rounds of such pilot work may be required in order to arrive at a final procedure.

Pilot work has other functions as well. These include training for the experimenter, help in selection of stimulus levels to cover the desired range with approximately equal spacing, guidance in final choice of design, and opportunity for a dress rehearsal on data analysis.

INDIVIDUAL ANALYSIS

Subjects should ordinarily be run individually, not in batches. Batch procedure, especially in a classroom situation, can be expected to yield low-grade data. Any moderate-sized group of subjects is likely to include some with low motivation and some with high carelessness. Such subjects can generally be monitored appropriately in individual sessions but not when they are part of a group. Moreover, group situations are susceptible to distractions that interfere with understanding of the instructions and with attention to the stimulus materials. The importance of attentional factors is illustrated in the work on primacy–recency (Section 3.3).

The disadvantage of individual sessions is their expense. To run 60 subjects in a single classroom batch is certainly cheaper than running 60 individuals in separate sessions. But group running is likely to be false economy when the meaningfulness of the data is at stake. The time spent in collecting data is typically only a small part of the total time of planning and designing the study, analyzing the data, and writing the report. Since the meaningfulness of the conclusions rests on the meaningfulness of the data, it is pennywise to scrimp at this basic level.

It may be useful to study the advantages of running two or three subjects at a time by incorporating group size as a methodological factor in substantive investigations. In many tasks this could be done without sacrificing the main advantages of individualized sessions. Running two subjects at a time will halve the required number of sessions, a savings of 50%. Further increases in group size show quickly diminishing returns; increasing the group from four subjects to five yields only a 5% savings. Since the disadvantages of group running would be expected to increase rapidly around this size, it seems generally unattractive to consider larger groups.

METHODOLOGY

Methodology is a bad word to many. Most investigators are truly concerned with methods, of course, but the term *methodology* suggests a dogmatic stance on standardization of procedure and correct data analysis. It connotes involvement in niceties, complexities of apparatus, and, especially, statistics that are generally barren, often useless digressions, and sometimes active hindrances to productive inquiry.

Properly considered, however, methodology is an organic part of substantive inquiry: necessarily so, for the validity of methods derives from the empirical results that they bring in. There are, of course, some properly methodological problems that must be studied in their own right, abstracted from particular applications. However, methodological studies are dangerously subject to involutional development that rapidly loses empirical relevance. Steady feedback from empirical substantive exploration is vital for minimizing wasted digressions and keeping methodology on useful and productive paths.

This view toward methodology fits with the inductive conception of scientific theory (Section 1.8.1). Knowledge is not divorced from the methods by which it was acquired; those methods themselves constitute an integral part of knowledge.

APPENDIX B*

Likableness L-values for Four Subsets of Thirty-two Trait Adjectives

H adjectives		M+ adjectives		M− adjectives		L adjectives	
Adjective	L-value	Adjective	L-value	Adjective	L-value	Adjective	L-value
truthful	545	persuasive	374	dependent	254	abusive	100
trustworthy	539	obedient	373	unsystematic	253	distrustful	99
intelligent	537	quick	373	self-conscious	249	intolerant	98
dependable	536	sophisticated	372	undecided	249	unforgiving	98
open-minded	530	thrifty	372	resigned	248	boring	97
thoughtful	529	sentimental	371	clownish	247	unethical	97
wise	528	objective	370	anxious	246	unreasonable	97
considerate	527	nonconforming	369	conforming	246	self-centered	96
good-natured	527	righteous	369	critical	243	snobbish	96
reliable	527	mathematical	367	conformist	241	unkindly	96
mature	522	meditative	366	radical	241	ill-mannered	95
warm	522	fearless	366	dissatisfied	239	ill-tempered	95
earnest	521	systematic	366	old-fashioned	239	unfriendly	92
kind	520	subtle	365	meek	238	hostile	91
friendly	519	normal	362	frivolous	237	dislikable	90
kind-hearted	514	daring	360	discontented	237	ultra-critical	90
happy	514	middleclass	360	troubled	235	offensive	88
clean	514	lucky	358	irreligious	234	belligerent	86
interesting	511	proud	358	overcautious	229	underhanded	86
unselfish	510	sensitive	357	silent	228	annoying	84
good-humored	507	moralistic	357	tough	228	disrespectful	83
honorable	507	talkative	352	ungraceful	228	loud-mouthed	83
humorous	505	excited	351	argumentative	227	selfish	82
responsible	505	moderate	351	withdrawing	227	narrow-minded	80
cheerful	504	satirical	351	uninquisitive	225	vulgar	79
trustful	504	prudent	348	forgetful	224	heartless	78
warm-hearted	504	reserved	348	inhibited	224	insolent	78
broad-minded	503	persistent	347	unskilled	224	thoughtless	77
gentle	503	meticulous	346	crafty	223	rude	76
well-spoken	501	unconventional	346	passive	223	conceited	74
educated	500	deliberate	345	immodest	222	greedy	72
reasonable	500	painstaking	345	unpopular	222	spiteful	72

*H, M+, M−, and L adjectives have frequently been mentioned in the text. These are exemplified by the following four subranges of 32 adjectives taken from the master list of 555 trait words (Anderson, 1968d).

References

Abelson, R. P., Aronson, E., McGuire, W. J., Newcomb, T. M., Rosenberg, M. J., & Tannenbaum, P. H. (Eds.). *Theories of cognitive consistency: A sourcebook.* Chicago, Illinois: Rand McNally, 1968.

Anderson, N. H. An analysis of sequential dependencies. In R. R. Bush & W. K. Estes (Eds.), *Studies in mathematical learning theory.* Stanford, California: Stanford University Press, 1959. (a)

Anderson, N. H. Test of a model for opinion change. *Journal of Abnormal and Social Psychology,* 1959, *59,* 371–381. (b)

Anderson, N. H. Scales and statistics: Parametric and nonparametric. *Psychological Bulletin,* 1961, *58,* 305–316. (a)

Anderson, N. H. Two learning models for responses measured on a continuous scale. *Psychometrika,* 1961, *26,* 391–403. (b)

Anderson, N. H. Application of an additive model to impression formation. *Science,* 1962, *138,* 817–818. (a)

Anderson, N. H. On the quantification of Miller's conflict theory. *Psychological Review,* 1962, *69,* 400–414. (b)

Anderson, N. H. *Application of an additive model to impression formation.* Paper presented at the third annual meeting of the Psychonomic Society. St. Louis, Missouri, August, 1962. (c)

Anderson, N. H. An evaluation of stimulus sampling theory: Comments on Prof. Estes' paper. In A. W. Melton (Ed.), *Categories of human learning.* New York: Academic Press, 1964. (a)

Anderson, N. H. Linear models for responses measured on a continuous scale. *Journal of Mathematical Psychology,* 1964, *1,* 121–142. (b)

Anderson, N. H. Note on weighted sum and linear operator models. *Psychonomic Science,* 1964, *1,* 189–190. (c)

Anderson, N. H. Test of a model for number-averaging behavior. *Psychonomic Science,* 1964, *1,* 191–192. (d)

Anderson, N. H. Averaging versus adding as a stimulus-combination rule in impression formation. *Journal of Experimental Psychology,* 1965, *70,* 394–400. (a)

Anderson, N. H. Primacy effects in personality impression formation using a generalized order effect paradigm. *Journal of Personality and Social Psychology*, 1965, *2*, 1–9. (b)

Anderson, N. H. Three methodological experiments on the positive context effect. Unpublished experiments, University of California, Los Angeles, 1965. (c)

Anderson, N. H. Component ratings in impression formation. *Psychonomic Science*, 1966, *6*, 279–280. (a)

Anderson, N. H. *A search task: Discussion of Professor Restle's paper*. Paper presented at conference on "Learning Processes and Thought," University of Pittsburgh, October 13–14, 1966. (b)

Anderson, N. H. Application of a weighted average model to a psychophysical averaging task. *Psychonomic Science*, 1967, *8*, 227–228. (a)

Anderson, N. H. Averaging model analysis of set-size effect in impression formation. *Journal of Experimental Psychology*, 1967, *75*, 158–165. (b)

Anderson, N. H. Application of a linear-serial model to a personality-impression task using serial presentation. *Journal of Personality and Social Psychology*, 1968, *10*, 354–362. (a)

Anderson, N. H. A simple model for information integration. In R. P. Abelson, E. Aronson, W. J. McGuire, T. M. Newcomb, M. J. Rosenberg, & P. H. Tannenbaum (Eds.), *Theories of cognitive consistency: A sourcebook*. Chicago, Illinois: Rand McNally, 1968. (b)

Anderson, N. H. Averaging of space and number stimuli with simultaneous presentation. *Journal of Experimental Psychology*, 1968, *77*, 383–392. (c)

Anderson, N. H. Likableness ratings of 555 personality-trait words. *Journal of Personality and Social Psychology*, 1968, *9*, 272–279. (d)

Anderson, N. H. Application of a model for numerical response to a probability learning situation. *Journal of Experimental Psychology*, 1969, *80*, 19–27. (a)

Anderson, N. H. A search task. In J. F. Voss (Ed.), *Approaches to thought*. Columbus, Ohio: Merrill, 1969. (b)

Anderson, N. H. Comment on "An analysis-of-variance model for the assessment of configural cue utilization in clinical judgment." *Psychological Bulletin*, 1969, *72*, 63–65. (c)

Anderson, N. H. Effects of choice and verbal feedback on preference values. *Journal of Experimental Psychology*, 1969, *79*, 77–84. (d)

Anderson, N. H. Averaging model applied to the size–weight illusion. *Perception & Psychophysics*, 1970, *8*, 1–4. (a)

Anderson, N. H. Functional measurement and psychophysical judgment. *Psychological Review*, 1970, *77*, 153–170. (b)

Anderson, N. H. Integration theory and attitude change. *Psychological Review*, 1971, *78*, 171–206. (a)

Anderson, N. H. Test of adaptation-level theory as an explanation of a recency effect in psychophysical integration. *Journal of Experimental Psychology*, 1971, *87*, 57–63. (b)

Anderson, N. H. Two more tests against change of meaning in adjective combinations. *Journal of Verbal Learning and Verbal Behavior*, 1971, *10*, 75–85. (c)

Anderson, N. H. An exchange on functional and conjoint measurement. *Psychological Review*, 1971, *78*, 457–458. (d)

Anderson, N. H. Cross-task validation of functional measurement. *Perception & Psychophysics*, 1972, *12*, 389–395. (a)

Anderson, N. H. Looking for configurality in clinical judgment. *Psychological Bulletin*, 1972, *78*, 93–102. (b)

Anderson, N. H. Comments on the articles of Hodges and of Schönemann, Cafferty, and Rotton. *Psychological Review*, 1973, *80*, 88–92. (a)

Anderson, N. H. Functional measurement of social desirability. *Sociometry*, 1973, *36*, 89–98. (b)

Anderson, N. H. Information integration theory applied to attitudes about U.S. presidents. *Journal of Educational Psychology*, 1973, *64*, 1–8. (c)

Anderson, N. H. Serial position curves in impression formation. *Journal of Experimental Psychology*, 1973, *97*, 8–12. (d)

Anderson, N. H. Algebraic models in perception. In E. C. Carterette & M. P. Friedman (Eds.), *Handbook of perception* (Vol. 2). New York: Academic Press, 1974. (a)

Anderson, N. H. Cognitive algebra. In L. Berkowitz (Ed.), *Advances in experimental social psychology* (Vol. 7). New York: Academic Press, 1974. (b)

Anderson, N. H. Cross-task validation of functional measurement using judgments of total magnitude. *Journal of Experimental Psychology*, 1974, *102*, 226–233. (c)

Anderson, N. H. Information integration theory: A brief survey. In D. H. Krantz, R. C. Atkinson, R. D. Luce, & P. Suppes (Eds.), *Contemporary developments in mathematical psychology* (Vol. 2). San Francisco: Freeman, 1974. (d)

Anderson, N. H. *The problem of change-of-meaning* (Tech. Rep. CHIP 42). La Jolla: Center for Human Information Processing, University of California, San Diego, June, 1974. (e)

Anderson, N. H. *Methods for studying information integration* (Tech. Rep. CHIP 43). La Jolla: Center for Human Information Processing, University of California, San Diego, June 1974. (f)

Anderson, N. H. *Basic experiments in person perception* (Tech. Rep. CHIP 44). La Jolla: Center for Human Information Processing, University of California, San Diego, June, 1974. (g)

Anderson, N. H. *Algebraic models for information integration* (Tech. Rep. CHIP 45). La Jolla: Center for Human Information Processing, University of California, San Diego, June 1974. (h)

Anderson, N. H. Unpublished experiments on psycholinguistic integration. University of California, San Diego, 1974. (i)

Anderson, N. H. On the role of context effects in psychophysical judgment. *Psychological Review*, 1975, *82*, 462–482. (a)

Anderson, N. H. Unpublished experiments on probability models. University of California, San Diego, 1975. (b)

Anderson, N. H. Equity judgments as information integration. *Journal of Personality and Social Psychology*, 1976, *33*, 291–299. (a)

Anderson, N. H. How functional measurement can yield validated interval scales of mental quantities. *Journal of Applied Psychology*, 1976, *61*, 677–692. (b)

Anderson, N. H. Integration theory, functional measurement and the psychophysical law. In H.-G. Geissler & Yu. M. Zabrodin (Eds.), *Advances in psychophysics*. Berlin: VEB Deutscher Verlag, 1976. (c)

Anderson, N. H. *Social perception and cognition* (Tech. Rep. CHIP 62). La Jolla: Center for Human Information Processing, University of California, San Diego, July 1976. (d)

Anderson, N. H. Failure of additivity in bisection of length. *Perception & Psychophysics*, 1977, *22*, 213–222. (a)

Anderson, N. H. Note on functional measurement and data analysis. *Perception & Psychophysics*, 1977, *21*, 201–215. (b)

Anderson, N. H. Some problems in using analysis of variance in balance theory. *Journal of Personality and Social Psychology*, 1977, *35*, 140–158. (c)

Anderson, N. H. Measurement of motivation and incentive. *Behavior Research Methods &*
    Instrumentation, 1978, *10*, 360–375. (a)

Anderson, N. H. Progress in cognitive algebra. In L. Berkowitz (Ed.), *Cognitive theories in*
    social psychology. New York: Academic Press, 1978. (b)

Anderson, N. H. Algebraic rules in psychological measurement. *American Scientist*, 1979,
    67, 555–563. (a)

Anderson, N. H. Indeterminate theory: Reply to Gollob. *Journal of Personality and Social*
    Psychology, 1979, *37*, 950–952. (b)

Anderson, N. H. Information integration theory in developmental psychology. In F.
    Wilkening, J. Becker, & T. Trabasso (Eds.), *Information integration by children*. Hills-
    dale, New Jersey: Erlbaum, 1980.

Anderson, N. H. Cognitive algebra and social psychophysics. In B. Wegener (Ed.), *Social*
    attitudes and psychophysical measurement. Hillsdale, New Jersey: Erlbaum, 1981. (a)

Anderson, N. H. Integration theory applied to cognitive responses and attitudes. In R.
    Petty, T. Ostrom, & T. Brock (Eds.), *Cognitive responses in persuasion*. Hillsdale, New
    Jersey: Erlbaum, 1981. (b)

Anderson, N. H. *Methods of information integration theory*. New York: Academic Press, in
    press.

Anderson, N. H., & Alexander, G. R. Choice test of the averaging hypothesis for informa-
    tion integration. *Cognitive Psychology*, 1971, *2*, 313–324.

Anderson, N. H., & Barrios, A. A. Primacy effects in personality impression formation.
    Journal of Abnormal and Social Psychology, 1961, *63*, 346–350.

Anderson, N. H., & Butzin, C. A. Performance = Motivation × Ability: An integration-
    theoretical analysis. *Journal of Personality and Social Psychology*, 1974, *30*, 598–604.

Anderson, N. H., & Butzin, C. A. Integration theory applied to children's judgments of
    equity. *Developmental Psychology*, 1978, *14*, 593–606.

Anderson, N. H., & Clavadetscher, J. E. Tests of a conditioning hypothesis with adjective
    combinations. *Journal of Experimental Psychology: Human Learning and Memory*, 1976,
    2, 11–20.

Anderson, N. H., & Cuneo, D. O. The Height + Width rule in children's judgments of
    quantity. *Journal of Experimental Psychology: General*, 1978, *107*, 335–378. (a)

Anderson, N. H., & Cuneo, D. O. The Height + Width rule seems solid: Reply to Bogartz.
    Journal of Experimental Psychology: General, 1978, *107*, 388–392. (b)

Anderson, N. H., & Farkas, A. J. New light on order effects in attitude change. *Journal of*
    Personality and Social Psychology, 1973, *28*, 88–93.

Anderson, N. H., & Farkas, A. J. Integration theory applied to models of inequity. *Person-*
    ality and Social Psychology Bulletin, 1975, *1*, 588–591.

Anderson, N. H., & Graesser, C. C. An information integration analysis of attitude change
    in group discussion. *Journal of Personality and Social Psychology*, 1976, *34*, 210–222.

Anderson, N. H., & Hovland, C. I. The representation of order effects in communication
    research. In C. I. Hovland (Ed.), *The order of presentation in persuasion*. New Haven,
    Connecticut: Yale University Press, 1957.

Anderson, N. H., & Hubert, S. Effects of concomitant verbal recall on order effects in
    personality impression formation. *Journal of Verbal Learning and Verbal Behavior*, 1963,
    2, 379–391.

Anderson, N. H., & Jacobson, A. Effect of stimulus inconsistency and discounting instruc-
    tions in personality impression formation. *Journal of Personality and Social Psychology*,
    1965, *2*, 531–539.

Anderson, N. H., & Jacobson, A. Further data on a weighted average model for judgment
    in a lifted weight task. *Perception & Psychophysics*, 1968, *4*, 81–84.

Anderson, N. H., & Lampel, A. K. Effect of context on ratings of personality traits. *Psychonomic Science*, 1965, *3*, 433–434.

Anderson, N. H., Lindner, R., & Lopes, L. L. Integration theory applied to judgments of group attractiveness. *Journal of Personality and Social Psychology*, 1973, *26*, 400–408.

Anderson, N. H., & Lopes, L. L. Some psycholinguistic aspects of person perception. *Memory & Cognition*, 1974, *2*, 67–74.

Anderson, N. H., & Norman, A. Order effects in impression formation in four classes of stimuli. *Journal of Abnormal and Social Psychology*, 1964, *69*, 467–471.

Anderson, N. H., Sawyers, B. K., & Farkas, A. J. President paragraphs. *Behavior Research Methods & Instrumentation*, 1972, *4*, 177–192.

Anderson, N. H., & Shanteau, J. C. Information integration in risky decision making. *Journal of Experimental Psychology*, 1970, *84*, 441–451.

Anderson, N. H., & Shanteau, J. Weak inference with linear models. *Psychological Bulletin*, 1977, *84*, 1155–1170.

Anderson, N. H., & Weiss, D. J. Test of a multiplying model for estimated area of rectangles. *American Journal of Psychology*, 1971, *84*, 543–548.

Anderson, T., & Birnbaum, M. H. Test of an additive model of social inference. *Journal of Personality and Social Psychology*, 1976, *33*, 655–662.

Argyle, M. *The psychology of interpersonal behavior*. Baltimore, Maryland: Penguin, 1967.

Asch, S. E. Forming impressions of personality. *Journal of Abnormal and Social Psychology*, 1946, *41*, 258–290.

Attneave, F. Perception and related areas. In S. Koch (Ed.), *Psychology: A study of a science* (Vol. 4). New York: McGraw-Hill, 1962.

Baron, P. H. Self-esteem, ingratiation, and evaluation of unknown others. *Journal of Personality and Social Psychology*, 1974, *30*, 104–109.

Bartlett, F. C. *Remembering: A study in experimental and social psychology*. London and New York: Cambridge University Press, 1932.

Beck, J., & Shaw, W. A. Ratio-estimations of loudness-intervals. *American Journal of Psychology*, 1967, *80*, 59–65.

Becker, L. A. *Meaning shifts in impression formation*. Paper presented at the meetings of the Midwestern Psychological Association, Detroit, Michigan, May 1971.

Bettman, J. R., Capon, N., & Lutz, R. J. Information processing in attitude formation and change. *Communication Research*, 1975, *2*, 267–278.

Berscheid, E., & Walster, E. Physical attractiveness. In L. Berkowitz (Ed.), *Advances in experimental social psychology* (Vol. 7). New York: Academic Press, 1974.

Birnbaum, M. H. Morality judgments: Tests of an averaging model. *Journal of Experimental Psychology*, 1972, *93*, 35–42.

Birnbaum, M. H. Morality judgment: Test of an averaging model with differential weights. *Journal of Experimental Psychology*, 1973, *99*, 395–399.

Birnbaum, M. H. The nonadditivity of personality impressions. *Journal of Experimental Psychology*, 1974, *102*, 543–561. (a)

Birnbaum, M. H. Using contextual effects to derive psychophysical scales. *Perception & Psychophysics*, 1974, *15*, 89–96. (b)

Birnbaum, M. H. Intuitive numerical prediction. *American Journal of Psychology*, 1976, *89*, 417–429.

Birnbaum, M. H. Differences and ratios in psychological measurement. In N. J. Castellan, Jr., & F. Restle (Eds.), *Cognitive theory* (Vol. 3). Hillsdale, New Jersey: Erlbaum, 1978.

Birnbaum, M. H., & Elmasian, R. Loudness "ratios" and "differences" involve the same psychophysical operation. *Perception & Psychophysics*, 1977, *22*, 383–391.

Birnbaum, M. H., Kobernick, M., & Veit, C. T. Subjective correlation and the size-numerosity illusion. *Journal of Experimental Psychology*, 1974, *102*, 537–539.

Birnbaum, M. H., Parducci, A., & Gifford, R. K. Contextual effects in information integration. *Journal of Experimental Psychology*, 1971, *88*, 158–170.

Birnbaum, M. H., & Stegner, S. E. Source credibility in social judgment: Bias, expertise, and the judge's point of view. *Journal of Personality and Social Psychology*, 1979, *37*, 48–74.

Birnbaum, M. H., Wong, R., & Wong, L. K. Combining information from sources that vary in credibility. *Memory & Cognition*, 1976, *4*, 330–336.

Björkman, M. Studies in predictive behavior: Explorations into predictive judgments based on functional learning and defined by estimation, categorization, and choice. *Scandinavian Journal of Psychology*, 1965, *6*, 129–156.

Blankenship, D. A., & Anderson, N. H. Subjective duration: A functional measurement analysis. *Perception & Psychophysics*, 1976, *20*, 168–172.

Bock, R. D., & Jones, L. V. *The measurement and prediction of judgment and choice*. San Francisco: Holden-Day, 1968.

Bogartz, R. S. On the meaning of statistical interactions. *Journal of Experimental Child Psychology*, 1976, *22*, 178–183.

Bogartz, R. S. Comments on Anderson and Cuneo's "The Height + Width rule in children's judgments of quantity." *Journal of Experimental Psychology: General*, 1978, *107*, 379–387.

Bogartz, R. S. Some functional measurement procedures for determining the psychophysical law. *Perception & Psychophysics*, 1980, *27*, 284–294.

Bogartz, R. S., & Wackwitz, J. H. Polynomial response scaling and functional measurement. *Journal of Mathematical Psychology*, 1971, *8*, 418–443.

Borges, M. A., & Sawyers, B. K. Common verbal quantifiers: Usage and interpretation. *Journal of Experimental Psychology*, 1974, *102*, 335–338.

Boring, E. G. *Sensation and perception in the history of experimental psychology*. New York: Appleton, 1942.

Boring, E. G. *A history of experimental psychology* (2nd ed.). New York: Appleton, 1950.

Bossart, P., & DiVesta, F. J. Effects of context, frequency, and order of presentation of evaluative assertions on impression formation. *Journal of Personality and Social Psychology*, 1966, *4*, 538–544.

Brehmer, B., & Slovic, P. Information integration in multiple-cue judgments. *Journal of Experimental Psychology: Human Perception and Performance*, 1980, *6*, 302–308.

Brigham, J. C. Ethnic stereotypes. *Psychological Bulletin*, 1971, *76*, 15–38.

Brink, J. H. Impression order effects as a function of the personal relevance of the object of description. *Memory & Cognition*, 1974, *2*, 561–565.

Bruner, J. S., Shapiro, D., & Tagiuri, R. The meaning of traits in isolation and in combination. In R. Tagiuri & L. Petrullo (Eds.), *Person perception and interpersonal behavior*. Stanford, California: Stanford University Press, 1958.

Brunswik, E. *Perception and the representative design of psychological experiments*. Berkeley, California: University of California Press, 1956.

Bryson, J. B., & Franco, L. B. Experimental differentiation of meaning change and averaging explanations for context effects. *Memory & Cognition*, 1976, *4*, 337–344.

Burnstein, E., & Vinokur, A. What a person thinks upon learning he has chosen differently from others: Nice evidence for the persuasive-arguments explanation of choice shifts. *Journal of Experimental Social Psychology*, 1975, *11*, 412–426.

Busemeyer, J. R. Importance of measurement theory, error theory, and experimental design for testing the significance of interactions. *Psychological Bulletin*, 1980, *88*, 237–244.

Bush, R. R., & Mosteller, F. *Stochastic models for learning*. New York: Wiley, 1955.

Butzin, C. A. *The effect of ulterior motive information on children's moral judgments*. Unpublished doctoral dissertation, University of California, San Diego, 1978.

Butzin, C. A., & Anderson, N. H. Functional measurement of children's judgments. *Child Development*, 1973, *44*, 529–537.

Byrne, D. Attitudes and attraction. In L. Berkowitz (Ed.), *Advances in experimental social psychology* (Vol. 4). New York: Academic Press, 1969.

Byrne, D. *The attraction paradigm*. New York: Academic Press, 1971.

Byrne, D., Clore, G. L., Griffitt, W., Lamberth, J., & Mitchell, H. E. When research paradigms converge: Confrontation or integration? *Journal of Personality and Social Psychology*, 1973, *28*, 313–320. (a)

Byrne, D., Clore, G. L., Griffitt, W., Lamberth, J., & Mitchell, H. E. One more time. *Journal of Personality and Social Psychology*, 1973, *28*, 323–324. (b)

Byrne, D., Lamberth, J., Palmer, J., & London, O. Sequential effects as a function of explicit and implicit interpolated attraction responses. *Journal of Personality and Social Psychology*, 1969, *13*, 70–78.

Campbell, N. R. *An account of the principles of measurement and calculation*. London: Longmans, Green, 1928.

Carterette, E. C., & Anderson, N. H. Bisection of loudness. *Perception & Psychophysics*, 1979, *26*, 265–280.

Cartwright, D. Determinants of scientific progress: The case of research on the risky shift. *American Psychologist*, 1973, *28*, 222–231.

Chalmers, D. K. Repetition and order effects in attitude formation. *Journal of Personality and Social Psychology*, 1971, *17*, 219–228.

Clavadetscher, J. E. *Two context processes in the Ebbinghaus illusion*. Unpublished doctoral dissertation, University of California, San Diego, 1977.

Clavadetscher, J. E., & Anderson, N. H. Comparative judgment: Tests of two theories using the Baldwin figure. *Journal of Experimental Psychology: Human Perception and Performance*, 1977, *3*, 119–135.

Cliff, N. Adverbs as multipliers. *Psychological Review*, 1959, *66*, 27–44.

Cliff, N. Scaling. *Annual Review of Psychology*, 1973, *24*, 473–506.

Clore, G. L., & Byrne, D. A reinforcement–affect model of attraction. In T. L. Huston (Ed.), *Foundations of interpersonal attraction*. New York: Academic Press, 1974.

Cofer, C. N. Constructive processes in memory. *American Scientist*, 1973, *61*, 537–543.

Coleman, B. J., Graf, R. G., & Alf, E. F. Assessing power function relationships in magnitude estimation. *Perception & Psychophysics*, 1981, *29*, 178–180.

Coombs, C. H. *A theory of data*. New York: Wiley, 1964.

Coombs, C. H. Thurstone's measurement of social values revisited forty years later. *Journal of Personality and Social Psychology*, 1967, *6*, 85–91.

Crano, W. D. Primacy versus recency in retention of information and opinion change. *Journal of Social Psychology*, 1977, *101*, 87–96.

Cross, D. V. *On judgments of magnitude*. Paper presented at the International Symposium on Social Psychophysics, University of Mannheim, October 1978. (In B. Wegener [Ed.], *Social attitudes and psychophysical measurement*. Hillsdale, New Jersey: Erlbaum, in press.)

Cuneo, D. O. *Children's judgments of numerical quantity: The role of length, density, and number cues*. Unpublished doctoral dissertation, University of California, San Diego, 1978.

Cuneo, D. O. A general strategy for judgments of quantity: The Height + Width rule. *Child Development*, 1980, *51*, 299–301.

Cuneo, D. O. Children's judgments of numerical quantity: A new view of early quantification. *Cognitive Psychology*, 1981, in press.

Curtis, D. W. Magnitude estimations and category judgments of brightness and brightness intervals: A two-stage interpretation. *Journal of Experimental Psychology*, 1970, *83*, 201–208.

Curtis, D. W., Attneave, F., & Harrington, T. L. A test of a two-stage model of magnitude judgment. *Perception & Psychophysics*, 1968, *3*, 25–31.

Curtis, D. W., & Mullin, L. C. Judgments of average magnitude: Analyses in terms of the functional measurement and two-stage models. *Perception & Psychophysics*, 1975, *18*, 299–308.

Curtis, D. W., & Rule, S. J. Magnitude judgments of brightness and brightness difference as a function of background reflectance. *Journal of Experimental Psychology*, 1972, *95*, 215–222.

Curtis, D. W., & Rule, S. J. Judgment of duration relations: Simultaneous and sequential presentation. *Perception & Psychophysics*, 1977, *22*, 578–584.

Curtis, D. W., & Rule, S. J. Judgments of average lightness and darkness: A further consideration of inverse attributes. *Perception & Psychophysics*, 1978, *24*, 343–348.

Dawson, W. E. Magnitude estimation of apparent sums and differences. *Perception & Psychophysics*, 1971, *9*, 368–374.

de Leeuw, J., Young, F. W., & Takane, Y. Additive structure in qualitative data: An alternating least squares method with optimal scaling features. *Psychometrika*, 1976, *41*, 471–503.

Dion, K., Berscheid, E., & Walster, E. What is beautiful is good. *Journal of Personality and Social Psychology*, 1972, *24*, 285–290.

Drake, S. *Galileo at work*. Chicago, Illinois: University of Chicago Press, 1978.

Dreben, E. K., Fiske, S. T., & Hastie, R. The independence of evaluative and item information: Impression and recall order effects in behavior-based impression formation. *Journal of Personality and Social Psychology*, 1979, *37*, 1758–1768.

Ebbesen, E. B., & Konečni, V. J. Decision making and information integration in the courts: The setting of bail. *Journal of Personality and Social Psychology*, 1975, *32*, 805–821.

Eiser, J. R. *Cognitive social psychology*. New York: McGraw-Hill, 1980.

Eisler, H. Similarity in the continuum of heaviness with some methodological and theoretical considerations. *Scandinavian Journal of Psychology*, 1960, *1*, 69–81.

Eisler, H. Magnitude scales, category scales, and Fechnerian integration. *Psychological Review*, 1963, *70*, 243–253.

Eisler, H. The connection between magnitude and discrimination scales and direct and indirect scaling methods. *Psychometrika*, 1965, *30*, 271–289.

Eisler, H. Subjective duration and psychophysics. *Psychological Review*, 1975, *82*, 429–450.

Eisler, H., Holm, S., & Montgomery, H. The general psychophysical differential equation: A comparison of three specifications. *Journal of Mathematical Psychology*, 1979, *20*, 16–34.

Ekman, G., & Sjöberg, L. Scaling. *Annual Review of Psychology*, 1965, *16*, 451–474.

Ellis, B. *Basic concepts of measurement*. London and New York: Cambridge University Press, 1966.

English, H. B., Welborn, E. L., & Killian, C. D. Studies in substance memorization. *Journal of General Psychology*, 1934, *11*, 233–260.

Estes, W. K. Some targets for mathematical psychology. *Journal of Mathematical Psychology*, 1975, *12*, 263–282.

Falmagne, J.-C. Random conjoint measurement and loudness summation. *Psychological Review*, 1976, *83*, 65–79.

Farkas, A. J. *A cognitive algebra for bystander judgments of interpersonal unfairness.* Unpublished doctoral dissertation, University of California, San Diego, 1977.

Farkas, A. J., & Anderson, N. H. Multidimensional input in equity theory. *Journal of Personality and Social Psychology*, 1979, 37, 879–896.

Farley, J., & Fantino, E. The symmetrical law of effect and the matching relation in choice behavior. *Journal of the Experimental Analysis of Behavior*, 1978, 29, 37–60.

Fechner, G. T. *In sachen der psychophysik.* Amsterdam: E. J. Bonset, 1968. (Originally published, 1877.)

Feldman, S. *Evaluative ratings of adjective–adjective combinations predicted from ratings of their components.* Unpublished doctoral dissertation, Yale University, New Haven, Connecticut, 1962.

Fishbein, M. A behavior theory approach to the relations between beliefs about an object and the attitude toward the object. In M. Fishbein (Ed.), *Readings in attitude theory and measurement.* New York: Wiley, 1967.

Fishbein, M., & Ajzen, I. Attitudes and opinions. *Annual Review of Psychology*, 1972, 23, 487–544.

Fishbein, M., & Ajzen, I. *Belief, attitude, intention and behavior.* Reading, Massachusetts: Addison-Wesley, 1975.

Fisher, R. A. *Statistical methods for research workers* (13th ed.). New York: Hafner, 1958.

Fiske, S. T. Attention and weight in person perception: The impact of negative and extreme behavior. *Journal of Personality and Social Psychology*, 1980, 38, 889–906.

Friedman, M. P., Carterette, E. C., & Anderson, N. H. Long-term probability learning with a random schedule of reinforcement. *Journal of Experimental Psychology*, 1968, 78, 442–455.

Garner, W. R. A technique and a scale for loudness measurement. *Journal of the Acoustical Society of America*, 1954, 26, 73–88. (a)

Garner, W. R. Context effects and the validity of loudness scales. *Journal of Experimental Psychology*, 1954, 48, 218–224. (b)

Garner, W. R. Advantages of the discriminability criterion for a loudness scale. *Journal of the Acoustical Society of America*, 1958, 30, 1005–1012.

Gollob, H. F. A reply to Norman H. Anderson's critique of the subject–verb–object approach to social cognition. *Journal of Personality and Social Psychology*, 1979, 37, 931–949.

Gollob, H. F., & Lugg, A. M. Effect of instruction and stimulus presentation on the occurrence of averaging responses in impression formation. *Journal of Experimental Psychology*, 1973, 98, 217–219.

Gollob, H. F., Rossman, B. B., & Abelson, R. P. Social inference as a function of the number of instances and consistency of information presented. *Journal of Personality and Social Psychology*, 1973, 27, 19–33.

Graesser, C. C. *A social averaging theorem for group decision making.* Unpublished doctoral dissertation, University of California, San Diego, 1977.

Graesser, C. C., & Anderson, N. H. Cognitive algebra of the equation: Gift size = Generosity × Income. *Journal of Experimental Psychology*, 1974, 103, 692–699.

Graham, C. H., & Ratoosh, P. Notes on some interrelations of sensory psychology, perception, and behavior. In S. Koch (Ed.), *Psychology: A study of a science* (Vol. 4). New York: McGraw-Hill, 1962.

Greenwald, A. G. Cognitive learning, cognitive response to persuasion, and attitude change. In A. G. Greenwald, T. C. Brock, & T. M. Ostrom (Eds.), *Psychological foundations of attitudes.* New York: Academic Press, 1968.

Gregson, R. A. M. *Psychometrics of similarity.* New York: Academic Press, 1975.

Griffitt, W., & Jackson, T. Context effects in impression formation as a function of context source. *Psychonomic Science*, 1970, *20*, 321–322.

Guilford, J. P. *Psychometric methods* (2nd ed.). New York: McGraw-Hill, 1954.

Gupta, M. *An information integration analysis of developmental trends in attribution of scholastic performance.* Unpublished doctoral dissertation, Indian Institute of Technology, Kanpur, 1978.

Hagiwara, S. Visual versus verbal information in impression formation. *Journal of Personality and Social Psychology*, 1975, *32*, 692–698.

Hamilton, D. L., & Huffman, L. J. Generality of impression-formation processes for evaluative and nonevaluative judgments. *Journal of Personality and Social Psychology*, 1971, *20*, 200–207.

Hamilton, D. L., & Zanna, M. P. Context effects in impression formation: Changes in connotative meaning. *Journal of Personality and Social Psychology*, 1974, *29*, 649–654.

Hammond, K. R. Probabilistic functioning and the clinical method. *Psychological Review*, 1955, *62*, 255–262.

Harris, R. J. The uncertain connection between verbal theories and research hypotheses in social psychology. *Journal of Experimental Social Psychology*, 1976, *12*, 210–219.

Hastie, R., Ostrom, T. M., Ebbesen, E. B., Wyer, R. S., Jr., Hamilton, D. L., & Carlston, D. E. *Person memory: The cognitive basis of social perception.* Hillsdale, New Jersey: Erlbaum, 1980.

Hays, W. L. An approach to the study of trait implication and trait similarity. In R. Tagiuri & L. Petrullo (Eds.), *Person perception and interpersonal behavior.* Stanford, California: Stanford University Press, 1958.

Helson, H. *Adaptation-level theory.* New York: Harper, 1964.

Hendrick, C. *Averaging versus summation in impression formation.* Unpublished doctoral dissertation, University of Missouri, Columbia, 1967.

Hendrick, C. Averaging vs. summation in impression formation. *Perceptual and Motor Skills*, 1968, *27*, 1295–1302. (a)

Hendrick, C. Measurement of trait implication. *Perceptual and Motor Skills*, 1968, *27*, 443–446. (b)

Hendrick, C. Asymmetry of the trait inference process in impression formation. *Perceptual and Motor Skills*, 1969, *28*, 715–720.

Hendrick, C. Effects of salience of stimulus inconsistency on impression formation. *Journal of Personality and Social Psychology*, 1972, *22*, 219–222.

Hendrick, C., & Costantini, A. F. Effects of varying trait inconsistency and response requirements on the primacy effect in impression formation. *Journal of Personality and Social Psychology*, 1970, *15*, 158–164. (a)

Hendrick, C., & Costantini, A. F. Number averaging behavior: A primacy effect. *Psychonomic Science*, 1970, *19*, 121–122. (b)

Hendrick, C., Costantini, A. F., McGarry, J., & McBride, K. Attention decrement, temporal variation, and the primacy effect in impression formation. *Memory & Cognition*, 1973, *1*, 193–195.

Herrnstein, R. J. On the law of effect. *Journal of the Experimental Analysis of Behavior*, 1970, *13*, 243–266.

Hewitt, J. Integration of information about others. *Psychological Reports*, 1972, *30*, 1007–1010.

Hicks, J. M., & Campbell, D. T. Zero-point scaling as affected by social object, scaling method, and context. *Journal of Personality and Social Psychology*, 1965, *2*, 793–808.

Higgins, E. T., & Rholes, W. S. Impression formation and role fulfillment: A "holistic reference" approach. *Journal of Experimental Social Psychology*, 1976, *12*, 422–435.

Himmelfarb, S. Effects of cue validity differences in weighting information. *Journal of Mathematical Psychology*, 1970, *7*, 531–539.

Himmelfarb, S. Integration and attribution theories in personality impression formation. *Journal of Personality and Social Psychology*, 1972, *23*, 309–313.

Himmelfarb, S. General test of a differential weighted averaging model of impression formation. *Journal of Experimental Social Psychology*, 1973, *9*, 379–390.

Himmelfarb, S. "Resistance" to persuasion induced by information integration. In S. Himmelfarb & A. H. Eagly (Eds.), *Readings in attitude change*. New York: Wiley, 1974.

Himmelfarb, S. On scale value and weight in the weighted averaging model of integration theory. *Personality and Social Psychology Bulletin*, 1975, *1*, 580–583.

Himmelfarb, S., & Anderson, N. H. Integration theory applied to opinion attribution. *Journal of Personality and Social Psychology*, 1975, *31*, 1064–1072.

Himmelfarb, S., & Senn, D. J. Forming impressions of social class: Two tests of an averaging model. *Journal of Personality and Social Psychology*, 1969, *12*, 38–51.

Hodges, B. H. Adding and averaging models for information integration. *Psychological Review*, 1973, *80*, 80–84.

Hodges, B. H. Effect of valence on relative weighting in impression formation. *Journal of Personality and Social Psychology*, 1974, *30*, 378–381.

Hoffman, P. J. The paramorphic representation of clinical judgment. *Psychological Bulletin*, 1960, *57*, 116–131.

Hoffman, P. J., Slovic, P., & Rorer, L. G. An analysis-of-variance model for the assessment of configural cue utilization in clinical judgment. *Psychological Bulletin*, 1968, *69*, 338–349.

Hommers, W. Information processing in children's choices among bets. In F. Wilkening, J. Becker, & T. Trabasso (Eds.), *Information integration by children*. Hillsdale, New Jersey: Erlbaum, 1980.

Hommers, W. *Die Entwicklungspsychologie der Delikts- und Geschäftsfähigkeit*. Unpublished Habilitationsschrift, Christian-Albrechts-Universität, Kiel, Federal Republic of Germany, February, 1981.

Hovland, C. I. (Ed.). *The order of presentation in persuasion*. New Haven, Connecticut: Yale University Press, 1957.

Hoyle, F. *Astronomy*. London: Rathbone, 1962.

Hull, C. L. *A behavior system*. New Haven, Connecticut: Yale University Press, 1952.

Hyman, R., & Well, A. Judgments of similarity and spatial models. *Perception & Psychophysics*, 1967, *2*, 233–248.

Insko, C. A. *Theories of attitude change*. New York: Appleton, 1967.

James, W. *The principles of psychology* (Vol. 1). New York: Dover, 1950. (Originally published, 1890.)

Johnson, D. M. *The psychology of thought and judgment*. New York: Harper, 1955.

Jones, E. E. The rocky road from acts to dispositions. *American Psychologist*, 1979, *34*, 107–117.

Jones, E. E., & Goethals, G. R. *Order effects in impression formation: Attribution context and the nature of the entity*. New York: General Learning Press, 1971.

Jones, F. N. Overview of psychophysical scaling methods. In E. C. Carterette & M. P. Friedman (Eds.), *Handbook of perception* (Vol. 2). New York: Academic Press, 1974.

Jones, L. V. Invariance of zero-point scaling under changes in stimulus context. *Psychological Bulletin*, 1967, *67*, 153–164.

Kahneman, D., & Tversky, A. Prospect theory: An analysis of decision under risk. *Econometrica*, 1979, *47*, 263–291.

Kanouse, D. E., & Hanson, L. R., Jr. *Negativity in evaluations.* New York: General Learning Press, 1972.

Kaplan, K. J. The effects of initial attitude and initial number of beliefs on subsequent attitude change: An exploration into underlying cognitive dynamics. *Dissertation Abstracts International*, 1969, *30* (1-A), 390.

Kaplan, K. J. From attitude formation to attitude change: Acceptance and impact as cognitive mediators. *Sociometry*, 1972, *35*, 448–467.

Kaplan, M. F. Context effects in impression formation: The weighted average versus the meaning-change formulation. *Journal of Personality and Social Psychology*, 1971, *19*, 92–99. (a)

Kaplan, M. F. Dispositional effects and weight of information in impression formation. *Journal of Personality and Social Psychology*, 1971, *18*, 279–284. (b)

Kaplan, M. F. The effect of judgmental dispositions on forming impressions of personality. *Canadian Journal of Behavioral Science*, 1971, *3*, 259–267. (c)

Kaplan, M. F. The effect of evaluative dispositions and amount and credibility of information on forming impressions of personality. *Psychonomic Science*, 1971, *24*, 174–176. (d)

Kaplan, M. F. The determination of trait redundancy in personality impression formation. *Psychonomic Science*, 1971, *23*, 280–282. (e)

Kaplan, M. F. Interpersonal attraction as a function of relatedness of similar and dissimilar attitudes. *Journal of Experimental Research in Personality*, 1972, *6*, 17–21. (a)

Kaplan, M. F. The modifying effect of stimulus information on the consistency of individual differences in impression formation. *Journal of Experimental Research in Personality*, 1972, *6*, 213–219. (b)

Kaplan, M. F. Stimulus inconsistency and response dispositions in forming judgments of other persons. *Journal of Personality and Social Psychology*, 1973, *25*, 58–64.

Kaplan, M. F. Evaluative judgments are based on evaluative information: Evidence against meaning change in evaluative context effects. *Memory & Cognition*, 1975, *3*, 375–380. (a)

Kaplan, M. F. Information integration in social judgment: Interaction of judge and informational components. In M. F. Kaplan & S. Schwartz (Eds.), *Human judgment and decision processes.* New York: Academic Press, 1975. (b)

Kaplan, M. F. Measurement and generality of response dispositions in person perception. *Journal of Personality*, 1976, *44*, 179–194.

Kaplan, M. F. Judgment by juries. In M. F. Kaplan & S. Schwartz (Eds.), *Human judgment and decision processes in applied settings.* New York: Academic Press, 1977. (a)

Kaplan, M. F. Discussion polarization effects in a modified jury decision paradigm: Informational influences. *Sociometry*, 1977, *40*, 262–271. (b)

Kaplan, M. F., & Anderson, N. H. Information integration theory and reinforcement theory as approaches to interpersonal attraction. *Journal of Personality and Social Psychology*, 1973, *28*, 301–312. (a)

Kaplan, M. F., & Anderson, N. H. Comment on "When research paradigms converge: Confrontation or integration?" *Journal of Personality and Social Psychology*, 1973, *28*, 321–322. (b)

Kaplan, M. F., & Major, G. *Will you like me at set-size 3 as you might at 6?: Amount of information and attraction.* Paper presented at the meetings of the Psychonomic Society, St. Louis, Missouri, November, 1973.

Kaplan, M. F., & Schersching, C. Reducing juror bias: An experimental approach. In P. D. Lipsitt & B. D. Sales (Eds.), *New directions in psycholegal research*. New York: Van Nostrand, 1980.

Kaplan, R. M., Bush, J. W., & Berry, C. C. Health status index: Category rating versus magnitude estimation for measuring levels of well-being. *Medical Care*, 1979, *17*, 501–523.

Kelley, H. H. The warm–cold variable in first impressions of persons. *Journal of Personality*, 1950, *18*, 431–439.

Kirk, L., & Burton, M. Meaning and context: A study of contextual shifts in meaning of Maasai personality descriptors. *American Ethnologist*, 1977, *4*, 734–761.

Klitzner, M. D. *Small animal fear: An integration-theoretical analysis*. Unpublished doctoral dissertation, University of California, San Diego, 1977.

Klitzner, M. D., & Anderson, N. H. Motivation × Expectancy × Value: A functional measurement approach. *Motivation and Emotion*, 1977, *1*, 347–365.

Krantz, D. H. A theory of magnitude estimation and cross-modality matching. *Journal of Mathematical Psychology*, 1972, *9*, 168–199. (a)

Krantz, D. H. Measurement structures and psychological laws. *Science*, 1972, *175*, 1427–1435. (b)

Krantz, D. H. Measurement theory and qualitative laws in psychophysics. In D. H. Krantz, R. C. Atkinson, R. D. Luce, & P. Suppes (Eds.), *Contemporary developments in mathematical psychology* (Vol. 2). San Francisco, California: Freeman, 1974.

Krantz, D. H., Luce, R. D., Suppes, P., & Tversky, A. *Foundations of measurement* (Vol. 1). New York: Academic Press, 1971.

Krantz, D. H., & Tversky, A. Conjoint-measurement analysis of composition rules in psychology. *Psychological Review*, 1971, *78*, 151–169. (a)

Krantz, D. H., & Tversky, A. An exchange on functional and conjoint measurement. *Psychological Review*, 1971, *78*, 457–458. (b)

Kruskal, J. B. Analysis of factorial experiments by estimating monotone transformations of the data. *Journal of the Royal Statistical Society* (B), 1965, *27*, 251–263.

Kun, A., Parsons, J. E., & Ruble, D. N. Development of integration processes using ability and effort information to predict outcome. *Developmental Psychology*, 1974, *10*, 721–732.

Lakoff, G. Hedges: A study in meaning criteria and the logic of fuzzy concepts. *Papers from the eighth regional meeting of the Chicago Linguistic Society*. Chicago, Illinois: University of Chicago Linguistics Department, 1972.

Lampel, A. K., & Anderson, N. H. Combining visual and verbal information in an impression-formation task. *Journal of Personality and Social Psychology*, 1968, *9*, 1–6.

Lane, J., & Anderson, N. H. Integration of intention and outcome in moral judgment. *Memory & Cognition*, 1976, *4*, 1–5.

Leach, C. The importance of instructions in assessing sequential effects in impression formation. *British Journal of Social and Clinical Psychology*, 1974, *13*, 151–156.

Lee, A. G., & Ostrom, T. M. Set size and impression certainty: Cue vs. extremity explanations. *Bulletin of the Psychonomic Society*, 1976, *8*, 371–373.

Leon, M. *Coordination of intent and consequence information in children's moral judgments*. Unpublished doctoral dissertation, University of California, San Diego, 1976.

Leon, M. *Coordination of intent and consequence information in children's moral judgments* (Tech. Rep. CHIP 72). La Jolla: Center for Human Information Processing, University of California, San Diego, August 1977.

Leon, M. Integration of intent and consequence information in children's moral judgments. In F. Wilkening, J. Becker, & T. Trabasso (Eds.), *Information integration by children*. Hillsdale, New Jersey: Erlbaum, 1980.

Leon, M. Rules mothers and sons use to evaluate intent and damage information in their moral judgments. *Child Development*, in press.

Leon, M., & Anderson, N. H. A ratio rule from integration theory applied to inference judgments. *Journal of Experimental Psychology*, 1974, *102*, 27–36.

Leon, M., Oden, G. C., & Anderson, N. H. Functional measurement of social values. *Journal of Personality and Social Psychology*, 1973, *27*, 301–310.

Levin, I. P. Information integration in numerical judgments and decision processes. *Journal of Experimental Psychology: General*, 1975, *104*, 39–53.

Levin, I. P. Processing of deviant information in inference and descriptive tasks with simultaneous and serial presentation. *Organizational Behavior and Human Performance*, 1976, *15*, 195–211.

Levin, I. P., Ims, J. R., & Vilmain, J. A. Information variability and reliability effects in evaluating student performance. *Journal of Educational Psychology*, 1980, *72*, 355–361.

Lichtenstein, S., Earle, T. C., & Slovic, P. Cue utilization in a numerical prediction task. *Journal of Experimental Psychology: Human Perception and Performance*, 1975, *1*, 77–85.

Loftus, E. F., Miller, D. G., & Burns, H. J. Semantic integration of verbal information into a visual memory. *Journal of Experimental Psychology: Human Learning and Memory*, 1978, *4*, 19–31.

Lopes, L. L. *Model-based decision and judgment in stud poker*. Unpublished doctoral dissertation, University of California, San Diego, 1974.

Lopes, L. L. Individual strategies in goal-setting. *Organizational Behavior and Human Performance*, 1976, *15*, 268–277. (a)

Lopes, L. L. Model-based decision and inference in stud poker. *Journal of Experimental Psychology: General*, 1976, *105*, 217–239. (b)

Lopes, L. L., & Ekberg, P.-H. S. Test of an ordering hypothesis in risky decision making. *Acta Psychologica*, 1980, *45*, 161–167.

Lopes, L. L., & Oden, G. C. *Judging similarity among kinship terms* (Tech. Rep. WHIPP 2). Madison: University of Wisconsin—Madison, October 1977.

Lopes, L. L., & Oden, G. C. Comparison of two models of similarity judgment. *Acta Psychologica*, 1980, *46*, 205–234.

Louviere, J. J. (Ed.). *Great Plains–Rocky Mountain Geographical Journal*, 1977, *6*, 1–107.

Luce, R. D. Remarks on the theory of the measurement and its relation to psychology. In *Les modèles et la formalisation du comportement*. Paris: Editions du Centre National de la Récherche Scientifique, 1967.

Luce, R. D. What sort of measurement is psychophysical measurement? *American Psychologist*, 1972, *27*, 96–106.

Luce, R. D., & Galanter, E. Psychophysical scaling. In R. D. Luce, R. R. Bush, & E. Galanter (Eds.), *Handbook of mathematical psychology* (Vol. 1). New York: Wiley, 1963.

Luce, R. D., & Tukey, J. W. Simultaneous conjoint measurement: A new type of fundamental measurement. *Journal of Mathematical Psychology*, 1964, *1*, 1–27.

Luchins, A. S. Experimental attempts to minimize the impact of first impressions. In C. I. Hovland (Ed.), *The order of presentation in persuasion*. New Haven, Connecticut: Yale University Press, 1957.

Luchins, A. S., & Luchins, E. H. The effects of order of presentation of information and explanatory models. *Journal of Social Psychology*, 1970, *80*, 63–70.

Lugg, A. M., & Gollob, H. F. An adding result in impression formation. *Memory & Cognition*, 1973, *1*, 356–360.

McGill, W. J. The slope of the loudness function: A puzzle. In H. Gulliksen & S. Messick (Eds.), *Psychological scaling: Theory and applications.* New York: Wiley, 1960.

McGuire, W. J. The nature of attitudes and attitude change. In G. Lindzey & E. Aronson (Eds.), *The handbook of social psychology* (Vol. 3, 2nd ed.). Reading, Massachusetts: Addison-Wesley, 1969.

McKillip, J. Credibility and impression formation. *Personality and Social Psychology Bulletin,* 1975, *1,* 521–524.

McKillip, J., Barrett, G., & DiMiceli, A. J. Trait ambiguity and impression formation: Sufficiency tests of the meaning change model. *Journal of General Psychology,* 1978, *98,* 161–171.

McKillip, J., & Edwards, J. D. Source characteristics and attitude change. *Personality and Social Psychology Bulletin,* 1975, *1,* 135–137.

Mandel, J. *The statistical analysis of experimental data.* New York: Wiley, 1964.

Mandler, G. *Mind and emotion.* New York: Wiley, 1975.

Manis, M., Gleason, T. C., & Dawes, R. M. The evaluation of complex social stimuli. *Journal of Personality and Social Psychology,* 1966, *3,* 404–419.

Marks, L. E. Stimulus-range, number of categories, and form of the category-scale. *American Journal of Psychology,* 1968, *81,* 467–479.

Marks, L. E. On scales of sensation. *Perception & Psychophysics,* 1974, *16,* 358–376. (a)

Marks, L. E. *Sensory processes.* New York: Academic Press, 1974. (b)

Marks, L. E. Translations and annotations concerning loudness scales and the processing of auditory intensity. In N. J. Castellan, Jr. & F. Restle (Eds.), *Cognitive theory* (Vol. 3). Hillsdale, New Jersey: Erlbaum, 1978. (a)

Marks, L. E. *Psychophysical measurement: Procedures, tasks, and scales.* Paper presented at the International Symposium on Social Psychophysics, University of Mannheim, October, 1978. (In B. Wegener [Ed.], *Social attitudes and psychophysical measurement.* Hillsdale, New Jersey: Erlbaum, in press.) (b)

Marks, L. E. A theory of loudness and loudness judgments. *Psychological Review,* 1979, *86,* 256–285. (a)

Marks, L. E. Sensory and cognitive factors in judgments of loudness. *Journal of Experimental Psychology: Human Perception and Performance,* 1979, *5,* 426–443. (b)

Marks, L. E., & Cain, W. S. Perception of intervals and magnitudes for three prothetic continua. *Journal of Experimental Psychology,* 1972, *94,* 6–17.

Massaro, D. W., & Anderson, N. H. A test of a perspective theory of geometrical illusions. *American Journal of Psychology,* 1970, *83,* 567–575.

Massaro, D. W., & Anderson, N. H. Judgmental model of the Ebbinghaus illusion. *Journal of Experimental Psychology,* 1971, *89,* 147–151.

Massaro, D. W., & Cohen, M. M. Voice onset time and fundamental frequency as cues to the /zi/-/si/ distinction. *Perception & Psychophysics,* 1977, *22,* 373–382.

Matthäus, W., Ernst, G., Kleinbeck, U., & Stoffer, T. Funktionales Messen und kognitive Algebra, Teil I. *Archiv für Psychologie,* 1976, *128,* 267–291.

Matthäus, W., Ernst, G., Kleinbeck, U., & Stoffer, T. Funktionales Messen und kognitive Algebra, Teil II. *Archiv für Psychologie,* 1977, *129,* 1–24.

Montgomery, H. *Intra- and interindividual variations in the form of psychophysical scales.* Paper presented at the International Symposium on Social Psychophysics, University of Mannheim, October, 1978. (In B. Wegener [Ed.], *Social attitudes and psychophysical measurement.* Hillsdale, New Jersey: Erlbaum, in press.)

Mosteller, F. The mystery of the missing corpus. *Psychometrika,* 1958, *23,* 279–289.

National Science Foundation Annual Report. Washington, D.C.: National Science Foundation, 1974.

Nisbett, R. E., & Bellows, N. Verbal reports about causal influences on social judgments: Private access versus public theories. *Journal of Personality and Social Psychology*, 1977, *35*, 613-624.

Nisbett, R. E., & Ross, L. *Human inference: Strategies and shortcomings of social judgment.* Englewood Cliffs, New Jersey: Prentice Hall, 1980.

Nisbett, R. E., & Wilson, T. D. Telling more than we can know: Verbal reports on mental processes. *Psychological Review*, 1977, *84*, 231-259. (a)

Nisbett, R. E., & Wilson, T. D. The halo effect: Evidence for unconscious alteration of judgments. *Journal of Personality and Social Psychology*, 1977, *35*, 250-256. (b)

Nisbett, R. E., Zukier, H., & Lemley, R. E. The dilution effect: Nondiagnostic information weakens the implications of diagnostic information. *Cognitive Psychology*, 1981, *13*, 248-277.

Norman, D. A., & Rumelhart, D. E. Reference and comprehension. In D. A. Norman & D. E. Rumelhart (Eds.), *Explorations in cognition*. San Francisco, California: Freeman, 1975.

Norman, K. L. A solution for weights and scale values in functional measurement. *Psychological Review*, 1976, *83*, 80-84. (a)

Norman, K. L. Effects of feedback on the weights and subjective values in an information integration model. *Organizational Behavior and Human Performance*, 1976, *17*, 367-387. (b)

Oden, G. C. *Semantic constraints and ambiguity resolution.* Unpublished doctoral dissertation, University of California, San Diego, 1974.

Oden, G. C. Fuzziness in semantic memory: Choosing exemplars of subjective categories. *Memory & Cognition*, 1977, *5*, 198-204. (a)

Oden, G. C. Integration of fuzzy logical information. *Journal of Experimental Psychology: Human Perception and Performance*, 1977, *3*, 565-575. (b)

Oden, G. C. Integration of place and voicing information in the identification of synthetic stop consonants. *Journal of Phonetics*, 1978, *6*, 83-93. (a)

Oden, G. C. Semantic constraints and judged preference for interpretations of ambiguous sentences. *Memory & Cognition*, 1978, *6*, 26-37. (b)

Oden, G. C. A fuzzy logical model of letter identification. *Journal of Experimental Psychology: Human Perception and Performance*, 1979, *5*, 336-352. (a)

Oden, G. C. Fuzzy propositional approach to psycholinguistic problems: An application of fuzzy set theory in cognitive science. In M. M. Gupta, R. K. Ragade, & R. R. Yager (Eds.), *Advances in fuzzy set theory and applications*. Amsterdam: North-Holland, 1979. (b)

Oden, G. C., & Anderson, N. H. Differential weighting in integration theory. *Journal of Experimental Psychology*, 1971, *89*, 152-161.

Oden, G. C., & Anderson, N. H. Integration of semantic constraints. *Journal of Verbal Learning and Verbal Behavior*, 1974, *13*, 138-148.

Oden, G. C., & Massaro, D. W. Integration of featural information in speech perception. *Psychological Review*, 1978, *85*, 172-191.

Örtendahl, M., & Sjöberg, L. Delay of outcome and preference for different courses of action. *Perceptual and Motor Skills*, 1979, *48*, 3-57. (Monograph Supplement I-V48).

Osgood, C. E., Suci, G. J., & Tannenbaum, P. H. *The measurement of meaning.* Urbana: University of Illinois Press, 1957.

Osgood, C. E., & Tannenbaum, P. H. The principle of congruity in the prediction of attitude change. *Psychological Review*, 1955, *62*, 42-55.

Ostrom, T. M. *Directionality of component ratings in impression formation* (Tech. Rep. 71-5). Columbus: Ohio State University, Computer and Information Science Research Center, 1971.

Ostrom, T. M. Between-theory and within-theory conflict in explaining context effects in impression formation. *Journal of Experimental Social Psychology*, 1977, *13*, 492–503.

Ostrom, T. M., & Davis, D. Idiosyncratic weighting of trait information in impression formation. *Journal of Personality and Social Psychology*, 1979, *37*, 2025–2043.

Ostrom, T. M., & Essex, D. W. *Meaning shift in impression formation.* Paper presented at the meetings of the Psychonomic Society, St. Louis, Missouri, 1972.

Ostrom, T. M., Werner, C., & Saks, M. J. An integration theory analysis of jurors' presumptions of guilt or innocence. *Journal of Personality and Social Psychology*, 1978, *36*, 436–450.

Parducci, A. Contextual effects: A range–frequency analysis. In E. C. Carterette & M. P. Friedman (Eds.), *Handbook of perception* (Vol. 2). New York: Academic Press, 1974.

Payne, J. W. Alternative approaches to decision making under risk: Moments versus risk dimensions. *Psychological Bulletin*, 1973, *80*, 439–453.

Person, H. B., & Barron, F. H. Polynomial psychophysics of group risk perception. *Acta Psychologica*, 1978, *42*, 421–428.

Phelps, R. H., & Shanteau, J. Livestock judges: How much information can an expert use? *Organizational Behavior and Human Performance*, 1978, *21*, 209–219.

Podell, J. E. A comparison of generalization and adaptation-level as theories of connotation. *Journal of Abnormal and Social Psychology*, 1961, *62*, 593–597.

Poulton, E. C. The new psychophysics: Six models for magnitude estimation. *Psychological Bulletin*, 1968, *69*, 1–19.

Poulton, E. C. Models for biases in judging sensory magnitude. *Psychological Bulletin*, 1979, *86*, 777–803.

Ramsay, J. O. Review of *Foundations of measurement* (Vol. 1). *Psychometrika*, 1975, *40*, 257–262.

Rapoport, A., & Wallsten, T. S. Individual decision behavior. *Annual Review of Psychology*, 1972, *23*, 131–176.

Richey, M. H., Richey, H. W., & Thieman, G. Negative salience in impression of character: Effects of new information on established relationships. *Psychonomic Science*, 1972, *28*, 65–67.

Rimoldi, H. J. A. Prediction of scale values for combined stimuli. *The British Journal of Statistical Psychology*, 1956, *9*, 29–40.

Rips, L. J., Shoben, E. J., & Smith, E. Semantic distance and the verification of semantic relations. *Journal of Verbal Learning and Verbal Behavior*, 1973, *12*, 1–20.

Riskey, D. R. Verbal memory processes in impression formation. *Journal of Experimental Psychology: Human Learning and Memory*, 1979, *5*, 271–281.

Riskey, D. R., & Birnbaum, M. H. Compensatory effects in moral judgment: Two rights don't make up for a wrong. *Journal of Experimental Psychology*, 1974, *103*, 171–173.

Rokeach, M., & Rothman, G. The principle of belief congruence and the congruity principle as models of cognitive interaction. *Psychological Review*, 1965, *72*, 128–142.

Rosch, E. H. On the internal structure of perceptual and semantic categories. In T. E. Moore (Ed.), *Cognitive development and the acquisition of language*. New York: Academic Press, 1973.

Rose, B. J., & Birnbaum, M. H. Judgments of differences and ratios of numerals. *Perception & Psychophysics*, 1975, *18*, 194–200.

Rosenbaum, M. E., & Levin, I. P. Impression formation as a function of source credibility and order of presentation of contradictory information. *Journal of Personality and Social Psychology*, 1968, *10*, 167–174.

Rosenbaum, M. E., & Levin, I. P. Impression formation as a function of source credibility and the polarity of information. *Journal of Personality and Social Psychology*, 1969, *12*, 34–37.

Rosenbaum, M. E., & Schmidt, C. F. Whither pooling models? Some additional variables. In R. P. Abelson, E. Aronson, W. J. McGuire, T. M. Newcomb, M. J. Rosenberg, & P. H. Tannenbaum (Eds.), *Theories of cognitive consistency: A sourcebook.* Chicago, Illinois: Rand McNally, 1968.

Rule, S. J., & Curtis, D. W. Converging power functions as a description of the size–weight illusion: A control experiment. *Bulletin of the Psychonomic Society,* 1976, *8,* 16–18.

Rule, S. J., & Curtis, D. W. The influence of the interaction of weight and volume on subjective heaviness. *Perception & Psychophysics,* 1977, *22,* 159–164.

Rule, S. J., & Curtis, D. W. *Levels of sensory and judgmental functioning: Strategies for the evaluation of a model.* Paper presented at the International Symposium on Social Psychophysics, University of Mannheim, October, 1978. (In B. Wegener [Ed.], *Social attitudes and psychophysical measurement.* Hillsdale, New Jersey: Erlbaum, in press.)

Rywick, T., & Schaye, P. Use of long-term memory in impression formation. *Psychological Reports,* 1974, *34,* 939–945.

Scheffé, H. *The analysis of variance.* New York: Wiley, 1959.

Schiffenbauer, A. Effect of observer's emotional state on judgments of the emotional state of others. *Journal of Personality and Social Psychology,* 1974, *30,* 31–35.

Schmidt, C. F., & Levin, I. P. Test of an averaging model of person preference: Effect of context. *Journal of Personality and Social Psychology,* 1972, *23,* 277–282.

Schneider, D. J., Hastorf, A. H., & Ellsworth, P. C. *Person perception* (2nd ed.). Reading, Massachusetts: Addison-Wesley, 1979.

Schönemann, P. H., Cafferty, T., & Rotton, J. A note on additive functional measurement. *Psychological Review,* 1973, *80,* 85–87.

Schümer, R. Context effects in impression formation as a function of the ambiguity of test traits. *European Journal of Social Psychology,* 1973, *3,* 333–338.

Shanteau, J. C. An additive model for sequential decision making. *Journal of Experimental Psychology,* 1970, *85,* 181–191. (a)

Shanteau, J. C. *Component processes in risky decision judgments.* Unpublished doctoral dissertation, University of California, San Diego, 1970. (b)

Shanteau, J. Descriptive versus normative models of sequential inference judgment. *Journal of Experimental Psychology,* 1972, *93,* 63–68.

Shanteau, J. Component processes in risky decision making. *Journal of Experimental Psychology,* 1974, *103,* 680–691.

Shanteau, J. An information-integration analysis of risky decision making. In M. F. Kaplan & S. Schwartz (Eds.), *Human judgment and decision processes.* New York: Academic Press, 1975. (a)

Shanteau, J. Averaging versus multiplying combination rules of inference judgment. *Acta Psychologica,* 1975, *39,* 83–89. (b)

Shanteau, J. *Repetition effects on memory-retrieval and decision-making time in problem solving.* Paper presented at meetings of Psychonomic Society, St. Louis, Missouri, 1976.

Shanteau, J. POLYLIN: A FORTRAN IV program for the analysis of multiplicative (multilinear) trend components of interactions. *Behavior Research Methods & Instrumentation,* 1977, *9,* 381–382.

Shanteau, J. C., & Anderson, N. H. Test of a conflict model for preference judgment. *Journal of Mathematical Psychology,* 1969, *6,* 312–325.

Shanteau, J., & Anderson, N. H. Integration theory applied to judgments of the value of information. *Journal of Experimental Psychology,* 1972, *92,* 266–275.

Shanteau, J., & McClelland, G. H. *Mental search processes in problem solving.* Paper presented at the mathematical psychology meetings, San Diego, California, 1972.

Shanteau, J., & McClelland, G. H. Mental search processes in problem solving. *Memory & Cognition*, 1975, *3*, 627–634.

Shanteau, J., & Nagy, G. Decisions made about other people: A human judgment analysis of dating choice. In J. S. Carroll & J. W. Payne (Eds.), *Cognition and social behavior*. Potomac, Maryland: Erlbaum, 1976.

Shanteau, J., & Nagy, G. F. Probability of acceptance in dating choice. *Journal of Personality and Social Psychology*, 1979, *37*, 522–533.

Shepard, R. N. Metric structures in ordinal data. *Journal of Mathematical Psychology*, 1966, *3*, 287–315.

Sherif, M., & Hovland, C. I. *Social judgment: Assimilation and contrast effects in communication and attitude change*. New Haven, Connecticut: Yale University Press, 1961.

Sidowski, J. B., & Anderson, N. H. Judgments of City–Occupation combinations. *Psychonomic Science*, 1967, *7*, 279–280.

Sigall, H., & Landy, D. Radiating beauty: Effects of having a physically attractive partner on person perception. *Journal of Personality and Social Psychology*, 1973, *28*, 218–224.

Simmonds, M. B. *Verbal factors modifying similarity judgments*. Unpublished doctoral dissertation. Christchurch, New Zealand: University of Canterbury, 1971.

Simpson, D. D., & Ostrom, T. M. Effect of snap and thoughtful judgments on person impressions. *European Journal of Social Psychology*, 1975, *5*, 197–208.

Singh, R., Gupta, M., & Dalal, A. K. Cultural difference in attribution of performance: An integration-theoretical analysis. *Journal of Personality and Social Psychology*, 1979, *37*, 1342–1351.

Singh, R., Sidana, U. R., & Saluja, S. K. Integration theory applied to judgments of personal happiness by children. *Journal of Social Psychology*, 1978, *105*, 27–31.

Singh, R., Sidana, U. R., & Srivastava, P. Averaging processes in children's judgment of happiness. *Journal of Social Psychology*, 1978, *104*, 123–132.

Sjöberg, L. Thurstonian methods in the measurement of learning. *Scandinavian Journal of Psychology*, 1965, *6*, 33–48.

Sjöberg, L. A method for sensation scaling based on an analogy between perception and judgment. *Perception & Psychophysics*, 1966, *1*, 131–136.

Sjöberg, L. Successive intervals scaling of paired comparisons. *Psychometrika*, 1967, *32*, 297–308.

Sjöberg, L. Studies of the rated favorableness of offers to gamble. *Scandinavian Journal of Psychology*, 1968, *9*, 257–273.

Sjöberg, L. Sensation scales in the size–weight illusion. *Scandinavian Journal of Psychology*, 1969, *10*, 109–112.

Sjöberg, L. Three models for the analysis of subjective ratios. *Scandinavian Journal of Psychology*, 1971, *12*, 217–240.

Sloan, L. R., & Ostrom, T. M. Amount of information and interpersonal judgment. *Journal of Personality and Social Psychology*, 1974, *29*, 23–29.

Smith, E. R., & Miller, F. D. Limits on perception of cognitive processes: A reply to Nisbett and Wilson. *Psychological Review*, 1978, *85*, 355–362.

Snyder, M., & Uranowitz, S. W. Reconstructing the past: Some cognitive consequences of person perception. *Journal of Personality and Social Psychology*, 1978, *36*, 941–950.

Spence, W., & Guilford, J. P. The affective value of combinations of odors. *American Journal of Psychology*, 1933, *45*, 495–501.

Sternberg, S. High-speed scanning in human memory. *Science*, 1966, *153*, 652–654.

Sternberg, S. The discovery of processing stages: Extensions of Donders' method. *Acta Psychologica*, 1969, *30*, 276–315.

Stevens, S. S. The direct estimation of sensory magnitudes—loudness. *American Journal of Psychology*, 1956, *69*, 1–25.

Stevens, S. S. On the psychophysical law. *Psychological Review*, 1957, *64*, 153–181.

Stevens, S. S. The psychophysics of sensory function. *American Scientist*, 1960, *48*, 226–253.

Stevens, S. S. The psychophysics of sensory function. In W. A. Rosenblith (Ed.), *Sensory communication*. Cambridge, Massachusetts: M.I.T. Press, 1961.

Stevens, S. S. A metric for the social consensus. *Science*, 1966, *151*, 530–541.

Stevens, S. S. Measurement, statistics, and the schemapiric view. *Science*, 1968, *161*, 849–856.

Stevens, S. S. Issues in psychophysical measurement. *Psychological Review*, 1971, *78*, 426–450.

Stevens, S. S. Perceptual magnitude and its measurement. In E. C. Carterette & M. P. Friedman (Eds.), *Handbook of perception* (Vol. 2). New York: Academic Press, 1974.

Stevens, S. S. *Psychophysics*. New York: Wiley, 1975.

Stewart, R. H. Effect of continuous responding on the order effect in personality impression formation. *Journal of Personality and Social Psychology*, 1965, *1*, 161–165.

Suppes, P., & Zinnes, J. L. Basic measurement theory. In R. D. Luce, R. R. Bush, & E. Galanter (Eds.), *Handbook of mathematical psychology* (Vol. 1). New York: Wiley, 1963.

Surber, C. F. Developmental processes in social inference: Averaging of intentions and consequences in moral judgment. *Developmental Psychology*, 1977, *13*, 654–665.

Svenson, O. A functional measurement approach to intuitive estimation as exemplified by estimated time savings. *Journal of Experimental Psychology*, 1970, *86*, 204–210.

Tagiuri, R. Person perception. In G. Lindzey & E. Aronson (Eds.), *Handbook of social psychology* (Vol. 3, 2nd ed.). Reading, Massachusetts: Addison-Wesley, 1969.

Takahashi, S. [Analysis of weighted averaging model on integration of informations in personality impression formation.] *Japanese Psychological Research*, 1970, *12*, 154–162.

Takahashi, S. Effect of inter-relatedness of informations on context effect in personality impression formation. *Japanese Psychological Research*, 1971, *13*, 167–175. (a)

Takahashi, S. Effect of the context upon personality-impression formation. *Japanese Journal of Psychology*, 1971, *41*, 307–313. (b)

Takahashi, S. Effect of cognitive complexity on the integration process of informations in personality impression formation. *Bulletin of the Faculty of Education, Hiroshima University*. 1971, *20*, 257–265. (c)

Tesser, A. Differential weighting and directed meaning as explanations of primacy in impression formation. *Psychonomic Science*, 1968, *11*, 299–300.

Thorndike, E. L. A constant error in psychological ratings. *Journal of Applied Psychology*, 1920, *4*, 25–29.

Thurstone, L. L. Psychophysical analysis. *American Journal of Psychology*, 1927, *38*, 368–389. (Reprinted in *The measurement of values*, 1959.) (a)

Thurstone, L. L. The method of paired comparisons for social values. *Journal of Abnormal and Social Psychology*, 1927, *21*, 384–400. (Reprinted in *The measurement of values*, 1959.) (b)

Thurstone, L. L. *The measurement of values*. Chicago, Illinois: University of Chicago Press, 1959.

Thurstone, L. L., & Jones, L. V. The rational origin for measuring subjective values. In L. L. Thurstone (Ed.), *The measurement of values*. Chicago, Illinois: University of Chicago Press, 1959.

Torgerson, W. S. *Theory and methods of scaling*. New York: Wiley, 1958.

Treisman, M. Sensory scaling and the psychophysical law. *Quarterly Journal of Experimental Psychology*, 1964, *16*, 11–22.

Triandis, H. C., & Fishbein, M. Cognitive interaction in person perception. *Journal of Abnormal and Social Psychology*, 1963, *67*, 446–453.

Troutman, C. M., & Shanteau, J. Inferences based on nondiagnostic information. *Organizational Behavior and Human Performance*, 1977, *19*, 43–55.

Tukey, J. W. One degree of freedom for non-additivity. *Biometrics*, 1949, *5*, 232–242.

Tukey, J. W. Analyzing data: Sanctification or detective work? *American Psychologist*, 1969, *24*, 83–102.

Tversky, A. Features of similarity. *Psychological Review*, 1977, *84*, 327–352.

Veit, C. T. Ratio and subtractive processes in psychophysical judgment. *Journal of Experimental Psychology: General*, 1978, *107*, 81–107.

Verdi, J. *Integrating duty and need information in moral judgments.* Unpublished paper. La Jolla: Center for Human Information Processing, University of California, San Diego, September 1979.

Verge, C. G., & Bogartz, R. S. A functional measurement analysis of the development of dimensional coordination in children. *Journal of Experimental Child Psychology*, 1978, *25*, 337–353.

Warr, P. Inference magnitude, range, and evaluative direction as factors affecting relative importance of cues in impression formation. *Journal of Personality and Social Psychology*, 1974, *30*, 191–197.

Warr, P., & Jackson, P. The importance of extremity. *Journal of Personality and Social Psychology*, 1975, *32*, 278–282.

Weiss, D. J. Averaging: An empirical validity criterion for magnitude estimation. *Perception & Psychophysics*, 1972, *12*, 385–388.

Weiss, D. J. *A functional measurement analysis of equisection.* Unpublished doctoral dissertation, University of California, San Diego, 1973. (a)

Weiss, D. J. FUNPOT: A FORTRAN program for finding a polynomial transformation to reduce any sources of variance in a factorial design. *Behavioral Science*, 1973, *18*, 150. (b)

Weiss, D. J. Quantifying private events: A functional measurement analysis of equisection. *Perception & Psychophysics*, 1975, *17*, 351–357.

Weiss, D. J., & Anderson, N. H. Subjective averaging of length with serial presentation. *Journal of Experimental Psychology*, 1969, *82*, 52–63.

Weiss, D. J., & Anderson, N. H. Use of rank order data in functional measurement. *Psychological Bulletin*, 1972, *78*, 64–69.

White, P. Limitations on verbal reports of internal events: A refutation of Nisbett and Wilson and of Bem. *Psychological Review*, 1980, *87*, 105–112.

Wilkening, F. Combining of stimulus dimensions in children's and adults' judgments of area: An information integration analysis. *Developmental Psychology*, 1979, *15*, 25–33.

Wilkening, F. Development of dimensional integration in children's perceptual judgment: Experiments with area, volume, and velocity. In F. Wilkening, J. Becker, & T. Trabasso (Eds.), *Information integration by children.* Hillsdale, New Jersey: Erlbaum, 1980.

Wilkening, F. Integrating velocity, time, and distance information: A developmental study. *Cognitive Psychology*, 1981, *13*, 231–247.

Wilkening, F. Children's knowledge about time, distance, and velocity interrelations. In W. J. Friedman (Ed.), *The developmental psychology of time.* New York: Academic Press, in press.

Wilkening, F., & Anderson, N. H. *Comparison of two rule assessment methodologies for studying cognitive development* (Tech. Rep. CHIP 94). La Jolla: Center for Human Information Processing, University of California, San Diego, June 1980.

Willis, R. H. Stimulus pooling and social perception. *Journal of Abnormal and Social Psychology*, 1960, *60*, 365–373.

Wilson, J. D. *The fortunes of Falstaff*. London and New York: Cambridge University Press, 1943.

Wishner, J. Reanalysis of "Impressions of personality." *Psychological Review*, 1960, *67*, 96–112.

Wong, R. *A theory for integration of two source-adjective combinations in a personality impression formation task*. Senior honors thesis, University of California, San Diego, June 1973.

Wong, L. *Combining source estimates and blue book values to judge the worth of a used car*. Senior honors thesis, University of California, San Diego, June 1973.

Wyer, R. S., Jr. A quantitative comparison of three models of impression formation. *Journal of Experimental Research in Personality*, 1969, *4*, 29–41.

Wyer, R. S., Jr. *Cognitive organization and change*. Potomac, Maryland: Erlbaum, 1974. (a)

Wyer, R. S., Jr. Changes in meaning and halo effects in personality impression formation. *Journal of Personality and Social Psychology*, 1974, *29*, 829–835. (b)

Wyer, R. S., Jr. Functional measurement methodology applied to a subjective probability model of cognitive functioning. *Journal of Personality and Social Psychology*, 1975, *31*, 94–100. (a)

Wyer, R. S., Jr. The role of probabilistic and syllogistic reasoning in cognitive organization and social inference. In M. F. Kaplan & S. Schwartz (Eds.), *Human judgment and decision processes*. New York: Academic Press, 1975. (b)

Wyer, R. S., Jr., & Dermer, M. Effect of context and instructional set upon evaluations of personality-trait adjectives. *Journal of Personality and Social Psychology*, 1968, *9*, 7–14.

Wyer, R. S., Jr., & Watson, S. F. Context effects in impression formation. *Journal of Personality and Social Psychology*, 1969, *12*, 22–33.

Young, F. W., de Leeuw, J., & Takane, Y. Regression with qualitative and quantitative variables: An alternating least squares method with optimal scaling features. *Psychometrika*, 1976, *41*, 505–529.

Youngblood, J., & Himmelfarb, S. The effects of prior neutral messages on resistance to evaluative communications. *Psychonomic Science*, 1972, *29*, 348–350.

Zadeh, L. A., Fu, K.-S., Tanaka, K., & Shimura, M. *Fuzzy sets and their applications to cognitive and decision processes*. New York: Academic Press, 1975.

Zalinski, J., & Anderson, N. H. Measurement of importance in multiattribute judgment models. In J. B. Sidowski (Ed.), *Conditioning, cognition, and methodology: Contemporary issues in experimental psychology*. Hillsdale, New Jersey: Erlbaum, in press.

Zanna, M. P., & Hamilton, D. L. Further evidence for meaning change in impression formation. *Journal of Experimental Social Psychology*, 1977, *13*, 224–238.

Zinnes, J. L. Scaling. *Annual Review of Psychology*, 1969, *20*, 447–478.

Index